Lecture Notes in Computer Science 2965

Edited by G. Goos, J. Hartmanis, and J. van Leeuwen

RENEWALS 458-4374

Springer
Berlin
Heidelberg
New York
Hong Kong
London
Milan
Paris
Tokyo

Maria Carla Calzarossa   Erol Gelenbe (Eds.)

Performance Tools
and Applications
to Networked Systems

Revised Tutorial Lectures

WITHDRAWN
UTSA LIBRARIES

Springer

Series Editors

Gerhard Goos, Karlsruhe University, Germany
Juris Hartmanis, Cornell University, NY, USA
Jan van Leeuwen, Utrecht University, The Netherlands

Volume Editors

Maria Carla Calzarossa
Università di Pavia
Dipartimento di Informatica e Sistemistica
via Ferrata 1, 27100 Pavia, Italy
E-mail: mcc@alice.unipv.it

Erol Gelenbe
Imperial College
Department of Electrical and Electronic Engineering
London SW7 2BT, UK
E-mail: e.gelenbe@imperial.ac.uk

Library
University of Texas
at San Antonio

Library of Congress Control Number: 2004104315

CR Subject Classification (1998): C.2, C.4, E.1, D.2

ISSN 0302-9743
ISBN 3-540-21945-5 Springer-Verlag Berlin Heidelberg New York

This work is subject to copyright. All rights are reserved, whether the whole or part of the material is concerned, specifically the rights of translation, reprinting, re-use of illustrations, recitation, broadcasting, reproduction on microfilms or in any other way, and storage in data banks. Duplication of this publication or parts thereof is permitted only under the provisions of the German Copyright Law of September 9, 1965, in its current version, and permission for use must always be obtained from Springer-Verlag. Violations are liable to prosecution under the German Copyright Law.

Springer-Verlag is a part of Springer Science+Business Media

springeronline.com

© Springer-Verlag Berlin Heidelberg 2004
Printed in Germany

Typesetting: Camera-ready by author, data conversion by PTP-Berlin, Protago-TeX-Production GmbH
Printed on acid-free paper   SPIN: 10998946   06/3142   5 4 3 2 1 0

Preface

This volume is dedicated largely to the performance-oriented design of modern computer networks, both wired and wireless. It is the consequence of the Tutorial Session which was held on 12th October 2003, preceeding the IEEE Computer Society's Symposium on Modeling, Analysis and Simulation of Computer and Telecommunication Systems, held in Orlando, Florida. In addition to the core tutorial presentations, which covered both advances in network quality of service (QoS) and in performance evaluation methodology, we felt that it would be useful to assemble a volume in the Lecture Notes in Computer Science series which would also include specific application areas of performance modeling and measurements. Thus the current volume includes three parts:

- a first part that specifically addresses performance and QoS of modern wired and wireless networks;
- a second one that discusses current advances in performance modeling and simulation; and
- a final part that addresses other specific applications of these methodologies.

The network-oriented portion of the volume itself comprises three complementary topics:

- a first group of chapters deals with novel designs and new issues related to broad-based network performance and QoS, without limitations concerning the connection technologies that are being used;
- a second group addresses the wireless context; and
- a final grouping of contributions considers the topic of wireless ad hoc networks more specifically.

The part of the volume dealing with methodologies discusses the software specification of models, the use of certain formalisms such as Petri nets, and some recent advances in simulation. Finally, the third part of the volume discusses other performance applications related to scheduling and to specific architectures.

The first paper in the part on general networking issues provides a useful survey of Content Delivery Networks which is of great current interest. The second paper also addresses a question of major importance concerning the modeling of computer virus propagation. Electronic mail still remains central to the role of networks, and the performance of mail systems is addressed in a third paper. The research represented in these three contributions was supported via the national Italian research project FIRB–PERF on computer and network performance. These papers are followed by a novel packet network protocol design, the Cognitive Packet Network (CPN), which provides QoS-driven routing in wired and wireless networks; this paper also gives detailed measurement results concerning user performance and QoS obtained in an experimental CPN testbed. Issues of network reliability and path restoration in mesh networks are discussed in the fifth paper.

The integration of wireless and wired networks and the use of novel wireless systems as they become available is discussed first in a contribution on the wireless internet and then also in a contribution on the performance of systems that exploit 2.5/3G wireless. The role of the IP protocol in the wireless context is also examined in a separate paper.

Ad hoc wireless networks are discussed in a set of three papers, the first of which examines the role of peer-to-peer computing in this context, followed by another contribution that studies the role of multipath routing. A review of ad hoc wireless algorithm designs is presented in the third paper covering this area.

The section on performance evaluation methodologies begins with a paper on performance management which discusses certain general issues related to QoS in systems. The combination of UML and Petri nets is discussed in a second paper, while a novel extension to the performance evaluation tool PEPA so as to include the possibility of evaluating computer networks is discussed in the third paper of this group. A contribution on the use of simulation within a visually "true" augmented reality environment completes the set of methodologically oriented papers.

This volume closes with two papers on significant scheduling problems, the first in the area of distributed systems, and the second on mass memories and disk arrays.

We believe that this volume constitutes a very useful tool for both the practitioner and the researcher. To the practitioner we offer a set of comprehensive pointers to performance and QoS issues and we feel that the didactic style and the numerous references provided in each paper can be of great use in understanding how the field can help solve practical problems. For the researcher, we have consciously selected a set of contributions not only for their research value but also for their novelty and use in identifying areas of active research where much further work can be done.

February 2004   Maria Carla Calzarossa
                                                          Erol Gelenbe

Table of Contents

A Walk through Content Delivery Networks

Novella Bartolini[1]*, Emiliano Casalicchio[2], and Salvatore Tucci[2]

[1] Università di Roma "La Sapienza", Via Salaria 113 - 00198 Roma, Italy,
novella@dsi.uniroma1.it
[2] Università di Roma "Tor Vergata", Via del Politecnico, 1 - 00133 Roma, Italy,
casalicchio@ing.uniroma2.it, tucci@uniroma2.it

Abstract. Content Delivery Networks (CDN) aim at overcoming the inherent limitations of the Internet. The main concept at the basis of this technology is the delivery at edge points of the network, in proximity to the request areas, to improve the user's perceived performance while limiting the costs. This paper focuses on the main research areas in the field of CDN, pointing out the motivations, and analyzing the existing strategies for replica placement and management, server measurement, best fit replica selection and request redirection.

1 Introduction

The commercial success of the Internet and e-services, together with the exploding use of complex media content online has paved the way for the birth and growing interest in Content Delivery Networks (CDN). Internet traffic often encounters performance difficulties characteristic of a non dedicated, best effort environment. The user's urgent request for guarantees on quality of service have brought about the need to study and develop new network architectures and technologies to improve the user's perceived performance while limiting the costs paid by providers. Many solutions have been proposed to alleviate the bottleneck problems and the most promising are based on the awareness of the content that has to be delivered. The traditional *"content-blind"* network infrastructures are not sufficient to ensure quality of service to all users in a dynamic and ever increasing traffic situation. New protocols and integrated solutions must be in place both on the network and on the server side to distribute, locate and download contents through the Internet.

The enhancement of computer networks by means of a content aware overlay creates the new architectural paradigm of the CDN. Today's CDN act upon the traditional network protocol stack at various levels, relying on dynamic and proactive content caching and on automatic application deployment and migration at the edge of the network, in proximity to the final users. Content replicas in a CDN are geographically distributed, to enable fast and reliable delivery to any end-user location: through CDN services, up-to-date content, can be retrieved by end-users locally rather than remotely.

* The work of Novella Bartolini has been funded by the WEB-MINDS project supported by the Italian MIUR under the FIRB program

© Springer-Verlag Berlin Heidelberg 2004

CDNs were born to distribute heavily requested contents from popular web servers, most of all image files. Nowadays, a CDN supports the delivery of any type of dynamic content, including various forms of interactive media streaming. CDN providers are companies devoted to hosting in their servers the content of third-party content providers, to mirroring or replicating such contents on several servers spread over the world, and to transparently redirecting the customers requests to the 'best replica' (e.g. the closest replica, or the one from which the customer would access content at the lowest latency). Designing a complete solution for CDN therefore requires addressing a number of technical issues: which kind of content should be hosted (if any) at a given CDN server (replica placement), how the content must be kept updated, which is the 'best replica' for a given customer, which mechanisms must be in place to transparently redirect the user to such replica. A proper placement of replica servers shortens the path from servers to clients thus lowering the risk of encountering bottlenecks in the non-dedicated environment of the Internet. A request redirection mechanism is provided at the access routers level to ensure that the best suited replica is selected to answer any given request of possibly different types of services with different quality of service agreements. The CDN architecture also relies on a measurement activity that is performed by cooperative access routers to evaluate the traffic conditions and the computational capacity and availability of each replica capable of serving the given request. Successfully implemented, a CDN can accelerate end user access to content, reduce network traffic, and reduce content provider hardware requirements.

This paper explores architectures, technologies and research issues in content delivery networks [53]. In section 2 we describe the core features of a CDN, discussing the motivations and how content delivery can alleviate internet performance problems. In section 3 we examine types of content and services that can beneficiate from content delivery techniques. Section 4 describes the architecture and working principles of a CDN. A detailed discussion on replica placement and management is provided in section 5. Section 6 introduces the problem of how measures can be taken to select the replica that can better fulfil an incoming request, while request redirection mechanisms are described and compared in section 7. Section 8 concludes the paper.

Let us point out that this paper is related to other papers contained in this volume. Issues related to QoS are discussed in several papers in this volume, including [8] [31] [42], while content delivery is related to peer-to-peer networking which is discussed in [39]. Content is often of multimedia nature, such as augmented reality [30] which will have high bandwidth and significant QoS needs, and the type of tools described in [32] can contribute simpler evaluation tools which are applicable to the systems we discuss.

2   Motivations for Content Delivery

Internet users commonly get frustrated by low performances and may decide to abandon a web site or to disconnect a multimedia session when experiencing performance difficulties, causing revenue to be lost. Though centralized models are still in place in the Internet today, these architectures are poor in terms of adaptivity and scalability.

If a provider establishes a content server in a single physical location from which it disseminates data, services, and information to all its users, the single server is likely to become overloaded and its links can easily be saturated. The speed at which users can access the site could become unpredictably higher than the maximum request rate the server and its links can tolerate. Since it is impossible, with this approach, to adapt to the exponential growth of the Internet traffic, the centralized model of content serving is inherently unscalable, incapable of adaptivity and produces performance losses when traffic bursts occur. Though this leads to the conclusion that a certain amount of servers must be adopted, a cluster of servers (also known as *server farm*, that is a multi-server localized architecture, is not necessarily a solution yet.

The server computational and link capacity is only the first source of performance difficulties that may be encountered while downloading content over the Internet. There are many other possible congestion causes that may lead to unacceptable user perceived quality. The non dedicated, best effort nature of the Internet is the inborn limit to the possibility of having any sort of performance guarantee while delivering content over it.

The Internet is a network of heterogeneous networks composed of thousands of different autonomous systems ranging from large backbone providers to small local ISPs. The autonomous systems connect to each other creating the global Internet. The communication between two networks is achieved through the connection of border routers in a peering session. Two peer routers periodically exchange routing information and forward the received packets to carry each packet to its correct destination. This structure of the Internet as an interconnection of individual networks is the key to its scalability but is not sufficient to guarantee that a quickly growing number of users, services and traffic do not create bottlenecks that, if left unaddressed, can slow down performance. Bottlenecks may occur at many points in the core Internet and most of all in correspondence to peering points and backbones.

The network capacity is determined by the capacity of its cables and routers and although cable capacity is not an issue, the strongest limit to the backbone capacity comes from the packet-forwarding hardware and software of the routers. Once a peering point has been installed, traffic may have grown beyond expectations, resulting in a saturated link, typically because a network provider purchases just enough capacity to handle current traffic levels, to maximize the link utilization. The practice of running links at full capacity is one of the major causes of traffic bottlenecks showing very high utilization but also high rates of packet loss and high latency. Further the capacity of long backbones cannot always be adapted to the sudden and fast increases of the Internet traffic.

2.1 Move the Content to the Edges: An Approach to Improve the Internet Performance

The current centralized or partially distributed model of Internet content distribution requires that all user requests and responses travel several subnetworks and, therefore, traverse many possibly congested links. The first solution adopted to distribute the content trough the Internet consisted in mirroring. This technique statically replicates the web content in many locations across the Internet. Users manually select, from a list of servers, the best suited replica. The replica selection mechanism was automated and became transparent to the end-users with the introduction of the distributed web server systems [11][10].

With the introduction of proxy caching techniques to disseminate the content across the Internet, the bottlenecks at the server level and at the peering points were considerably reduced, though not ensuring a complete controllability of those systems by the content provider due to the absence of an intelligent and automated layer to perform server measures and request redirection. Proxy caching is only a semi-transparent mechanisms: the users, aware of the presence of a proxy server in their network, can/must configure their browser to use it, while the ISPs transparently manage their proxy caches. Large ISP proxy caches may also transparently cooperate with each other in a semi hierarchical structure. Proxy caches may experiment performance losses becoming themselves a bottleneck if there are frequent cache misses or cache inconsistencies. Besides this, proxy caches serve all requests independently of the required content, and do not prioritize users and QoS requirements.

In a CDN, by moving the content from multiple servers located at the edge of the Internet, a much more scalable model of distributing information and services to end-users is obtained, that is the so called edge delivery. In other words, a user would be able to find all requested content on a server within its home network. In this solution the requested content doesn't cross all the network before reaching its final destination, but only traverses the network part between the edge and the end-user. Further, cooperative access routers can be endowed with measure and server selection capabilities to perform a tradeoff solution between load balancing among the available servers and choosing the best suited replica to fulfil the agreements on quality of service.

2.2 The Features of a CDN

The design of a CDN requires, together with the distribution of replica servers at the edge of the network, a set of supporting services and capabilities. In order to be efficient for a significant number of users and for a considerably wide area, the edge servers must be deployed in thousands of networks, at different geographically spread locations. Optimal performance and reliability depend on the granularity of the distribution of the edge servers. The establishment of a CDN requires the design of some important features.

- **Replica placement mechanisms** are needed to decide the replica server locations and to adaptively fill them with the proper content prior to the request arrival (pre-fetching). Thus servers are not filled upon request like in traditional proxy caching, but are pro-actively updated, causing a one time offloading overhead that is not repeated for every access to the origin server. Adaptivity in replica placement is required to cope with changing traffic condition and is not related to a pull behavior like in traditional caching.

– **Content update mechanisms** must be provided to automatically check the host site for changes and retrieve updated content for delivery to the edges of the network, thus ensuring content freshness. Standard mechanisms adopted in proxy caching do not guarantee content freshness since content stored on standard cache servers does not change as the source content changes.

– **Active measurement mechanisms** must be added to cooperative access routers to have immediate access to a real-time picture of the Internet traffic, in order to recognize the fastest route from the requesting users to the replica servers in any type of traffic situations, especially in presence of "flash crowds", that is sudden heavy demand, expected or not, for a single site. A measurement activity is at the basis of the replica selection mechanism.

– **Replica selection mechanisms** must be added to cooperative access routers to accurately locate the closest and most available edge server from which the end users can retrieve the required content. A robust service must also keep its servers from getting overloaded by means of access control and load balancing.

– **Re-routing mechanisms** must be able to quickly re-route content requests in response to traffic bursts and congestion as revealed by the measurement activity.

Also, the CDN infrastructure, must allow the service providers to access directly the caches and control their consistency and to get the statistics information about the accesses to the site, available from the cooperative access routers.

3   Types of Content and Services in a CDN

CDN providers host third party contents to fasten the delivery of any type of digital content, e.g. audio/video streaming media, html pages, images, formatted documents or applications. The content sources could be media companies, large enterprises, broadcasters, web/Internet service provider. Due to the heterogeneous nature of the content to be delivered, various architectures and technologies can be adopted to design and develop a CDN. We now analyze the characteristics of the content and of the applications that most likely take advantages of a CDN architecture.

– **Static web based services.** Used to access static content (static html pages, images, document, software patches, audio and/or video files) or content that change with low frequency or timely (volatile web pages, stock quote exchange). All CDN provider (Akamai Inc., Speedera Inc., AT&T inc., Globix Inc. just to mention some) support this type of content delivery. This type of content can easily be cached and its freshness maintained at the edge using traditional content caching technologies.

- **Web storage services**. Essentially, this application can be based on the same techniques used for static content delivery. Additional features to manage logging and secure file transfer should be added. This type of application can require processing at the origin site or at the edge.
- **File transfer services**. World wide software distribution (patch, virus definition, etc.), e-learning material from an enterprise to all their global employees, movies-on-demand from a large media company, highly detailed medical images that are shared between doctors and hospitals, etc. All these content types are essentially static and can be maintained using the same techniques adopted for static web services.
- **E-commerce services**. The semantic of the query used in browsing a product catalogue is not complex, so frequent query results can be successfully cached using traditional DB query caching techniques[29][33]. Shopping charts can be stored and maintained at the replica server and also orders and credit card transactions can be processed at the edge: this requires trusted transaction-enabled replica servers. In [9] the authors propose a framework for enabling dynamic content caching for e-commerce site.
- **Web application**. Web transactions, data processing, database access, calendars, work schedules, all these services are typically characterized by an application logic that elaborates the client requests producing as results a dynamic web page. A partial solution to the employment of a CDN infrastructure in presence of dynamic pages is to fill the replica servers with the content that most frequently composes the dynamically generated web pages, and maintaining the application and its processing activity that produces the dynamic pages at the origin server. Another approach is to replicate both the application (or a portion of it) and the content at the edge server. In this way all the content generation process (application logic and content retrieval) are handled by the replica server thus offloading the origin server.
- **Directory services**. Used for access to database servers. For example, in the case of a LDAP server, frequent query results or a subsets of directories can be cached at the edge. Traditional DB query caching techniques [29] may be adopted.
- **Live or on-demand streaming**. In this case the edge server must have streaming capability. See section 3.1 for details.

Streaming media and application delivery are a challenge in CDN. A more detailed description of the solutions adopted for media streaming and dynamic contents can be found in the following subsections.

3.1   Streaming Media Content

Streaming media can be *live* and *on-demand*, thus a CDN needs to be able to deliver media in both these two modes. *Live* means that the content is delivered "instantly" from the encoder to the media server, and then onto the media client.

This is typically used for live events such as concerts or broadcasts. The end-to-end delay is at a minimum 20 seconds with today's technologies, so "live mode" is effectively "semi real-time". In *on-demand*, the content is encoded and then stored as streaming media files on media servers. The content is then available for request by media clients. This is typically used for content such as video or audio clips for later replay, e.g., video-on-demand, music clips, etc. A specialized server, called a media server, usually serves the digitalized and encoded content. The media server generally consists of media server software that runs on a general-purpose server. When a media client wishes to request a certain content, the media server responds to the query with the specific video or audio clip. The current product implementations of streaming servers are generally proprietary and demand that the encoder, server, and player all belong to the same vendor. Streaming servers also use specialized protocols (such as RTSP, RTP and MMS) for delivery of the content across the IP network. In [55] a typical streaming media CDN architecture is described. In streaming media CDNs a replica server must have, at least, the additional functionalities listed below.

- The ability to serve *live content* such as newscasts, concerts, or meetings etc. either in Multicast or Unicast mode.
- Support for delivery of stored or *on-demand content* such as training, archived meetings, news clips, etc.
- *Caching capability* of streaming media. Caching a large media file is unproductive, so typically media files are split in segment. Neighbor replica must be capable to share and exchange segment to minimize the network load and cache occupancy.
- *Peering capability* to exchange and retrieve content from the neighbor streaming cache in case of cache miss. Streaming cache node can be organized in a hierarchy.
- *Media transcoding functionality*, to adapt media streams for different client capabilities, e.g., low quality/bandwidth content to dial-up users, high quality/bandwidth to xDSL users.
- *Streaming session handoff* capability. The typically long life of a streaming session, in presence of user's mobility, causes the need for midstream handovers of streaming session between replica servers [5,49].

3.2 Web Application

Accessing dynamic content and other computer applications is one of the major challenges in CDN. CDN supporting this kind of content and services are also called Application Content Delivery Networks (ACDN). Some providers like AppStream Inc. and PIVIA Inc., implement ACDN using the so called "fat client" solution: the application is partitioned in "streamlets" or special applets and sent to the client. The client receives enough code to start the application and execute it, the other parts of the application are sent on demand. These solution use patented and proprietary technologies. Another approach is to migrate the application to the edge server using general utility such as Ajasent[1] and vMatrix[4]. However application replication may be expensive especially if performed on demand. A completely different solution is to automatically deploy the application at the replica server. In [47] the authors define an ACDN architecture relying on standard technologies such as HTTP protocol, web servers, CGI/FastCGI scripts or servlets. Rabinovich et al. define the additional capabilities of an ACDN in terms of: an *application distribution framework* capable to dynamically deploy the application at the edge and to keep the replica consistent, a *content placement mechanism* to decide where and when to deploy the application, a *request distribution mechanism* aware of the location of the involved applications.

4   Content Delivery Networks Architecture

The main goal of server replication in a CDN is to avoid large amounts of data repeatedly traversing possibly congested links on the Internet. As Figure 1 shows, there are a variety of ways and scale (local area or wide area networks) in which content networks may be implemented. Local solutions are web clusters, that typically hosts single site, and web farms, typically used to host multiple sites. Wide area solutions include: distributed web server systems, used to host single or multiple sites; cooperative proxy cache networks (a service infrastructure to reduce latency in downloading web objects) and content delivery networks [53] that are the focus of this paper.

A typical server farm is a group of servers, ranging from two to thousands, that makes use of a so-called cooperative dispatcher, working at OSI layers 4 and/or 7, to hide the distributed nature of the system, thus appearing as a single origin site. A layer 4 web switch dispatches the requests, among a group of servers, on the basis of network layer information such as IP address and TCP port. A content switch, working at the application layer, examines the content of requests and dispatches them among a group of servers. The goals of a server cluster/farm include: load-balancing of requests across all servers in the group; automatic routing of requests away from servers that fail; routing all requests

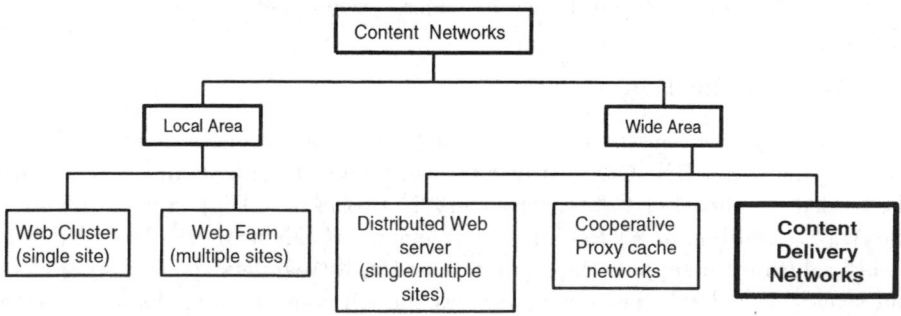

Fig. 1. Taxonomy of Content Networks

for a particular user agent's session to the same server, if necessary to preserve session state.

A type of content network that has been in use for several years is a caching proxy deployment. Such a network might typically be employed by an ISP for the benefit of narrow bandwidth users accessing the Internet. In order to improve performance and reduce bandwidth utilization, caching proxies are deployed close to the users. These users are encouraged to send their web requests through the caches rather than directly to origin servers, by configuring their browsers to do so. When this configuration is properly done, the user's entire browsing session goes through a specific caching proxy. This way the proxy cache would contain the hot portion of content that is being viewed by all the users of that caching proxy. A provider that deploys caches in many geographically locations may also deploy regional parent caches to further aggregate user requests thus creating an architecture known as hierarchical caching. This may provide additional performance improvements and bandwidth savings. Using rich parenting protocols, redundant parents may be deployed such that a failure in a primary parent is detected and a backup is used instead. Using similar parenting protocols, requests may be partitioned such that requests for certain content domains are sent to a specific primary parent. This can help to maximize the efficient use of caching proxy resources. Clients may also be able to communicate directly with multiple caching proxies.

Though certainly showing better scalability than a single origin server, both hierarchical caching and server farms have their limits. In these architectures, the replica servers are typically deployed in proximity to the origin server, therefore they do not introduce a significant improvement to the performance difficulties that are due to the network congestion. Caching proxies can improve performance difficulties due to congestion (since they are located in proximity to the final users) but they cache objects reactively to the client demand. Reactive caching based on client demand performs poorly if the requests for a given object, while numerous in aggregate, are spread among many different caching proxies.

To address these limitations, CDNs employ a solution based on proactive rather than on reactive caching, where the content is prefetched from the origin server and not cached on demand. In a CDN, multiple replicas host the same content. A request from a browser for a single content item is directed to the replica that is considered the best suited at the moment of the request arrival, and the item is served to the client in a shorter time than the one it would have taken to fetch it from its origin server. Since static information about geographic locations and network connectivity are not sufficient to choose the best replica, a CDN typically incorporates dynamic information about network conditions and load on the replicas, to redirect requests and balance the load among the servers. Operating a CDN is therefore a complex and expensive activity. For this reason a CDN is typically built and operated by a network/service provider that offers a content distribution service to several content providers.

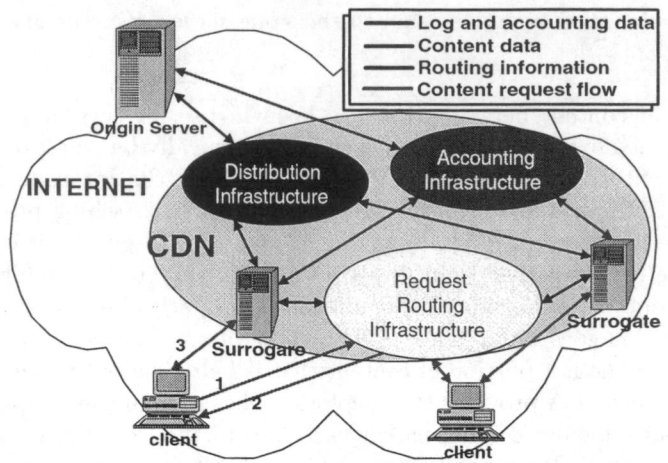

Fig. 2. Infrastructure components of a Content Delivery Network

A content delivery architecture consists of a set of **surrogate servers** that deliver copies of content to the users while combining different activities (see figure 2).

- the **request-routing** infrastructure consists of mechanisms to redirect content requests from a client to a suitable surrogate.
- the **distribution** infrastructure consists of mechanisms to move contents from the origin server to the surrogates.
- the **accounting** infrastructure tracks and collects data on request-routing, distribution, and delivery functions within the CDN creating logs and reports of distribution and delivery activities.

The **origin server** (hosting the content to be delivered) interacts with the CDN in two ways (see figure 2):

- it pushes new content to the replica servers, (the replica themselves request content updates from the origin server through the distribution infrastructure);
- it requests logs and other accounting data from the CDN or the CDN itself provides this data to the origin server through the accounting infrastructure.

The **clients** interact with the CDN through the request routing infrastructure and surrogate servers. Figure 2 shows one of the possible scenarios of interaction between the clients, the access routers, the replica servers and the origin server.

The user agent sends (1) a content request to the routing infrastructure, that redirects (2) the client request to a surrogate server, to which the client subsequently asks (3) the desired content.

5 Replica Placement and Management

5.1 Content Caching Techniques

The proactive caching infrastructure must be transparent to the end-users that must see no difference with being served directly by the central server. Proactive caching to the edges offers better delivery to the client because the content is located in their proximity. Therefore the requesting users perceive a lower latency, higher availability and lower load on the network links. Such architecture is also inherently protected from sudden burst that can be distributed among many servers so that no single device has to cope with a massive load. This close-to-the-client deployment mode is commonly known as forward proxy caching. Forward proxy implementations can reduce wide area network traffic by 30 to 50 percent (results vary based on the "cacheability" of the requested content). A Web cache monitors Internet traffic, intercepts requests for Web objects and then fulfils those requests from the set of objects it stores (*cache hit*). If the requested object is not in the cache (*cache miss*), the cache forwards the request to the origin server, that sends a copy of the object back to the cache. The cache store the object and sends it back to the requester. Caches in CDN cooperate interacting through the Internet Cache Protocol (ICP). ICP is typically used to build cache clusters or child-parent relationships in hierarchical caching[15] [44]. A cache can react to a cache miss inquiring other cooperative caches, in spite of the origin server, in order to retrieve the content from a closer location. Caching also acts as a point of control and security. Today's caches frequently include support for content filtering, anti-virus, access control and bandwidth management. Anti-virus and content filtering give users an extra level of security across the network. The access control and bandwidth management further assists in the reduction of the overall network utilization by making sure that only approved users get access to bandwidth, and that the bandwidth is being allocated in a way that ensures the best adherence to the signed agreements on quality. Caching activity in a CDN may involve different types of contents and therefore different functionalities.

Static Caching: to cache and replicate static content, such as html pages, images, documents, audio/video file etc.

Dynamic Caching: to cache and replicate dynamically generated content. This include application delivery and replication.

Streaming Media Caching: to store streaming media objects, as well as to serve streaming media to clients. Essentially, the cache acts as a streaming media server, storing media clips for later use.

Live Splitting: to cache replicated live streams, so that only one copy is pulled down from the upstream server and is then distributed to the subscribing clients.

5.2 Replica Placement

An hot topics in content delivery design is the replica placement problem: where and how the replica could be distributed across the Internet to minimize the user latency, the number of replica, and the bandwidth used for replica management?

The majority of the schemes presented in the literature tackle the problem of static replica placement that can be formulated as follows. Given a network topology, a set of CDN servers and a given request traffic pattern, decide where content has to be replicated so that some objective function is optimized while meeting constraints on the system resources. The solutions so far proposed typically try to either maximize the user perceived quality given an existing infrastructure, or to minimize the CDN infrastructure cost while meeting a specified user perceived performance. Examples of constraints taken into account are limits on the servers storage, on the servers sustainable load, on the maximum delay tolerable by the users etc.

A thorough survey of the different objective functions and constraints considered in the literature can be found in [37]. For the static case, simple efficient greedy solutions have been proposed in [46], [35] and [48]. In [46] Qiu et al. formulate the static replica placement problem as a minimum K median problem, in which K replicas have to be selected so that the sum of the distances between the users and their 'best replica' is minimized. In [35] and [48] Jamin et al. and Radoslavov et al. propose fan-out based heuristics in which replicas are placed at the nodes with the highest fan-out irrespective of the actual cost function. The rationale is that such nodes are likely to be in strategic places, closest (on average) to all other nodes, and therefore suitable for replica location. In [48] a performance evaluation based on real-world router-level topology shows that the fan-out based heuristic has behavior close to the greedy heuristic in terms of the average client latency. In both [18] and [41] the authors consider the problem of placing replicas for one origin server on a tree topology. In [40] the authors also consider very simple topology like rings and lines and tree topologies while considering the placement of intercepting proxies inside the network to reduce download time. In [36] the problem of optimally replicating objects in CDN servers is analyzed. All these solutions lack in considering the dynamics of the system (e.g. changes in the requests traffic pattern, network topology, replica sites).

In [6] a different approach is proposed and a dynamic allocation strategy is considered which explicitly takes into account the system dynamics as well as the costs of modifying the replica placement. By assuming the users requests dynamics to obey to a Markovian model a formulation of the dynamic replica placement problem as a Markovian decision process is obtained. Albeit this model may not accurately capture the user dynamics and can be numerically solved only for limited sized CDNs, it allows us to identify an optimal policy for dynamic replica placement that can be used as a benchmark for heuristics evaluation and provides insights on the formulation of a conservative placement heuristic.

The solution of replica placement for heavy contents, like video streaming, makes it impossible storing the entire content of several long streams because it would exhaust the capacity of a conventional cache. In this case, not only the content must be replicated and distributed to replicas, but also in a way that avoids to overload servers. To address this problem, in [51,54] the authors propose a prefix caching technique whereby a proxy stores the initial frames of popular clips. Upon receiving request for the stream, the proxy initiates transmission to the client and simultaneously requests the remaining frames from the server. In addition to hiding the delay, throughput and loss effects of a weaker service model between the server and the proxy, this caching technique aids the proxies in performing work ahead smoothing into the client playback buffer, by transmitting large frames in advance of each burst. The prefix caching techniques reduces the peak and variability of the network resources requirements along the path from the proxy to the client.

5.3 Cache Consistency and Content Flow

One of the important problems in CDNs is how to manage the consistency of content at replicas with that at the origin server, especially for those documents changing dynamically. Cached objects typically have associated expiration times after which they are considered stale and must be validated with a remote server (origin or another cache) before they can be sent to a client. Sometimes, a considerable fraction of cache hits involve stale copies that turned out to be current. These validations of current objects have small message size, but nonetheless, they often induce latency comparable to cache misses. Thus the functionality of caches as latency-reducing mechanism highly depends not only on content availability but also on its freshness.

A technique to achieve cache consistency consists in pre-populating, or pushing, content to the cache before requests arrive. When automatically pushing a new, or updated, Web object to a cache, the content in the cache is guaranteed to be always fresh and there is no reason for the cache to initiate a freshness check with the side effect that this technique often generates a large amount of traffic.

In [21] the authors propose policies for populating caches to proactively validate selected objects as they become stale, and thus allow for more client requests to be processed locally. Pre-populating content takes on even more importance with broadband content. The size of rich content files can be huge and increasing every day. Compression technologies have been invented specifically for these new and emerging types of media. The load limits for servers need to be established by means of load testing of the specific environment. However, with content pre-populating a high resolution file can be pushed across low speed lines directly to the Web cache in the branch office and then serve that streaming file at the top speed of your LAN. In the traditional propagation approach the updated version of a document is delivered to all replicas whenever a change is made to the document at the origin server. It may generate significant levels of unnecessary traffic if documents are updated more frequently than accessed.

Another approach is invalidation, in which an invalidation message is sent to all replicas when a document is changed at the origin server. This approach doesn't make full use of the distribution networks for content delivery and each replica needs to fetch an updated version individually at a later time. This can also lead to inefficiency in managing consistency at replicas.

In [27] the author propose a hybrid approach that generates less traffic than the propagation and the invalidation approach. The origin server makes the decision of using either propagation or invalidation method for each document, based on the statistics about the update frequency at the origin server and the request rated collected by replicas. They develop a technique that can reduce the burden of request rate collection at replicas and avoid the implosion problem when replicas send the statistics to the origin server.

Another main focus in CDNs research activity is how content at the origin servers have to be delivered to replicas. Two common approaches to this problem are to deliver data over N unicast channels, or over an application-level (tunnelled) multicast tree that connects the replicas [16,17]. Basically they run an auto-configuration protocol to establish a delivery structure of tunnelled topology among participating members. These approaches consist in building a mesh topology first and running the spanning tree algorithm to select a delivery tree. Intuitively, the unicast approach wastes network bandwidth and can cause congestion at bottleneck links, while application-level multicast approach is more efficient in delivery (although not as efficient as native IP multicast).

Another issue in cache management is the propagation of changes (adds and deletes of content). When content is added to the source-tree, it must also become available throughout the CDN. If there are delays involved in propagating the content, through all the CDN replicas, the content providers must be aware of that there may be periods in which contents may be inconsistent and even unavailable. For example, if a film studio wishes to make a new movie available to their global customer base, they need to know how long time it takes to make it globally available. This way, they can make sure that they do not start selling it until everyone can access it. Similarly, when the customer deletes content from their file area, it should also be expired from the caches and devices within the CDN.

6   Measurements Techniques for Request Routing

Request routing systems can use a variety of metrics in order to determine the best surrogate that can serve a client's request. The decentralized nature of the Internet makes quantitative assessment of network performance very difficult. Collecting network statistics directly on network devices (router and server) could be more expensive in terms of system performance. So typically, the acquisition of network statistics relies on the use of a combination of active network probing methods, passive traffic monitoring and feedback from surrogate servers. For deeper details of how measures can be inferred and network tomography can be done see also [7,20]. In CDN networks it is possible to combine multiple metrics using both proximity concept and surrogate feedback for best surrogate selection. Performance measurement is often a component of network management systems and offers the ability to monitor, understand, and project end-to-end performance of the CDN. Moreover one would need to measure both the internal performance, as well as the performance from the customer perspective. Typical parameters that would be useful to measure are: packet-loss and latency for all type of content and in particular for streaming content average bandwidth, startup time and frame rate. By deploying hardware-based or software probes, strategically throughout the network, one could correlate the information collected by the probes with the cache and server logs to determine delivery and QoS statistics. The most useful place to put the probes is at the edges of the network, thus measuring the performance as perceived by the end-users throughout the CDN. Network and geographical proximity measurements can be used by the request routing system to direct users to the "closest" surrogate. Furthermore, proximity measurements can be exchanged between surrogates and the requesting entity. In many cases, proximity measurements are "one-way" in that they measure either the forward or reverse path of packets from the surrogate to the requesting entity. This is important as many paths in the Internet are asymmetric. In order to obtain a set of proximity measurements, a network may employ active probing techniques and/or passive measurement techniques. The request-routing system can use also feedback from surrogates in order to select a "least-loaded" delivery node. Feedback can be delivered from each surrogate or can be aggregated by site or by location. We now discuss in detail about passive measurement, active probing and feedback information

Passive Measurement. Passive measurements could be obtained when a client performs data transfers to or from a surrogate. Once the client connects, the actual performance of the transfer is measured. This data is then fed back into the request routing system. An example of passive measurement is to watch the packet loss from a client to a surrogate, or the user perceived latency by observing TCP behavior. Basically, a good mechanism is needed to ensure that not every surrogate is tested per client in order to obtain the data. In [52] the authors proposed a system based on passive measurements of the network performance to be used with adaptive applications.

Active Probing. Active probing is when past or possible requesting entities are probed using one or more techniques to determine one or more metrics from each surrogate or set of surrogates. An example of a probing technique is an ICMP ECHO request that is periodically sent from each surrogate or set of surrogates to a potential requesting entity. Active probing techniques are limited for some reasons. Measurements can only be taken periodically and should not have a perceivable load since it cannot influence the measured traffic, firewalls and NATs disallow probes, and last, probes often cause security alarms to be triggered on intrusion detection systems.

Feedback information. These information may be obtained by periodically probing a surrogate by issuing application specific requests (e.g. HTTP) and taking related measures. The problems with probing for surrogate information is that it is difficult to obtain "real-time" information and the non-real-time information are sometimes inaccurate and obsolete. Consequently, feedback information can be obtained by agents that reside on surrogates that can communicate a variety of metrics about their nodes. There are two methods to obtain feedback information: static, in which the route that minimize the number of hops or to optimize other static parameters is selected [25][50]; dynamic probing (Real Time probing) allow to compute round trip time or other QoS parameters in "real time" [25][22]. Hybrid methods are also used to obtain other useful feedback information [25][26][43].

6.1   Metrics for Request Redirection

Replica server selection is performed with the goal to minimize some performance parameters perceived by the end-users. In this section we give a classification of the metrics that can be adopted to measure the network and systems performance in a CDN to decide where to redirect client's requests.

Geographical proximity is often used to redirect all users within a certain region to the same POP.

A measure of the *network proximity* is typically derived through active probing of the BGP routing table.

The ability to select the POP that shows the lowest latency can be obtained by enhancing the request redirection systems with active probing and passive measurement mechanisms, to maintain knowledge of *response time*.

The *server load state* can be computed, using SNMP or feedback agents on the server side, on the basis of the load state of server components (CPU, disk, memory, network interfaces) or on the basis of some aggregate performance measure, such as the server throughput or server response time.

All these measures may be relevant information to feed the server selection mechanism, combined with the knowledge of the *users identity*, that is intended to classify the user priority in accessing contents and services. As an example, paying customer may get access to better service than non-paying. The identity of paying users can be revealed by means of a cookie retrieved from the client system, or through an authentication process.

Figure 3 summarizes the above performance metrics classification.

7   Request-Redirection Mechanisms

A key challenge in CDN design is to realize efficient request-redirection service that tracks the servers where data objects are replicated and assigns each client request to a server that can offer the "best" service, this process is called server selection. Server selection algorithms include criteria like network topology, server availability and server load [13]. The ability to quickly locate replicas and perform request distribution has critical implications for the user-perceived response time. Figure 4 illustrates a high-level view of the request-redirection process: 1) the client requests a given content, e.g. a streaming file, residing at www.site.com; 2) since site.com doesn't host the requested file, but uses cdn.com as its CDN provider, the request is redirected to the cdn.com site. 3-3') By using

Metrics	Goals	Measurement Techniques
Latency	Select replica with lowest delay	Active Probing / Passive Measurement
Packet loss	Select path with lowest error rate (useful for streaming traffic)	Active Probing / Passive Measurement (TCP header info)
Network proximity	Select the shortest path	Active Probing
Avg. Bandwidth	Select the best path for streaming traffic	
Sturtup Time		
Frame Rate		
Geographical proximity	Redirect requests from a region to the same POP	IP header information, bind information
CPU load, net. interface load, active connection, storage I/O load	Select the server with the aggregated least load	feedback agents/active probing

Fig. 3. Metrics used in replica selection

some redirection algorithm, the media client gets redirected to the most appropriate replica. If the client has a CDN replica directly placed at its ISP network that is capable of guaranteeing the fulfillment of the SLA, this replica is selected (3) first (e.g. CDN cache-2), otherwise another one (e.g. CDN cache-1) is selected (3'). 4-4') The selected CDN replica serves the content to the client.

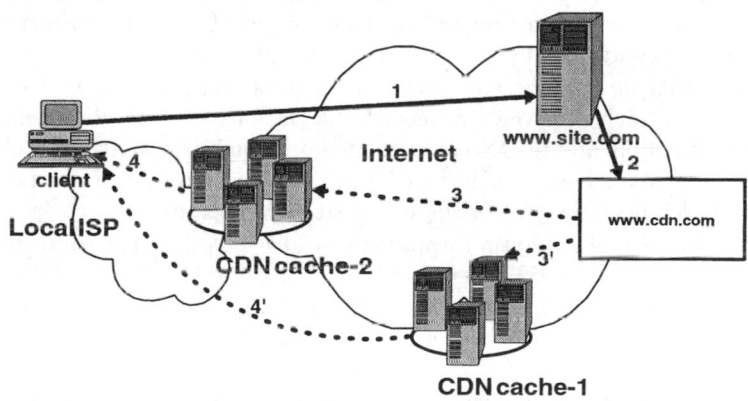

Fig. 4. The request-redirection process

Various schemes have been proposed in research literature to perform redirection at the *IP level* through some address packet rewriting scheme [24,34], at the *DNS level* through mapping of URL-name to IP-address of one of the servers in a cluster [19,3,56], or at the *application level* [2] using the redirection features of the HTTP protocol. In [23] the authors propose a hybrid scheme based on adaptive replication of the entries of the location directory that provides the redirection service. Network level support enables client requests to be redirected to appropriate servers in a quick and efficient manner. In [12] the authors focus on an alternative architecture that integrates DNS dispatching mechanism with a HTTP redirection technique carried out by Web servers. In the following sections we describe in details the main request-redirection mechanisms.

7.1   DNS Based Request Routing

Today commercial content distribution services rely on modified Domain Name System servers to dynamically redirect clients to the appropriate content server. This is the simplest form of redirection, according to which a domain name, e.g. www.cdn.com has multiple IP records attached to it. When a client requests the IP address of the domain, any one of the IP records from the pool will be selected based on the DNS action.

The popularity of DNS based request-routing techniques is mainly due to the ubiquity of DNS as a directory service. They mainly consist in inserting a specialized DNS server in the name resolution process. This specialized server is capable of returning a different set of A, NS or CNAME records based on user defined policies, metrics, or a combination of both.

In the **single reply** approach, the DNS server returns the IP address of the best surrogate in an A record to the requesting DNS server (client site DNS server's). The IP address of the surrogate could also be a virtual IP address of the best set of surrogates for requesting DNS server. The best surrogate will be selected, in a second step, by a server switching mechanism.

In the **multiple replies** approach, the request-routing DNS server returns multiple replies such as several A records for various surrogates. Common implementations of client site DNS servers cycle through the multiple replies in a round robin fashion. The order in which the records are returned can be used to direct multiple clients using a single client site DNS server.

A **multiple-level** resolution approach is also possible according to which multiple request-routing DNS servers can be involved in a single DNS resolution, thus demanding complex decisions from a single server to multiple, more specialized, request-routing DNS servers, disseminated in different points of the Internet.

The most common mechanisms used to insert multiple request-routing DNS servers in a single DNS resolution is the use of NS and CNAME records: NS records allow the DNS server to redirect the authority of the next level domain to another request-routing DNS server; CNAME records allow the DNS server to redirect the resolution request to an entirely new domain.

There are three main drawbacks of using NS records. First the number of request-routing DNS servers are limited by the number of parts in the DNS name; second the last DNS server can determine the Time To Live (TTL) of the entire resolution process. The client will cache the returned NS record and use it for further request resolutions until it expires. As a third drawback, a delay is added in the resolution process due to the use of multiple DNS servers.

Request-routing based on CNAME record has the advantage to redirect the resolution process to another domain, and the number of request-routing DNS servers is independent of the format of the domain name. The main disadvantage is the introduction of an additional overhead in resolving the new domain name.

The basic limitations of DNS based request-routing techniques can be summarized as follows below:

- DNS allows resolution only at the domain level. However, an ideal request resolution system should serve requests at object granularity (preserving sessions if needed).
- A short TTL of DNS entry allows to react quickly to network outages. This in return may increase the volume of requests to DNS servers. Therefore many DNS implementations do not honor the DNS TTL field.
- DNS request-routing does not take into account the IP address of the clients. Only the Internet location of the client DNS server is known: this limits the ability of the request-routing system to determine a client's proximity to the surrogate.
- Users that share a single client site DNS server will be redirected to the same set of IP addresses during the TTL interval. This might lead to overloading the surrogate during a flash crowd.

7.2   Transport Layer Request Routing

At the transport-layer, finer levels of granularity can be achieved by means of a closer inspection of client's requests. This level provides information about the client's IP address, TCP port, and other layer 4 header information. These data could be used in conjunction with other load state metrics to select the surrogate that is better suited to serve a given request. In general, the forward-flow traffic (client to newly selected surrogate) will flow through the request routing server or through a first step surrogate originally chosen by the DNS. The reverse-flow traffic (surrogate to client), which normally transfers much more data than the forward-flow, would typically take the direct path from the servant surrogate. The overhead associated with transport-layer request-routing makes it better suited for long-lived sessions such as file transfer (FTP), secure transactions (TLS, SSL), streaming (RTP, RTSP). However, transport-layer request-routing could also be used to redirect clients away from overloaded surrogates. A first course grain replica selection is operated by a DNS request router. The selected server, may operate a more accurate refinement of replica selection, based on local and server side information, using the transport-layer request-routing mechanism.

7.3   Application-Layer Request Routing

With application layer request-routing a fine-grained control, at the level of individual objects composing the multimedia content, can be achieved. The process could be performed, in real time, when the object request reaches the content server or a switching element. In most cases the header of the client's packet contains enough information to perform request routing. Some application level protocols such as HTTP, RTSP, and SSL provide necessary information in the initial portion of the session about how the client request must be directed. Application level protocols such as HTTP and RTSP may describe the requested content by means of its URL, other redirection information come from other parts of the MIME request header such as Cookies. In many cases the URL is sufficient to disambiguate the content and suitably redirect the request.

Header Inspection. This approach is based on the inspection of the header of client requests. In a first solution, also known as *302 redirection* [28], the client's request is first resolved to a virtual surrogate that returns an application-specific code, such as the 302 in the case of HTTP or RTSP, to redirect the client to the delegate content delivery node. The application layer redirection is relatively simple to implement. Nevertheless, this approach introduces an additional latency involved in sending the redirect message back to the client. Another approach, known as the *In-Path element* considers a network element, in the forwarding path of the client's request, that provides transparent interception of the transport connection. This In-Path network element, establishes a connection with the client, examines the client's content request and performs request-routing decisions. Then, the client connection is joined to a connection with the appropriate content delivery node. Drawbacks are the delay introduced for URL-parsing and the possible bottleneck introduced by the In-Path element.

Content modification. This technique enables a content provider to take direct control over request-routing decisions without the need for specific switching devices or directory services in the path between the client and the origin server. Basically, a content provider can directly communicate to the client the best surrogate that can serve the request. Decisions about the best surrogate can be made on a per-object basis or it can depend on a set of metrics. In general, the method takes advantage of content objects that consist of basic structure that includes references to additional, embedded objects. For example, most web pages, consist of an HTML document that contains plain text together with some embedded objects (e.g. GIF, JPEG images or PDF documents) referenced using embedded HTML directives. In general embedded objects are retrieved from the origin server. A content provider can now modify references to embedded objects such that they could be fetched from the best surrogate.

Pro-active URL Rewriting. According to this scheme, a content provider formulates the embedded URLs of a main html page before the content is loaded on the origin server. In this case, URL rewriting can be done either manually or by using software tools that parse the content and replace embedded URLs. Since these scheme consists in rewriting URLs in a proactive way, it cannot take into consideration client specific information while performing request routing. However, it can be used in combination with DNS request routing to direct related DNS queries into the domain name space of the service provider. Dynamic request routing based on client specifics are then done using the DNS approach.

Reactive URL Rewriting. This dynamic scheme consists in rewriting the embedded URLs of a html page when the client request reaches the origin server. In spite of the previous scheme, this one has the possibility to consider the identity of the client when rewriting the embedded URLs. In particular, an automated process can determine, on-demand, which surrogate would serve the requesting client best. The embedded URLs can then be rewritten to redirect the client to retrieve the objects from the surrogate that can fulfill the client request better than the other, with a consideration of the specific location and priority of the considered client.

A drawback of content modification based request-routing is that the first request, from a client to a specific site, must be served from the origin server. Besides, content that has been modified to include references to nearby surrogates rather than to the origin server should be marked as non-cacheable. To reduce this limitation, such pages can be marked to be cacheable only for a relatively short period of time. However, rewritten URLs on cached pages can cause problems, because they can get outdated and point to surrogates that are no longer available or no longer the best choice.

7.4   Anycast

This solution aims at solving the request-routing problem at the IP packet routing level. The base principle is that a group of servers providing the same service can be addressed using an anycast name and an anycast address[38,14]. A user willing to access some service, e.g., a given content, from any of the (equivalent) servers issues a request with the anycast name. This is mapped to the anycast address and the request is sent into the network with the anycast address as destination. The role of the anycast service, is to redirect these request to one of the servers, thus selecting the server which will serve the user request. Since the redirection system has the obvious goal of improving clients performance, redirection is based upon some performance criteria, e.g. minimize the user perceived response time. The redirection is therefore performed by a cooperative access router that is capable of selecting the best suited replica from the anycast table. Anycasting is a more elaborate location mechanism that targets network-wide replication of the servers over potentially heterogeneous platforms. A mechanism for request redirection that is based on the anycasting concept must allow for the maintenance of information about the servers state and performance. An anycast service can be implemented at different levels in the network protocol stack. At the network layer, anycasting mechanisms consist in associating a common IP anycast address with the group of replicated servers. The routing protocol routes datagrams to the closest server, using the routing distance metric. Standard intra-domain unicast routing protocols can accomplish this, assuming each server advertises the common IP address. In [45] the authors propose a solution to the server selection problem by introducing the idea of anycasting at the network layer. This work established the semantics of anycasting service within the

Internet. At the network level the implementation of an anycast service entails the following mechanisms:

- Anycast request interception, that maybe dealt at the network level by the edge routers. Edge routers are set to filter packets with anycast destination address.
- Anycast name to Anycast address translation mechanism.
- Server Selection that maybe dealt by an edge router module (or by an application interacting with the router). This module interacts with the measurement module and keeps and updates anycast server performance metrics.
- Anycast address to IP address translation that maybe dealt at the network level by the edge routers. Upon receiving an anycast address, edge routers perform redirection by translating them into a selected unicast address.
- Measurements collection to be used for the selection process itself.

Application layer anycast implementation is also proposed in [27,57]. The authors claim that at the application layer a better flexibility can be achieved if compared to the network layer implementation. At the application level the resolution of the anycast address is performed by means of a hierarchy of anycast resolvers that map the anycast domain name (AND) onto an IP address. To perform the mapping the resolvers maintain two types of information: 1) the list of IP addresses that form particular anycast groups, and 2) a metric database information associated with each member of the anycast group, while authoritative resolvers maintain the definitive list of IP addresses for a group, whereas local resolvers cache this information.

8   Conclusions

In this paper we have introduced the design principles at the basis of Content Delivery Networks (CDN). CDNs rely on a proactive distribution of content replicas to geographically distributed servers close to the edge of the network, in proximity to the end users. The redirection of a request to the best suited replica is performed by cooperative access routers that are capable of taking measures regarding the performance of the available replica servers, and performing the replica selection and the request redirection to the selected replica. CDNs are therefore complex systems. Their understanding and design involves knowledge in many research fields: internet traffic measurement, content caching, request rerouting at various communication protocol levels, request admission control, load balancing and more, depending on the type of content and application considered. We gave a high level description of all the mechanisms and technologies used in CDN. We first introduced the main motivations for content delivery and explored what types of content and services may beneficiate from the introduction of a CDN architecture. Then we focused on the main research problems related to the CDN design, in particular, the problems of replica placement and management, server selection and request redirection have been analyzed in more details.

References

1. Utility computing: Solutions for the next generation it infrastructure. www.ajasent.com/technology/whitepapers2.html.
2. Winddance networks corporation. http://www.winddancenet.com.
3. D. Andersen, T. Yang, V. Holmedahl, and O. Ibarra. Sweb: Toward a scalable world wide web server on multicomputers. In *Proceedings of International Parallel Processing Symposium*, April 1996.
4. A. Awadallah and M. Rosenblum. The vmatrix: A network of virtual machine monitors for dynamic content distribution, 2002.
5. N. Bartolini, E. Casalicchio, and S. Tucci. Mobility-aware admission control in content delivery networks. In *Proceedings of IEEE/ACM MASCOTS*, Orlando, Florida, October 2003.
6. N. Bartolini, F. Lo Presti, and C. Petrioli. Optimal dynamic replica placement in content delivery networks. In *Proc of the 11th IEEE Int. Conference on Networks (ICON)*, Sydney, Australia, Sept. 2003.
7. T. Bu, N. Duffield, F. Lo Presti, and D. Towsley. Network tomography on general topologies. In *Proc of ACM SIGMETRICS*, Marina del Rey, California, June 2002.
8. M. Calzarossa. Performance Evaluation of Mail Systems. In M. Calzarossa and E. Gelenbe, editors, *Performance Tools and Applications to Networked Systems*, volume 2965 of *Lecture Notes in Computer Science*. Springer, 2004.
9. K. S. Candan, W.-S. Li, Q. Luo, W.-P. Hsiung, and D. Agrawal. Enabling dynamic content caching for database-driven web sites. In *Proc. of ACM/SIGMOD Conference 2001, Santa Barbare, California, USA*, May 2001.
10. V. Cardellini, E. Casalicchio, M. Colajanni, and P. Yu. The state of the art in locally distributed web-server systems. *ACM Computing Surveys*, 34(2):263–311, 2002.
11. V. Cardellini, M. Colajanni, and P. Yu. Dynamic load balancing on web-server systems. *IEEE Internet Computing*, 3(3):28–39, May-June 1999.
12. V. Cardellini, M. Colajanni, and P. Yu. Redirection algorithms for load sharing in distributed web server systems. In *Proc. IEEE 19th Int'l Conf. on Distributed Computing Systems (ICDCS)*, 1999.
13. R. L. Carter and M. E. Crovella. On the network impact of dynamic server selection. *Computer Network*, 1999.
14. M. Castro, P. Druschel, A.-M. Kermarrec, and A. Rowstron. Scalable application-level anycast for highly dynamic groups.
15. A. Chankhunthod, P. B. Dansig, C. Neerdaels, M. F. Schwartz, and K. J. Worrel. A hierarchical internet object cache. In *Proceedings of USENIX*, January 1996.
16. Y. Chawathe, S. McCanne, and E. A. Brewer. An architecture for internet content distribution as an infrstructure service. not published
17. Y. Chu, S. Rao, and H. Zhang. A case for end system multicast. In *Proceedings of ACM Sigmetrics*, June 2000.
18. I. Cidon, S. Kutten, and R. Soffer. Optimal allocation of electronic content. In *Proceedings of IEEE Infocom*, 2001.
19. Cisco. Cisco's distributed director. http://www.cisco.com/warp/public/cc/pd/cxsr/dd/index.shtml.
20. M. Coates, A. O. Hero, R. Novak, and B. Yu. Internet tomography. *IEEE Signal Processing Magazine*, May 2002.
21. E. Cohen and H. Kaplan. Refreshment policies for web content caches. In *Proceedings of IEEE Infocom*, 2001.

22. M. E. Crovella and R. L. Carter. Dynamic server selection in the internet. In *Proceedings of the Third IEEE Workshop on the Architecture and implementation of High Performance Communication Subsystems, HPCS*, 1995.
23. K. Dasgupta and K. Kalpakis. Maintaining replicated redirection services in web-based information systems. In *Proceedings of the 2nd IEEE Workshop on Internet Applications*, 2001.
24. D. M. Dias, W. Kish, R. Mukherhee, and R. Tewari. A scalable and high available web server. In *Proceedings of the 41st IEEE Computer Society International Conference*, 1996.
25. S. G. Dykes, K. A. Robbins, and C. L. Jeffery. An empirical evaluation of client-side server selection algorithm. In *Proceedings of IEEE Infocom*, 2000.
26. EICE. Internet qos measurement for the server selection. Technical report, Technical Report of EICE, CQ2000-48, 2000.
27. Z. Fei, S. Bhattacharjee, E. Zegura, and M. H. Ammar. A novel server selection technique for improving the response time of a replicated service. In *Proceedings of Infocom'98*, 1998.
28. R. Fielding, J. Gettys, J. Mogul, H. Frystyk, L. Masinter, P. Leach, and T. Berners-Lee. Rfc2616: Hypertext transfer protocol – http/1.1. Rfc, June 1999.
29. D. Florescu, A. Y. Levy, and A. O. Mendelzon. Database techniques for the world-wide web: A survey. *SIGMOD Record*, 27(3):59–74, 1998.
30. E. Gelenbe, K. Hussain, and V. Kaptan. Enabling Simulation with Augmented Reality. In M. Calzarossa and E. Gelenbe, editors, *Performance Tools and Applications to Networked Systems*, volume 2965 of *Lecture Notes in Computer Science*. Springer, 2004.
31. E. Gelenbe, R. Lent, M. Gellman, P. Liu, and P. Su. CPN and QoS Driven Smart Routing in Wired and Wireless Networks. In M. Calzarossa and E. Gelenbe, editors, *Performance Tools and Applications to Networked Systems*, volume 2965 of *Lecture Notes in Computer Science*. Springer, 2004.
32. S. Gilmore, J. Hillston, and L. Kloul. PEPA Nets. In M. Calzarossa and E. Gelenbe, editors, *Performance Tools and Applications to Networked Systems*, volume 2965 of *Lecture Notes in Computer Science*. Springer, 2004.
33. R. Goldman and J. Widom. WSQ/DSQ: A practical approach for combined querying of databases and the web. In *SIGMOD Conference*, pages 285–296, 2000.
34. G. Hunt, G. Goldzmidt, R.P.King, and R. Mukherjee. Network dispatcher: A connection router for scalable internet services. In *Proceedings of 7th International World Wide Web Conference*, April 1998.
35. S. Jamin, C. Jin, A. R. Kurc, D. Raz, and Y. Shavitt. Constrained mirror placement on the internet. In *INFOCOM*, pages 31–40, 2001.
36. J. Kangasharju, J. Roberts, and K. W. Ross. Object replication strategies in content distribution networks. *Computer Communications*, 25(4):367–383, March 2002.
37. M. Karlsson, C. Karamanolis, and M. Mahalingam. A unified framework for evaluating replica placement algorithms. *Technical Report HPL-2002, Hewlett Packard Laboratories*.
38. D. Katabi and J. Wroclawski. A framework for scalable global IP-anycast (GIA). In *SIGCOMM*, pages 3–15, 2000.
39. A. Klemm, C. Lindemann, and O. Waldhorst. Peer–to–Peer Computing in Mobile Ad Hoc Networks. In M. Calzarossa and E. Gelenbe, editors, *Performance Tools and Applications to Networked Systems*, volume 2965 of *Lecture Notes in Computer Science*. Springer, 2004.

40. P. Krishan, D. Raz, and Y. Shavitt. The cache location problem. *IEEE/ACM Transactions on Networking*, October 2000.

41. B. Li, J. Golin, G. F. Italiano, and X. Deng. On the optimal placement of web proxies in the internet. In *Proceedings of IEEE Infocom*, 1999.

42. P. Lorenz. IP–Oriented QoS in the Next Generation Networks: Application to Wireless Networks. In M. Calzarossa and E. Gelenbe, editors, *Performance Tools and Applications to Networked Systems*, volume 2965 of *Lecture Notes in Computer Science*. Springer, 2004.

43. K. Mase, A. Tsuno, Y. Toyama, and N. Karasawa. A web server selection algorithm using qos measurement. In *Proceedings of International Conference on Communication*, 2001.

44. S. Michel, K. Nguyen, A. Rosenstein, L. Zhang, S. Floyd, and V. Jacobson. Adaptive web caching: Towards a new caching architecture. *Computer Network and ISDN Systems*, November 1998.

45. C. Partridge, T. Mendez, and W. Milliken. Host anycasting service. http://rfc.sunsite.dk/rfc/rfc1546.html.

46. L. Qiu, V. N. Padmanabhan, and G. M. Voelker. On the placement of web server replicas. In *INFOCOM*, pages 1587–1596, 2001.

47. M. Rabinovich, Z. Xiao, and A. Aggarwal. Computing on the edge: A platform for replicating internet applications. In *Proc. of Int. Workshop on Web Content Caching and Distribution*, Hawthorne, NY USA, Sept. 2003.

48. P. Radoslavov, R. Govindan, and D. Estrin. Topology-informed internet replica placement. *Proceedings of WCW'01: Web Caching and Content Distribution Workshop, Boston, MA, June 2001.*

49. S. Roy, B. Shen, V. Sundaram, and R. Kumar. Application level hand-off support for mobile media transcoding sessions. In *Proceedings of the 12th international workshop on Network and operating systems support for digital audio and video*, pages 95–104. ACM Press, 2002.

50. M. Sayal, Y. Breithart, P. Scheuermann, and R. Vingralek. Selection algorithms for replicated web servers. 26(3), December 1998.

51. S. Sen, J. Rexford, and D. Towsley. Proxy prefix caching for multimedia streams. In *Proceedings of IEEE Infocom*, 1999.

52. M. Stemm, R. Katz, and S. Seshan. A network measurement architecture for adaptive applications. In *Proceedings of IEEE Infocom*, 2001.

53. D. C. Verma. *Content Distribution Networks: An engineering approach.* Wiley Inter-Science, 2002.

54. B. Wang, S. Sen, M. Adler, and D. Towsley. Optimal proxy cache allocation for efficient streaming media distribution. In *Proceedings of Infocom*, 2002.

55. S. Wee, J. Apostolopoulos, W. Tan, and S. Roy. Research and design of a mobile streaming media content delivery network. In *Proceedings of IEEE International Conference on Multimedia & Expo*, 2003.

56. P. Yu and D. M. Dias. Analysis of task assignment policies in scalable distributed web-server systems. *IEEE Transaction on Parallel and Distributed Systems*, June 1998.

57. E. Zegura, M. Ammar, Z. Fei, and S. Bhattacharjee. Application-layer anycasting: a server selection architecture and use in a replicated web service. *IEEE/ACM Transaction on Networking*, 8(4), August 2000.

Computer Virus Propagation Models

Giuseppe Serazzi and Stefano Zanero*

Dipartimento di Elettronica e Informazione, Politecnico di Milano,
Via Ponzio 34/5, 20133 Milano, Italy,
{giuseppe.serazzi,stefano.zanero}@polimi.it

Abstract. The availability of reliable models of computer virus propagation would prove useful in a number of ways, in order both to predict future threats, and to develop new containment measures. In this paper, we review the most popular models of virus propagation, analyzing the underlying assumptions of each of them, their strengths and their weaknesses. We also introduce a new model, which extends the Random Constant Spread modeling technique, allowing us to draw some conclusions about the behavior of the Internet infrastructure in presence of a self-replicating worm. A comparison of the results of the model with the actual behavior of the infrastructure during recent worm outbreaks is also presented.

1   Introduction

The concept of a computer virus is relatively old, in the young and expanding field of information security. First developed by Cohen in [1] and [2], the concept of "self-replicating code" has been presented and studied in many researches, both in the academic world and in the so-called "underground". The spread of computer viruses still accounts for a significant share of the financial losses that large organizations suffer for computer security problems [3].

While many researches deal with concept such as the creation of new viruses, enhanced worms and new viral vectors, or the development of new techniques for detection and containment, some effort has been done also in the area of modeling viral code replication and propagation behavior. The importance of this work, and the shortcomings of many existing models, are described in [4].

Creating reliable models of virus and worm propagation is beneficial for many reasons. First, it allows researchers to better understand the threat posed by new attack vector and new propagation techniques. For instance, the use of conceptual models of worm propagation allowed researchers to predict the behavior of future malware, and later to verify that their predictions were substantially correct [5].

In second place, using such models, researchers can develop and test new and improved models for containment and disinfection of viruses without resorting

---

* Work partially supported by IEIIT-CNR institute, and COFIN 2001-Web and FIRB-Perf projects.

to risky "in vitro" experimentation of zoo virus release and cleanup on testbed networks [6].

Finally, if these models are combined with good load modeling techniques such as the queueing networks, we can use them to predict failures of the global network infrastructure when exposed to worm attacks. Moreover, we can individuate and describe characteristic symptoms of worm activity, and use them as an early detection mechanism.

In order to be useful, however, such a model must exhibit some well-known characteristics: it must be accurate in its predictions and it must be as general as possible, while remaining as simple and as low-cost as possible.

Although this has not yet been widely recognized, the issue of virus infection via networks can be viewed as a part of network quality-of-service, and should be considered in studies of mail system performance as discussed in [7], while techniques discussed in [8] can be potentially used to route traffic in a manner which avoids virus infection.

In this paper we present a critical review of most of the existing models of virus propagation, showing the underlying assumptions of each of them, and their strengths and weaknesses. We also introduce a new model, based on the same foundations of an existing technique, which allows us to draw some conclusions about the stability of the Internet infrastructure in presence of a self-replicating worm. We compare our modeling results with actual behavior of the infrastructure during recent worm crises and show that our model can accurately describe some effects observed during fast worms propagation.

2   A Survey of Existing Modeling Techniques

Viral code propagation vectors have evolved over the years. In the beginning of the virus era, the most common vector of propagation was the exchange of files via floppy disks and similar supports. The pathogens were viruses, in the strictest sense: they propagated by appending their code to a host program, which had to be executed in order to spread the infection, and to execute the payload if present. This applies, with some modification, also to the so-called "boot sector viruses" which infected the boot loader of the operating system, and spread by infecting the boot sector of floppy disks, which would run whenever a disk was unintentionally left in the machine at boot time. The same concept, in more recent times, has been extended to macro languages embedded in office automation suites, generating the so-called "macro viruses".

The concept of a worm, i.e. a self-contained, self-propagating program which did not require an host program to be carried around, was also developed, but was somehow neglected for a long time. In 1988, however, the *Internet Worm* [9] changed the landscape of the threats. The Internet Worm was the first successful example of a self-propagating program which did not infect host files, but was self contained. Moreover, it was the first really successful example of an active network worm, which propagated on the Internet by using well-known vulnerabilities of the UNIX operating system. Other worms used open network shares, or exploited vulnerabilities in operating systems and server software to propagate.

With the widespread adoption of the Internet, mass-mailing worms began to appear. The damage caused by *Melissa* virus in 1999, *Love Letter* in 2000 and *Sircam* in 2001 demonstrated that tricking users into executing the worm code attached to an e-mail, or exploiting a vulnerability in a common e-mail client to automatically launch it, is a successful way to propagate viral code.

Each of these different propagation vectors has inspired various propagation models. In the next sections, we will review the most interesting and successful models for each class: file, macro and boot sector viruses, e-mail based worms, "active" (or "scanning") Internet worms.

It is important to note that modern viruses often use a mix of these techniques to spread (for instance, Sircam uses both mass mailing and open network shares, while *Nimda* uses four different mechanisms to propagate). We are not aware, however, of any existing model which takes into account multi-vector viruses and worms. So, we will follow a traditional taxonomy in presenting the existing modeling techniques.

2.1   Modeling Traditional Viruses

The first complete application of mathematical models to computer virus propagation appeared in [10]. The basic intuitions of this work still provide the fundamental assumptions of most computer epidemiological models.

Epidemiological models abstract from the individuals, and consider them units of a population. Each unit can only belong to a limited number of states (e.g. "susceptible" or "infected"; see Table 1 for additional states): usually, the chain of these states gives the name to the model, e.g., a model where the Susceptible population becomes Infected, and then Recovers, is called a SIR model, whereas a model with a Susceptible population which becomes Infected, and then goes back to a Susceptible state is called SIS.

Another typical simplification consists in avoiding a detailed analysis of virus transmission mechanics, translating them into a probability that an individual will infect another individual (with some parameters). In a similar way, transitions between other states of the model are described by simple probabilities. Such probabilities could be calculated directly by the details of the infection mechanism or, more likely, they can be inferred by fitting the model to actual propagation data. An excellent analysis of mathematics for infectious diseases in the biological world is available in [11].

Most epidemiological models, however, share two important shortcomings: they are *homogeneous*, i.e. an infected individual is equally likely to infect any other individual; and they are *symmetric*, which means that there is no privileged direction of transmission of the virus. The former makes these models unappropriate for illnesses that require a non-casual contact for transmission; the latter constitutes a problem, for instance, in the case of sexually-transmitted diseases.

Table 1. Typical states for an epidemiological model

M	Passive immunity
S	Susceptible state
E	Exposed to infection
I	Infective
R	Recovered

In the case of computer viruses, however, both problems are often grievous. For example, most individuals exchange programs and documents (by means of e-mails or diskettes) in almost closed groups, and thus an homogeneous model may not be appropriate. Furthermore, there are also "sources" of information and programs (e.g. computer dealers and software distributors) and "sinks" (final users): that makes asymmetry a key factor of data exchange.

In [10] both these shortcomings are addressed by transferring a traditional SIS model onto a directed random graph, and the important effects of the topology of the graph on propagation speed are analyzed. The authors describe the behavior of virus infections on *sparse* and *local* graphs. In a sparse graph, each node has a small, constant average degree; on the contrary, in a local graph, the probability of having a vertex between nodes B and C is significantly higher if both have a vertex connected to the same node A. The authors discuss that in the landscape of the beginnings of the 90s the latter situation approximated very well the interaction between computer users. Among other results, it is shown that the more sparse a graph is, the slower is the spread of an infection on it; and the higher is the probability that an epidemic condition does not occur at all, which means that sparseness helps in containing *global* virus spread (while local spread is unhindered). Further elaborations on this type of model can be found in [12].

These findings are useful and interesting. However, it must be noted that often a SIR model, in which a "cured" system is not susceptible any more, could approximate better the behavior of many real cases of propagation when a patch or antivirus signature is available. Also, the introduction of the Internet as a convenient and immediate way for software and data exchange has arguably made the assumptions of locality and sparseness of the graph no longer valid.

2.2 Modeling E-mail Based Worms

In a technical report [13] Zou et al. describe a model of e-mail worm propagation. The authors model the Internet e-mail service as an undirected graph of relationship between people (i.e. if user A has user B's e-mail address in his address book, B has probably A's address in her contacts also). In order to build a simulation of this graph, they assume that each node degree is distributed on a power-law probability function.

They draw this assumption from the analysis of "Yahoo!" discussion group sizes, which result to be heavy-tailed. Since once a user puts a discussion group address in his contact book he actually adds an edge toward all the group members, the node degree should be heavy-tailed too. It is unclear if this distribution reflects also the true distribution of contacts (i.e. not considering e-mail lists) among Internet users. However, considering that nowadays most Internet discussion groups employ content filtering or ban attachments altogether, they are not a very viable virus propagation vector. The construction method, based on a "small world" network topology, seems to ignore completely the existence of interest groups and organizations, which naturally create clusters of densely connected vertices.

Furthermore, the authors assume that each user "opens" an incoming virus attachment with a fixed probability P_i, a function of the user but constant in time. This does not describe very well the typical behavior of users. Indeed, most experienced users avoid virus attachments altogether, while unexperienced users open them every time, thus making this approximation misleading.

They model e-mail checking time T_i as either an exponentially or Erlang distributed random variable. The means of these distributions, $T = E[T_i]$, and $P = E[P_i]$ are assumed to be independently distributed gaussians.

An interesting observation the authors make is that since the user e-mail checking time is much larger than the average e-mail transmission time, the latter can be disregarded in the model. The authors proceed in considering a "reinfection" model, where a user will send out copies of the e-mail virus each time he reopens an infected attachment, as opposed to a "non-reinfection" model where this happens just once. Neither model is very realistic: in many cases, e-mail viruses install themselves as startup services on the system, and spread themselves at each opportunity.

These observations suggest that the results of the simulation should be considered as qualitative, rather than quantitative, indications. A couple of interesting points can be drawn from this study. Firstly, infecting users with high "degrees", i.e. an high number of contacts, in the early phase of infection speeds up the process considerably (and conversely, making these nodes immune to the virus helps defending against it). A second observation is that the overall spread rate of viruses gets higher as the variability of users' e-mail checking times increases.

2.3 Modeling a Scanning Worm: The Random Constant Spread Model

The Random Constant Spread (RCS) model [5] was developed by Staniford, Paxson and Weaver using empirical data derived from the outbreak of the *Code Red* worm. This worm was released in its first version (which we will call CRv1) onto the Internet on July 13th 2001, according to the initial analysis from eEye Digital Security [14]. The CRv1 worm was disassembled and studied quickly [15], making its diffusion mechanism very clear.

Code Red propagates by using the *.ida vulnerability* discovered by eEye itself on June 18th 2001 [16], thus infecting vulnerable web servers running Microsoft IIS version 4.0 and 5.0. When Code Red infects an host, it spreads by launching 99 threads, which randomly generate IP addresses (excluding subnets 127.0.0.0/8, loopback, and 224.0.0.0/8, multicast) and try to compromise the hosts at those addresses using the same vulnerability.

CRv1, however, had a fundamental design flaw: the random number generator of each thread was initialized with a fixed seed, so all the copies of the worm in a particular thread, on all infected hosts, generated the same sequence of target IP addresses, attempting to compromise them in that fixed order. The thread identifier is used as part of the seed, so each instance of the worm has 99 different sequences, but these sequences were the same for all the instances of the worm. For this reason, CRv1 was not very effective nor dangerous: since all the instances of the worm scanned the same sequence of IP addresses, the worm spread was only linear.

A particularity of this worm is that it does not reside on the file system of the target machine, but it is carried over the network as the shellcode of the buffer overflow attack it uses (for a definition of buffer overflows, see [17]). When it infects an host, it resides only in memory: thus a simple reboot eliminates the worm, but does not avoid reinfection. Applying a patch to fix the IIS server or using temporary workarounds (e.g. activating a firewall, or shutting down the web server) makes instead the machine completely invulnerable to the infection.

A "version 2" (CRv2) of the same worm "fixes" this bug by randomizing the seed of each thread. It also adds a nasty subroutine to attempt a DDoS attack against www1.whitehouse.gov on the days between the 20th and the 28th of each month, then reactivating on the 1st of the following month. CRv2 should **not** be confused with the so-named *Code Red II* virus, which in spite of the name is a completely different virus (we will address it later).

The RCS model actually describes CRv2, since it assumes that the worm has a good random number generator that is properly seeded. Let N be the total number of vulnerable servers which can be potentially compromised *from* the Internet. The model here makes two approximations: it ignores that systems can be patched, powered and shut down, deployed or disconnected. Also, it ignores any sudden spread of the worm behind firewalls on private intranets, which could be misleading. In other words, in the model the pool of vulnerable targets is considered to be constant.

An additional, more crucial approximation, is that the Internet topology is considered an undirected complete graph. In truth, the Internet being (as S. Breidbart defined it) "the largest equivalence class in the reflexive, transitive, symmetric closure of the relationship *can be reached by an IP packet from*", it is all but completely connected. In fact, recent researches [18] show that as much as the 5% of the routed (and used) address space is not reachable by various portions of the network, due to misconfiguration, aggressive filtering, or even commercial disputes between carriers. Intuitively, however, this does not fundamentally alter the conclusions of the study.

Let K be the average initial compromise rate, i.e. the number of vulnerable hosts that an infected host can compromise per unit of time at the beginning of

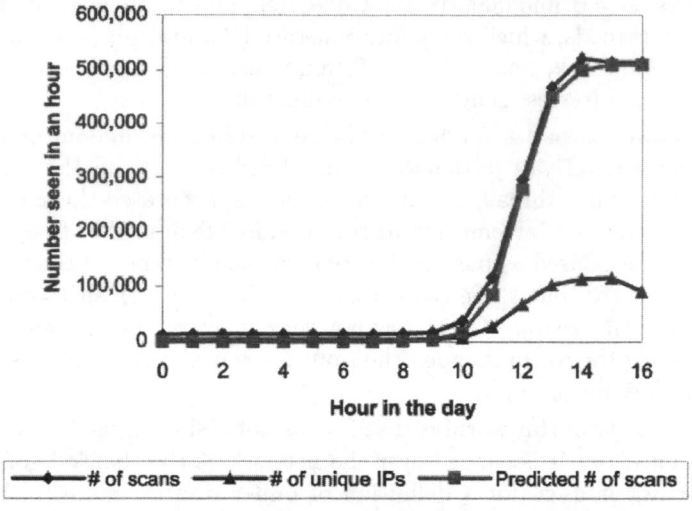

Fig. 1. Comparison between predicted and detected number of scans by Code Red, in data offered by Chemical Abstracts Service; originally appeared in [5]

the outbreak. The model assumes that K is constant, averaging out the differences in processor speed, network bandwidth and location of the infected host. The model also assumes that a machine cannot be compromised multiple times. If $a(t)$ is the proportion of vulnerable machines which have been compromised at the instant t, $N \cdot a(t)$ is the number of infected hosts, each of which scans other vulnerable machines at a rate K per unit of time. But since a portion $a(t)$ of the vulnerable machines is already infected, only $K \cdot (1 - a(t))$ new infections will be generated by each infected host, per unit of time. The number n of machines that will be compromised in the interval of time dt (in which we assume a to be constant) is thus given by:

$$n = (Na) \cdot K(1 - a)dt \tag{1}$$

We are obviously considering that, being 2^{32} a very large address space, and since CRv2 target list is truly random, the chance that two different instances of the worm simultaneously try to infect a single target is negligible. Now, under the hypothesis that N is constant, $n = d(Na) = Nda$, we can also write:

$$Nda = (Na) \cdot K(1 - a)dt \tag{2}$$

From this, it follows the simple differential equation:

$$\frac{da}{dt} = Ka(1 - a) \tag{3}$$

The solution of this equation is a *logistic curve*:

$$a = \frac{e^{K(t-T)}}{1 + e^{K(t-T)}} \tag{4}$$

where T is a time parameter representing the point of maximum increase in the growth. In [5] the authors fit their model to the "scan rate", or the total number of scans seen at a single site, instead than using the number of distinct attacker IP addresses. We show this in Figure 1, where the logistic curve has parameters $K = 1.6$ and $T = 11.9$. The scan rate is directly proportional to the total number of infected IPs on the Internet, since each infected host has a fixed probability to scan the observation point in the current time interval. On the contrary, the number of distinct attacker addresses seen at a single site is evidently distorted (as can be seen in Figure 1), since each given worm copy takes some random amount of time before it scans a particular site. If the site covers just a small address space, the delay makes the variable "number of distinct IP addresses of attackers" to lag behind the actual rate of infection.

Researchers from CAIDA also published data on the Code Red outbreak [19]; their monitoring technique is based on the usage of a "network telescope" [20], i.e. a large address-space block, routed but with no actual hosts connected. Three of such telescope datasets (one observed from a /8 network, and two from /16 networks respectively) were merged to generate the data presented in the paper. On such a large portion of IP space, the "distortion" is less evident, as we can see in Figure 2. In Figure 3 the cumulative total of "attacker" IP addresses seen by the telescopes is plotted on a log-log scale and fitted against the predictions of the RCS model, on a logistic curve with parameter $K = 1.8$ and $T = 16$. CAIDA data are expressed in the UTC timezone, while Chemical Abstracts Service data were expressed in CDT timezone: this accounts for the different T parameter.

In Figure 4 instead we can see the deactivation rate of the worm, considering as "deactivated" an host which did not attempt to spread the infection anymore. The worm was built to deactivate its propagation routine on midnight of July 20, UTC time (to begin the Denial of Service process). This is clearly visible in the graphic.

As we can see from Figures 3 and 1, at that time the worm was approaching saturation. A total of about 359.000 hosts were infected by CRv2 in about 14 hours of activity (corresponding to the plateau in Figure 2).

In Figure 5 the hourly probe rate detected at the Chemical Abstracts Service on day August 1st 2001 is compared to a fitting curve. On that day CRv2 reactivated after the denial of service cycle, as discussed before. CAIDA observes that at peak 275.000 hosts were infected. The lower number is probably due to ongoing patching activity during the 10-days grace period.

Other authors [21] propose the AAWP discrete time model, in the hope to better capture the discrete time behavior of a worm. However, a continuous model is appropriate for such large scale models, and the epidemiological literature is clear in this direction. The assumptions on which the AAWP model is based are not completely correct, but it is enough to note that the benefits of using a discrete time model seem to be very limited.

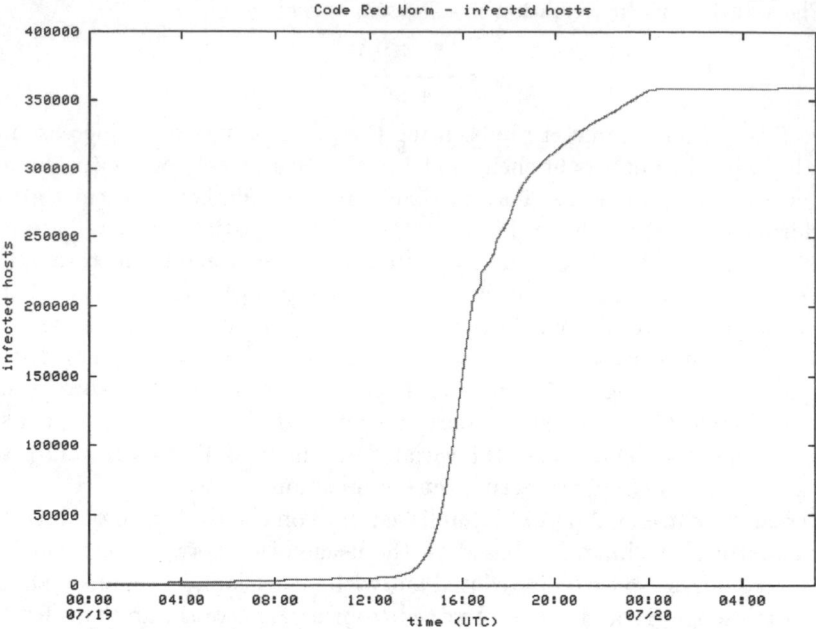

Fig. 2. The number of distinct IP addresses infected by Code Red v2 during its first outbreak, drawn from [19]

3   A New Model for the Propagation of Worms on the Internet

3.1   Slammer, or the Crisis of Traditional Models

As described in section 2, a number of models have been developed for the propagation of viruses and worms. However, most of them have critical shortcomings when dealing with new, aggressive types of worms, called "flash" or "Warhol" worms.

On Saturday, January 25th, 2003, slightly before 05:30 UTC, the *Sapphire Worm* (also known as *SQ-Hell* or *Slammer*) was released onto the Internet. Sapphire propagated by exploiting a buffer overflow vulnerability in computers on the Internet running Microsoft's SQL Server or MSDE 2000 (Microsoft SQL Server Desktop Engine). The vulnerability had been discovered in July 2002, and a patch for it was actually available even before the vulnerability was announced.

The characteristic which made this worm so different from the previous ones was its speed: it effectively showed a doubling time of 8.5($\pm$1) seconds, infecting more than 90 percent of vulnerable hosts within the first 10 minutes. It was thus a lot faster than Code Red, which had a doubling time of about 37 minutes. At least 75.000 hosts were infected by Sapphire.

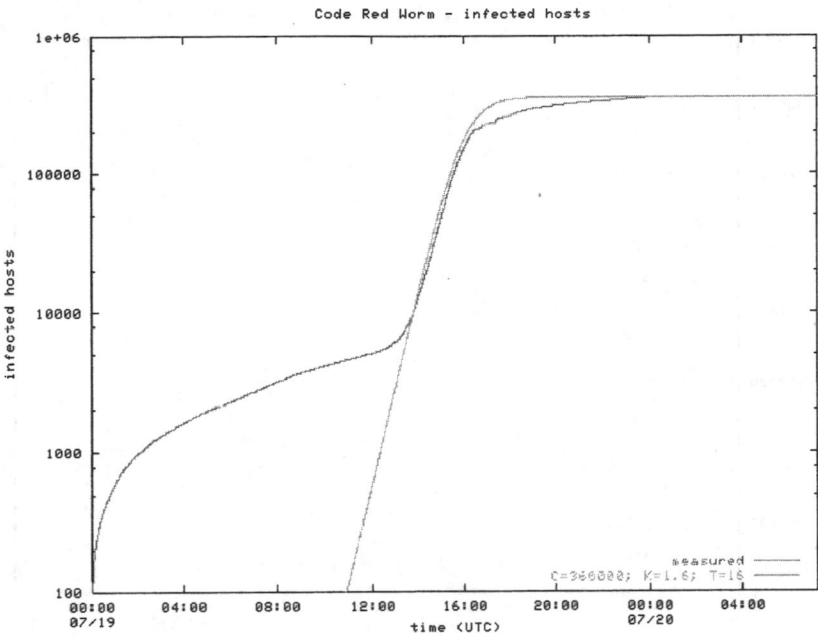

Fig. 3. The number of distinct IP addresses infected by Code Red v2 during its first outbreak, plotted on a log-log scale and fitted against the RCS model, drawn from [19]

Sapphire's spreading strategy is based on random scanning, like Code Red. Thus, the same RCS model that described CRv2 should fit also Sapphire's growth. However, as it appears from Figure 6, the model fits well only for the initial stage of growth. Then, suddenly, there is an abrupt difference between the model and the real data.

We must remember that this data shows the total number of *scans*, not the actual number of infected machines. After approximately 3 minutes from the beginning of the infection, the worm achieved its full scanning rate of more than 55 million scans per second; after this point, the rate of growth slowed down somewhat. The common explanation for this phenomenon is that significant portions of the network did not have enough bandwidth to support the propagation of the worm at its full speed: in other words, the worm saturated the network bandwidth before saturating the number of infectable hosts.

Why was Sapphire so deadly efficient, when compared to Code Red? The difference relies mainly in the transmission mechanism: the exploit used by Sapphire was based on UDP, while the exploit of Code Red was based on TCP. So, Code Red had to establish a connection before actually exploiting the vulnerability: having to complete the three-way handshake, waiting for answers, it was

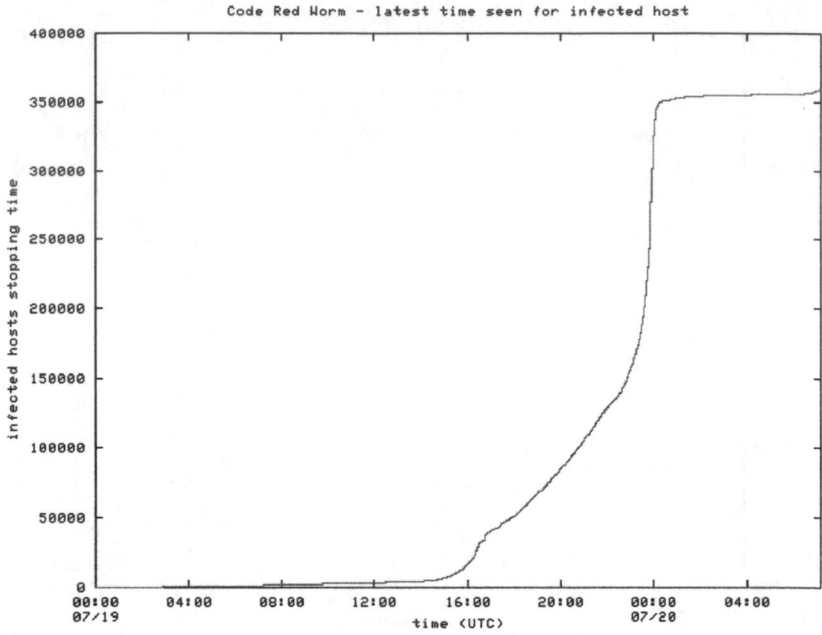

Fig. 4. Rate of "deactivation" of infected hosts, drawn from [19]

latency limited. Sapphire, on the contrary, could scan at the full speed allowed by the network bandwidth available, so it was *network limited.*

In order to properly model such a worm, the bandwidth between nodes must be taken into account: this means that most of the proposed models are not applicable in this situation because they use the "global reachability" property of the Internet as a simplifying assumption.

3.2   Building a Compartment-Based Model

Modeling the whole Internet as a graph, with each node representing an host, is unfeasible. Even modeling the communication infrastructure, representing routers as nodes of the graph and links as edges, is an almost impossible task.

Luckily, we do not need such granularity. The Internet can be macroscopically thought of as the interconnection of a number of *Autonomous Systems.* An AS is a subnetwork which is administered by a single authority. Usually, the bottlenecks of Internet performance are located in the inter-AS connections (i.e. the peering networks and the NAP connections), not in the intra-AS connections. However, some AS are instead very large entities, which comprise densely connected regions and bottleneck links: in this case we could split these ASs in smaller regions that satisfy the property.

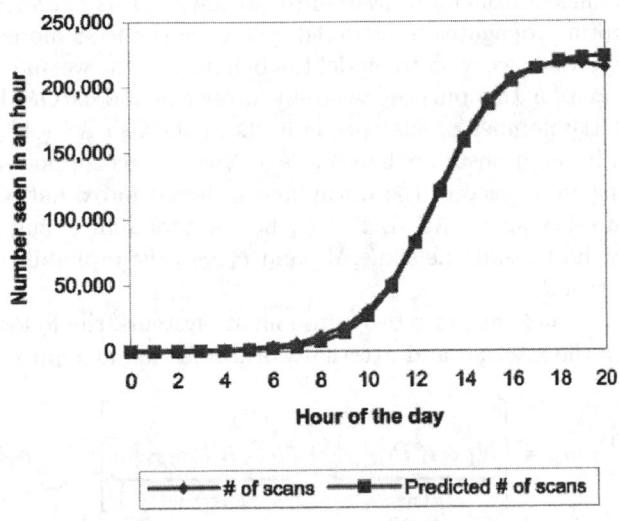

Fig. 5. The real and predicted scan rate of Code Red v2 during the second outbreak on August 1st, as it appears in the CAS [5] dataset. The time of day is expressed in Central US Time.

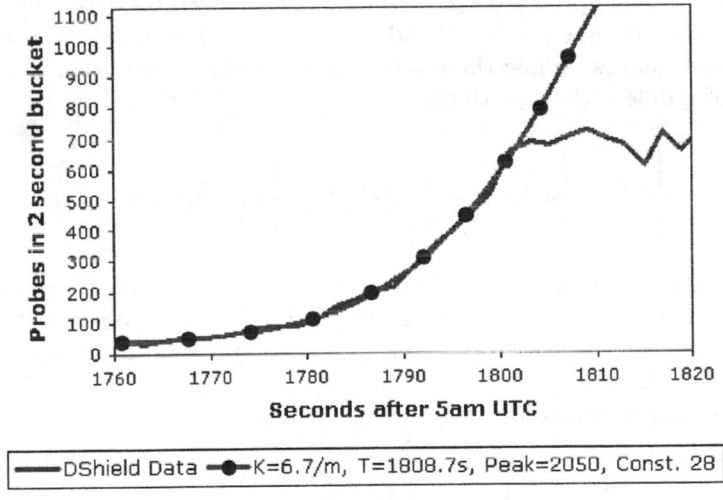

Fig. 6. The growth of Slammer scans, as seen from Dshield.org, fitted against the RCS model; K, T and Const are the parameters of the fitting curve

For this reason, we propose a compartment-based model, in which we suppose that inside a single autonomous system (or inside a densely connected region of an AS) the worm propagates unhindered, following the RCS model described in Section 2.3. However, we wish to model the behavior of the worm in the intra-AS propagation, and for this purpose we need to rewrite and extend Equation 3.

Let N_i be the number of susceptible hosts in the i-th AS (AS_i), and a_i the proportion of infected hosts in the same AS. Now, let us suppose that K is the average propagation speed of the worm, and in first approximation let us say it is constant in every single AS. Let $P_{IN,i}$ be the probability that a host inside AS_i targets an host inside the same AS, and $P_{OUT,i}$ the probability that instead it attacks another AS.

In a simple model with just two autonomous systems, the following equation describes both the internal and external worm infection attempts on AS_1:

$$N_1 da_1 = \left[\underbrace{N_1 a_1 K P_{IN,1} dt}_{Internal} + \underbrace{N_2 a_2 K P_{OUT,2} dt}_{External} \right] (1 - a_1)$$

A similar equation can obviously be drawn for AS_2 simply by switching the terms. We thus have a system of two differential equations:

$$\begin{cases} \frac{da_1}{dt} = \left[a_1 K P_{IN,1} + \frac{N_2}{N_1} a_2 K P_{OUT,2} \right] (1 - a_1) \\ \frac{da_2}{dt} = \left[a_2 K P_{IN,2} + \frac{N_1}{N_2} a_1 K P_{OUT,1} \right] (1 - a_2) \end{cases}$$

Under the assumption that the worm randomly generates the target IP addesses, it follows that $P_{IN,1} = N_1/N$ and $P_{OUT,1} = 1 - P_{IN,1} = N_2/N$. Substituting these values, and extending the result to a set of n ASs, we obtain the following system of n differential equations:

$$\begin{cases} \frac{da_i}{dt} = \left[a_i K \frac{N_i}{N} + \sum_{\substack{j=1 \\ j \neq i}}^{n} \frac{N_j}{N_i} a_j K \frac{N_i}{N} \right] (1 - a_i) \ 1 \leq i \leq n \end{cases} \tag{5}$$

We can think of the result of the integration of each equation as a logistic function (similar to the one generated by the RCS model), somehow "forced" in its growth by the second additive term (which represents the attacks incoming from outside the AS).

Simplifying the equation we obtain:

$$\begin{cases} \frac{da_i}{dt} = \left[a_i K \frac{N_i}{N} + \underbrace{\sum_{\substack{j=1 \\ j \neq i}}^{n} \frac{N_j}{N} a_j K}_{incoming \ attacks} \right] (1 - a_i) \end{cases} \tag{6}$$

in which we left in evidence the term describing the incoming attack rate, but we can further reduce the equations to the following:

$$\left\{\frac{da_i}{dt} = \left[\sum_{j=1}^{n} N_j a_j\right](1 - a_i)\frac{K}{N}\right. \tag{7}$$

This is a nonlinear system of differential equations. It can be easily shown that the results of equation 7 are a solution also for this model, with the same K and N parameters. Considering that $a = \frac{\sum_{i=1}^{n} N_i a_i}{N}$, we have:

$$\frac{da}{dt} = \frac{d}{dt}\left[\frac{\sum_{i=1}^{n} N_i a_i}{N}\right] = \frac{1}{N}\sum_{i=1}^{n} N_i \frac{da_i}{dt}$$

and, from equation 7:

$$\frac{da}{dt} = \frac{1}{N}\sum_{i=1}^{n}\left\{N_i\left[\sum_{j=1}^{n} N_j a_j\right](1 - a_i)\frac{K}{N}\right\} = \frac{K}{N^2}\sum_{i=1}^{n} N_i N a (1 - a_i) =$$

$$= \frac{aK}{N}\sum_{i=1}^{n} N_i (1 - a_i) = \frac{aK}{N}\left[\sum_{i=1}^{n} N_i - \sum_{i=1}^{n} N_i a_i\right] = \frac{aK}{N}[N - N a] =$$

$$= aK(1 - a)$$

Thus we obtain equation 3.

We can explore the solutions of a linearization of the system, in the neighborhood of the unstable equilibrium point in $a_j = 0$, $\forall j$. With the convention of using the newtonian notation (denoting the first derivative with an upper dot), and using the traditional substitution of $a_i = (\overline{a_i} + \delta a_i)$, we obtain:

$$\left\{(\overline{a_i} + \delta \dot{a_i}) = \left(\left[\sum_j N_j a_j\right](1 - a_i)\frac{K}{N}\right)\Bigg|_{\overline{a}} + \frac{\partial\left(\left[\sum_j N_j a_j\right](1 - a_i)\frac{K}{N}\right)}{\partial a_i}\Bigg|_{\overline{a}}\delta a_i\right.$$

$$\left\{\dot{\delta a_i} = \left(\left[\sum_j N_j \overline{a_j}\right](1 - \overline{a_i})\frac{K}{N}\right) + \left(-\left[\sum_j N_j \overline{a_j}\right] + N_i(1 - \overline{a_i})\right)\frac{K}{N}\delta a_i\right.$$

Now we can inject a small number of worms in the k-th autonomous system, in a model initially free of virus infections: $\overline{a_i} = 0$; $\delta a_k(0) = 0$, $\forall i \neq k$; $\delta a_k(0) = \varepsilon > 0$. The system initially behaves in this way:

$$\left\{ \dot{\delta a_i} = N_i \frac{K}{N} \delta a_i \right.$$

Thus, in the k-th autonomous system, the worm begins to grow according to the RCS model, while the other AS are temporarily at the equilibrium. This state, however, lasts only until the worm is first "shot" outside the AS, which happens a few moments after the first infection.

We can now calculate analytically the bandwidth consumed by incoming attacks on a "leaf" AS, AS_i, connected to the network via a single connection (a *single homed* AS). Let s be the size of the worm, r_j the number of attacks generated in a time unit by AS_j. Let T describe the total number of systems present on the Internet, and T_i the number of systems in AS_i. The rate of attacks that a single infected host performs in average is $R \cong K\frac{T}{N}$ (since K is the rate of successful outgoing attacks). The incoming bandwidth $b_{i,incoming}$ wasted by the incoming worms on the link is therefore described by:

$$b_{i,incoming} = s \sum_{\substack{j=1 \\ j \neq i}}^{n} r_j \frac{T_i}{T} = s \sum_{\substack{j=1 \\ j \neq i}}^{n} a_j N_j R \frac{T_i}{T}$$

$$= s \sum_{\substack{j=1 \\ j \neq i}}^{n} a_j N_j \frac{K T}{N} \frac{T_i}{T} = s K \frac{T_i}{N} \sum_{\substack{j=1 \\ j \neq i}}^{n} a_j N_j \tag{8}$$

If we compare equations 8 and 6, we can see the structural analogy:

$$b_{i,incoming} = s\, T_i \underbrace{\sum_{\substack{j=1 \\ j \neq i}}^{n} \frac{N_j}{N} a_j K}_{incoming\ attacks} \tag{9}$$

We must also consider the *outgoing* attack rate, which equals the generated attacks minus the attacks directed against the AS itself:

$$b_{i,outgoing} = s r_i \left[1 - \frac{T_i}{T}\right] = s\, a_i N_i R \left[1 - \frac{T_i}{T}\right] = s\, a_i N_i K \frac{T}{N} \left[\frac{T - T_i}{T}\right]$$

Also in this case we can see a structural analogy with equation 6:

$$b_{i,outgoing} = s\,(T - T_i) \underbrace{a_i \frac{N_i}{N} K}_{outgoing\ attacks} \tag{10}$$

Adding equation 9 to equation 10 we can thus obtain the amount of bandwidth the worm would waste on AS_i if unconstrained:

$$b_i = s\,(T - T_i)\, a_i \frac{N_i}{N} K + s\, T \sum_{\substack{j=1 \\ j \neq i}}^{n} \frac{N_j}{N} a_j K \tag{11}$$

Considering that:

$$a_i \frac{N_i}{N} K + \sum_{j \neq i} \frac{N_j}{N} a_j K = \sum_j \frac{N_j}{N} a_j K$$

we can easily see that:

$$b_i = s(T - T_i) a_i \frac{N_i}{N} K + s T \left[\sum_j \frac{N_j}{N} a_j K - a_i \frac{N_i}{N} K \right] =$$

$$= s T \sum_j \frac{N_j}{N} a_j K - s T_i a_i \frac{N_i}{N} K = \frac{s K}{N} \left[T \sum_j N_j a_j - T_i a_i N_i \right]$$

We could also extend this result to a multi-homed *leaf* AS, that is, an AS with multiple connections to the Internet, but which does not carry traffic from one peer to another for policy reasons (as it is the case for many end-user sites). We must simply divide this equation, using a different equation for each link, each carrying the sum of the AS that are reachable through that link. We should also rewrite equation 10 to split up the outgoing attacks depending on the links. This would not change the overall structure of the equations.

It would be a lot more complex to model a *non-leaf* AS, because we should take into account also the incoming and outgoing traffic on each link that is being forwarded from a neighbor AS to another. The complexity lies in describing in a mathematically tractable way the paths on a complex network. However, we can ignore, in first approximation, non-leaf AS, because they tend to be *carriers* of traffic, not containing hosts susceptible to the worm: they can thus be considered as a part of the intra-AS connection links.

Let us go back to the leaf, single-homed model, and let us suppose now that there is a structural limit to the available bandwidth on the link, B_i. In the simplest possible model, we will see that only a fraction $Q_i, 0 < Q_i \leq 1$ of packets will be allowed to pass through the i-th link, such that $Q_i b_i \leq B_i$. The actual behavior of the system under an increasing load is not known a priori, but we can suppose that there exists a simple relation expressing the saturation of the link, such as:

$$B_i \left(1 - e^{-\lambda \frac{b_i}{B_i}} \right) = Q_i b_i \Rightarrow Q_i = \frac{B_i \left(1 - e^{-\lambda \frac{b_i}{B_i}} \right)}{b_i}$$

This is justified by thinking that:

$$\lim_{b_i \to 0} B_i \left(1 - e^{-\lambda \frac{b_i}{B_i}} \right) = 0 \quad \lim_{b_i \to +\infty} B_i \left(1 - e^{-\lambda \frac{b_i}{B_i}} \right) = B_i$$

Resubstituting this:

$$Q_i b_i = Q_i \left[s (T - T_i) a_i \frac{N_i}{N} K + s T \sum_{j \neq i} \frac{N_j}{N} a_j K \right] =$$

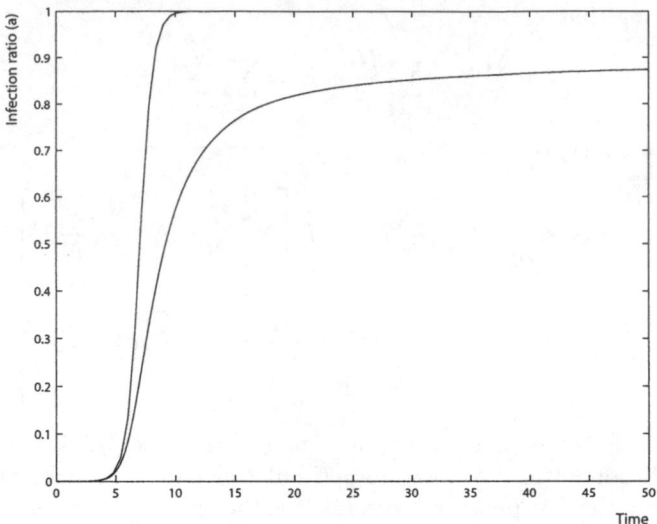

Fig. 7. A comparison between the unrestricted growth predicted by an RCS model and the growth restricted by bandwidth constraints

$$= s\left(T - T_i\right)\left[\underbrace{Q_i a_i \frac{N_i}{N} K}_{reduced\,outgoing\,rate}\right] + s\,T\left[\underbrace{Q_i \sum_{j\neq i} \frac{N_j}{N} a_j K}_{reduced\,incoming\,rate}\right]$$

As we see, in order to reduce the bandwidth to the value $Q_i b_i$, and under the hypothesis that the incoming and outgoing stream of data are curtailed by the same factor, the incoming and outgoing attack rate must be decreased of this same factor Q_i. We can now substitute this reduced attack rate into the system of equations 6 (remembering that the inner worm propagation will be unaffected), and thus obtain:

$$\left\{\frac{da_i}{dt} = \left[a_i K \frac{N_i}{N} + Q_i \sum_{j\neq i} Q_j \frac{N_j}{N} a_j K\right](1 - a_i)\right. \tag{12}$$

Equation 12 expresses the same model, but with a limit on the dataflow rate of the links between different ASs. We have plotted this equation using Simulink and Matlab, obtaining the result shown in Figure 7. Here we compare the different results of equations 6 and 12. They are built with the same parameters, and thus their initial growth is totally symmetric. However, as soon as the links begin to saturate, the growth of equation 12 slows down.

We can then insert into the Simulink model an additional component in order to simulate the disruption in the Internet links caused by traffic overload. In

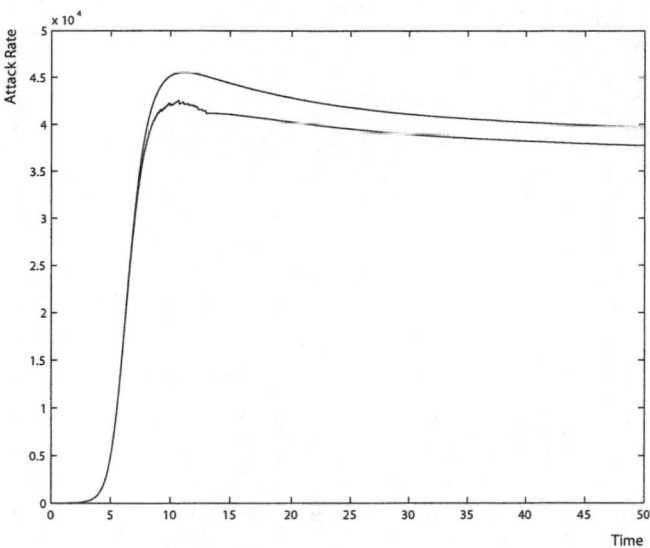

Fig. 8. Attack rates observed on a single link, under the hypotheses that links hold (upper curve) or randomly fail (lower curve)

particular, we suppose that a proportion p of the links will shut down when they are flooded by worm traffic (i.e. when the worm generates a flow of packets a lot above B_i). In Figure 8 we see the comparison on the traffic observed on a single link with or without this additional detail. We can observe the small oscillations that are generated, even in this static and simplified model of network failure.

We repeated our simulations for a variety of parameters of the model, concluding that as the bandwidth limit increases the small peak of traffic seen on the link is less and less evident, up to the point where the model behaves exactly as if the bandwidths were unlimited. Increasing the number of failing links, instead, increases the oscillations after the peak.

In Figure 9 we instead plot the number of attacks seen on a large subset of the links. This is very similar to the behavior observed by DShield during the Slammer outbreak (see Figure 6): DShield, in fact, monitors a large subset of different links. However, as can be seen in Figure 7, the actual growth of the worm is only slowed down, not stopped at all, by the vanishing links.

In our opinion, then, the sudden and strange stop in the increase of observed attacks can indeed be explained by the disruption of Internet links, as hypothesized in various previous works, but this does not imply a similar slow down in the growth of the worm.

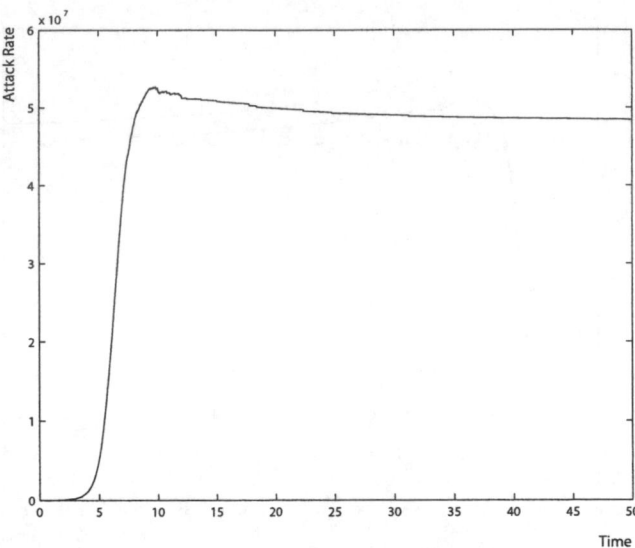

Fig. 9. The number of attack rates seen by a global network telescope, under the hypothesis that some links fail during the outbreak

4   A Discussion of Proposed Countermeasures

4.1   Monitoring and Early Warning

In [22] the authors use the models of active worm propagation to describe a monitoring and alerting system, based on distributed *ingress* and *egress* sensors for worm activity. Ingress sensors detect as possible worm activity any incoming scan trying to contact unused addresses on the network (with a principle similar to the one of the network telescopes discussed in section 2.3). Egress sensors instead try to capture outgoing worm activity from the network.

In order to create a global early warning distributed sensor network, the authors propose a data collection engine, capable of correcting the statistical biases responsible for the distortions described in section 2.3. They propose the use of a Kalman filter for estimating parameters such as K, N and a from the observations, and thus have a detailed understanding of how much damage the spreading worm could generate. In addition, using some properties of the filter, it can be used to generate and early warning of worm activity as early as when $1\% \leq a \leq 2\%$.

The authors also show that this early warning method works well also with fast spreading worms, and even if an hit-list startup strategy is used.

4.2   Modeling Removal and Disinfection of Hosts

Models such as RCS purposefully avoid to take into account the dynamics of countermeasures deployed to stop or contain virus outbreaks, considering worm propagation to be too quick to be influenced by human response.

A study by Zou et al. [23], focused on slower propagating worms such as Code Red, builds upon the RCS equations, incorporating the Kermack-Mckendrick model which accounts for the removal of infectious hosts, and extending it to the removal of susceptible hosts as well.

Additionally, the authors propose that the infection rate K should be considered a function of time: $K = K(t)$, because of intervening network saturation and router collapse. Basically they rewrite the model as:

$$\frac{da}{dt} = K(t)\,a\,(1 - a - q - r) - \frac{dr}{dt} \qquad (13)$$

Where $q(t)$ is the proportion of susceptible hosts that are immunized at time t, and $r(t)$ is the proportion of infected hosts that are cured and immunized at time t. This model is called the *two-factor worm model*. In order to completely describe the model, the authors make some assumptions on $q(t)$ and $r(t)$. In particular, they hypothesize that a constant portion of the infected machines are cured on a unit of time:

$$\frac{dr}{dt} = \gamma a$$

Additionally, with an hypothesis close to the kill signal theory described by Wang et al. in [24], they describe the patching process as a diffusion similar to the one of the worm:

$$\frac{dq}{dt} = \mu(1 - a - q - r)(a + r)$$

it is unclear, however, how the authors chose this particular equation, and how the variables have been chosen. Also, as it has been pointed out in comments to the paper by Wang et al., it is far from evident that the kill signal propagation and the worm propagation follow the same parameters, or even the same topology of network. However the simulation of this model yields interesting analogies with the real data of the Code Red outbreak.

A model by Wang et al. [25] shows the interdependence between the timing parameters of propagation and removal, and their influence on the worm propagation.

4.3   Quarantine: The World's Oldest Defense

In [26] the authors study a dynamic preventive quarantine system, which places suspiciously behaving hosts under quarantine for a fixed interval of time. We omit many details of their analysis, but their conclusion is as follows. Let $1/\lambda_1$ be the mean time before an infected host is detected and quarantined, $1/\lambda_2$ be the mean time before a false positive occurs, i.e. a non-infected host is wrongly quarantined, and T be the quarantine time.

The probability that an infectious host is correctly quarantined is

$$p_1 = \frac{\lambda_1 T}{1 + \lambda_1 T}$$

and the probability of a false positive conversely is:

$$p_2 = \frac{\lambda_2 T}{1 + \lambda_2 T}$$

So the RCS model may be applied, by discounting the infection rate K in order to take into account the effects of quarantine:

$$K' = (1 - p_1)(1 - p_2)K$$

An extension of the Kermack-Mckendrick model, omitted here for brevity, is also presented, and the results of simulation runs on both these models are discussed.

It should be noted that such a dynamic quarantine system would be difficult to implement, because each host cannot be trusted to auto-quarantine itself. Practically, on most networks, the number of remotely manageable enforcement points (i.e. firewalls and intelligent network switches) is limited. Entire blocks of network would need to be isolated at once, uncontrollably increasing the factor p_2. This could help to stop the warm, but with a steep price, since p_2 represents the probability that innocent hosts will be harmed by quarantine.

In addition, as shown by the model presented in 3.2, the virus spread is not stopped but only slowed down inside each quarantined block. Moreover, it should be considered that the "kill signal" effect (i.e. the distribution of anti-virus signatures and patches) would be hampered by aggressive quarantine policies.

On the same topic Moore et al. [27] simulated various containment strategies (namely content filtering and blacklisting), deriving lower and upper bounds of efficacy for each. Albeit interesting, the results on blacklisting share the same weakness pointed out before: it's not realistic to think about a global blacklisting engine.

Real-world isolation techniques are far less efficient. On a LAN, an intelligent network switch could be used to selectively shut down the ports of infected hosts, or to cut off an entire sensitive segment. Network firewalls and perimeter routers can be used to shut down the affected services. Reactive IDSs (the so-called "intrusion prevention systems") can be used to selectively kill worm connections on the base of attack signatures.

Automatic reaction policies are intrinsically dangerous. False positives and the possibility of fooling a prevention system into activating a denial-of-service are dangerous enough to make most network administrators wary.

4.4   Immunization

In [24] the effect of selective immunization of computers on a network is discussed. The dynamics of infection and the choice of immunization targets are examined for two network topologies: a hierarchical, tree-like topology (which is obviously not realistic for modeling the Internet), and a cluster topology. The results are interesting, but the exact meaning of "node immunization" is left open.

If it means the deployment of anti-virus software, as we discussed before, it consists of a reactive technology which cannot prevent the spread of malicious code. If it means the accurate deployment of patches, the study could be used to prioritize the process, in order to patch sooner the most critical systems.

4.5   Honeypots and Tarpits

Honeypots are fake computer system and networks, used as a decoy to delude intruders. They are installed on dedicated machines, and left as a bait so that aggressors will lose time attacking them and trigger an alert. Since honeypots are not used for any production purpose, any request directed to the honeypot is at least suspect. Honeypots can be made up of real sacrificial systems, or of simulated hosts and services (created using Honeyd by Niels Provos, for example).

An honeypot could be used to detect the aggressive pattern of a worm (either by attack signatures, or by using a technique such as the one described above). When a worm is detected, all the traffic incoming from it can be captured at the gateway level and routed to a fake version of the real network.

Using signatures has the usual disadvantage that they may not be readily available for an exploding, unknown worm. Using anomaly detection filters is prone to false positives, and could send legitimate traffic into the fake honeypot.

Once a worm has entered a honeypot, its payload and replication behaviors can be easily studied, without risk. An important note is that hosts on the honeypot must be quarantined and made unable to actually attack the real hosts outside. By using sacrificial unprotected machines, copies of the worm can be captured and studied; sometimes, even using honeyd with some small tricks is sufficient in order to capture copies of the worm.

As an additional possibility, an honeypot can be actually used to slow down the worm propagation, particularly in the case of TCP based worms. By delaying the answers to the worm connections, a honeypot may be able to slow down its propagation: when a copy of the worm hits the honeypot, it sees a simulated open TCP port, and thus it is forced to attack the fake host, losing time in the process.

This technique is used in the Labrea "tarpit" tool. LaBrea can reply to any connection incoming on any unused IP address of a network, and simulate a TCP session with the possible aggressor. Afterward it slows down the connection: when data transfer begins to occur, the TCP window size is set to zero, so no data can be transferred. The connection is kept open, and any request to close the connection is ignored. This means that the worm will have to wait for a timeout in order to disconnect, since it uses the standard TCP stack of the host machine which follows RFC standards. A worm won't be able to detect this slowdown, and if enough fake targets are present, its growth will be slowed down. Obviously, a multi-threaded worm will be less affected by this technique.

4.6   Counterattacks and Good Worms

Counter-attack may seem a viable cure to worms. When an host A sees an incoming worm attack from host B, it knows that host B must be vulnerable to the particular exploit that the worm uses to propagate, unless the worm itself removed that vulnerability. By using the same type of exploit, host A can automatically take control of host B and try to cure it from infection and patch it.

The first important thing to note is that, fascinating as the concept may seem, this is not legal, unless host B is under the control of the same administrator of host A. Additionally, automatically patching a remote host is always a dangerous thing, which can cause considerable unintended damage (e.g. breaking services and applications that rely on the patched component).

Another solution which actually proves to be worse than the illness is the release of a so-called "good" or "healing" worm, which automatically propagates in the same way the bad worm does, but carries a payload which patches the vulnerability. A good example of just how dangerous such things may be is the *Welchia* worm, which was meant to be a cure for *Blaster*, but actually caused devastating harm to the networks.

5   Conclusions and Future Work

In this paper, we reviewed existing modeling techniques for computer virus propagation, presenting their underlying assumptions, and discussing whether or not these assumptions can still be considered valid. We also presented a new model, which extends the Random Constant Spread model, which allows us to derive some conclusions about the behavior of the Internet infrastructure in presence of a self-replicating worm. We compared our modeling results with data collected during the outbreak of the Slammer worm and proposed an explanation for some observed effects. We also discussed briefly countermeasures for fighting a self-replicating worm, along with their strengths and weaknesses. As a future extension of this work, we will try to model these countermeasures in order to assess their value in protecting the Internet infrastructure.

Acknowledgments.   We thank David Moore of CAIDA, and Stuart Staniford of Silicon Defense, for allowing us to reproduce their measurements of Code Red v2 expansion. We also wish to thank Sergio Savaresi, Giuliano Casale and Paolo Albini for reading a preliminary version of the equations in section 3.2 and providing helpful suggestions for the improvement of the model.

References

1. Cohen, F.: Computer Viruses. PhD thesis, University of Southern California (1985)
2. Cohen, F.: Computer viruses – theory and experiments. Computers & Security **6** (1987) 22 35
3. Power, R.: 2003 csi/fbi computer crime and security survey. In: Computer Security Issues & Trends. Volume VIII. Computer Security Institute (2002)
4. White, S.R.: Open problems in computer virus research. In: Proceedings of the Virus Bulletin Conference. (1998)
5. Staniford, S., Paxson, V., Weaver, N.: How to 0wn the internet in your spare time. In: Proceedings of the 11th USENIX Security Symposium (Security '02). (2002)
6. Whalley, I., Arnold, B., Chess, D., Morar, J., Segal, A., Swimmer, M.: An environment for controlled worm replication and analysis. In: Proceedings of the Virus Bulletin Conference. (2000)
7. Calzarossa, M.C.: Performance Evaluation of Mail Systems. In: Calzarossa, M. and Gelenbe, E., editors, Performance Tools and Applications to Networked Systems. Volume 2965 of Lecture Notes in Computer Science. Springer. (2004)
8. Gelenbe, E., Lent, R., Gellman, M., Liu, P., Su, P.: CPN and QoS Driven Smart Routing in Wired and Wireless Networks. In: Calzarossa, M. and Gelenbe, E., editors, Performance Tools and Applications to Networked Systems. Volume 2965 of Lecture Notes in Computer Science. Springer. (2004)
9. Spafford, E.H.: Crisis and aftermath. Communications of the ACM **32** (1989) 678–687
10. Kephart, J.O., White, S.R.: Directed-graph epidemiological models of computer viruses. In: IEEE Symposium on Security and Privacy. (1991) 343–361
11. Hethcote, H.W.: The mathematics of infectious diseases. SIAM Review **42** (2000) 599–653
12. Billings, L., Spears, W.M., Schwartz, I.B.: A unified prediction of computer virus spread in connected networks. Physics Letters A (2002) 261–266
13. Zou, C.C., Towsley, D., Gong, W.: Email virus propagation modeling and analysis. Technical Report TR-CSE-03-04, (University of Massachussets, Amherst)
14. Permeh, R., Maiffret, M.: .ida 'code red' worm. Advisory AL20010717 (2001)
15. Permeh, R., Maiffret, M.: Code red disassembly. Assembly code and research paper (2001)
16. Permeh, R., Hassell, R.: Microsoft i.i.s. remote buffer overflow. Advisory AD20010618 (2001)
17. Levy, E.A.: Smashing the stack for fun and profit. Phrack magazine **7** (1996)
18. Craig Labovitz, A.A., Bailey, M.: Shining light on dark address space. Technical report, Arbor networks (2001)
19. Moore, D., Shannon, C., Brown, J.: Code-red: a case study on the spread and victims of an internet worm. In: Proceedings of the ACM SIGCOMM/USENIX Internet Measurement Workshop. (2002)
20. Moore, D.: Network telescopes: Observing small or distant security events. In: Proceedings of the 11th USENIX Security Symposium. (2002)
21. Chen, Z., Gao, L., Kwiat, K.: Modeling the spread of active worms. In: Proceedings of IEEE INFOCOM 2003. (2003)
22. Zou, C.C., Gao, L., Gong, W., Towsley, D.: Monitoring and early warning for internet worms. In: Proceedings of the 10th ACM conference on Computer and communication security, ACM Press (2003) 190–199

23. Zou, C.C., Gong, W., Towsley, D.: Code red worm propagation modeling and analysis. In: Proceedings of the 9th ACM conference on Computer and communications security, ACM Press (2002) 138–147
24. Wang, C., Knight, J.C., Elder, M.C.: On computer viral infection and the effect of immunization. In: ACSAC. (2000) 246–256
25. Wang, Y., Wang, C.: Modelling the effects of timing parameters on virus propagation. In: Proceedings of the ACM CCS Workshop on Rapid Malcode (WORM'03). (2003)
26. Zou, C.C., Gong, W., Towsley, D.: Worm propagation modeling and analysis under dynamic quarantine defense. In: Proceedings of the ACM CCS Workshop on Rapid Malcode (WORM'03). (2003)
27. Moore, D., Shannon, C., Voelker, G.M., Savage, S.: Internet quarantine: Requirements for containing self-propagating code. In: INFOCOM. (2003)

Performance Evaluation of Mail Systems*

Maria Carla Calzarossa

Dipartimento di Informatica e Sistemistica
Università di Pavia
I-27100 Pavia, Italy
mcc@unipv.it

Abstract. Electronic mail is the core of modern communications. People rely on email to conduct their business and stay in touch with families and friends. The ubiquity and popularity of email make its QoS an important issue. Performance of mail systems is the result of the interactions between their hardware and software components and the user behavior, that is, how users exploit and exercise these components. This paper addresses the performance issues of mail systems by focusing on the characterization of their workloads and on benchmarking. The mail systems considered in this paper rely on Internet standard messaging protocols.

1 Introduction

Electronic mail is one of the earliest and most popular Internet applications that has had major societal and economical impacts [16], [27]. Email is an asynchronous communication medium that connects people across time zones and culture. Email is the core of modern communications and plays a central role for both personal and professional lives. It is used to stay in touch with families and friends and conduct business. The ubiquity of email availability is one of the strengths of the Internet whose phenomenal growth and popularity are also due to the popularity of email. Email is fast, cheap and easy to distribute and is now as necessary as telephone and other communication media or even more necessary. Email services have become more and more elaborate and powerful and continue to evolve [24]. Messages are no longer limited to textual contents; they can include any type of multimedia contents, that is, images, sound, video and other applications such as those described in [12]. Many people rely on email to get work done and information communicated across an enterprise or a research institution. Hence, they expect mail services with a good QoS. Service providers have to meet user requirements by designing and delivering mail systems characterized by high availability and reliability and good performance.

There is a variety of commercial products available for supporting email services. Most of them rely on Internet standard protocols to format, transport

---

* This work has been supported by the Italian Ministry of Education, Universities and Research (MIUR) under the FIRB and Cofin Programmes.

M.C. Calzarossa and E. Gelenbe (Eds.): MASCOTS 2003, LNCS 2965, pp. 51–67, 2004.
© Springer-Verlag Berlin Heidelberg 2004

and retrieve messages, whereas only few use their own protocols, message formats and transports.

The performance of mail systems is the result of a complex set of interactions between their hardware and software components and of the user behavior, that is, how users exercise these components. Hence, it is important to assess and predict the performance of mail systems both at the design stage and during their entire life cycle.

Despite the importance of these topics, few papers deal with the design of mail systems and the analysis of their performance. In [23], the behavior of Lotus Notes email clients is modeled with the objective of balancing the workload on mail servers. The model is based on the distribution of the work generated by users. The characterization of the workload associated with Internet standard mail protocols is addressed in [3]. The models obtained in this study are used describe the input of benchmark experiments. A high–availability high–performance email cluster is presented in [20]. The design of the system is based on the analysis of user needs and behavior. Benchmarking is applied to test the system under heavy load conditions. Benchmarking of mail servers is also addressed in [4] and a prototype of a benchmark tool is presented.

The purpose of this paper is to present a survey of the performance issues typical of mail systems that rely on Internet standard protocols. We first introduce the architecture of mail systems by describing the Internet mail model and the corresponding protocols. We then focus on the performance of mail systems by presenting a methodology for the analysis of their workloads. Benchmarking is then introduced as a technique to assess and compare the performance of different mail systems.

The paper is organized as follows. Section 2 presents the Internet mail model and the protocols used to transport and access messages. Section 3 focuses on workload characterization of mail systems. Section 4 discusses the benchmarking techniques in the framework of mail systems. Finally, Section 5 summarizes the paper and outlines some open issues.

2   Internet Mail Model

Mail systems rely on a client/server architecture consisting of a large number of hardware and software components interacting together. The mail server is the component that provides mail services. The mail clients are the components used to access these services. The network is the communication medium between clients and server. The architecture of mail systems is organized around the Internet mail model, namely, a collection of standard components whose goal is to provide a framework for carrying electronic messages between users. One main characteristic of these components is their interoperability, that is, their ability to operate on different platforms.

The framework of the Internet mail model consists of agents, mailstores and standards [14], [19]. The standards are the rules that define how to handle messages, that is, how to format, encode, transport and access the messages. The agents are the software programs responsible to perform a certain task on behalf of a human user or of another program. The mailstore is the filing cabinet of a mail system. User mailboxes are stored in the portion of the mailstore that contains the messages addressed to the specific user.

Mail messages are managed by three different types of agents, namely:

- Mail User Agent (MUA)
- Mail Transport Agent (MTA)
- Mail Delivery Agent (MDA).

The Mail User Agents, known also as mail readers, are the programs that run on the user clients, e.g., PCs, workstations, PDAs, and allow users to read, reply, save and compose messages. Current MUAs are provided with graphical interfaces and a rich set of features that allow users to send and view multimedia messages and attachments and to address security issues. The most popular MUAs are Microsoft Outlook, Netscape Messenger and Qualcomm Eudora.

The Mail Transport Agent is the program that sends and receives messages between mail servers and is responsible to know how to route them. The Mail Delivery Agent is the program that places in the mailstore the messages handed by the MTA.

Figure 1 sketches the hardware and software components and the main functions of the mail architecture. The Mail User Agent is a client/server program

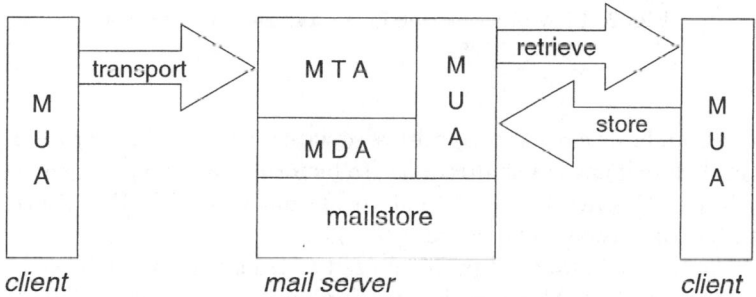

Fig. 1. Components of a mail architecture.

whose client component runs on the user client and whose server component runs on the mail server. Both the MTA and the MDA run on the mail server. An MTA is a client/server program that acts as a client when it sends messages to another MTA and as a server when it receives messages from another MTA. Note that both components of the MTA run on every mail server. The mailstore is located in the mail server. A mailstore can be organized as a traditional flat file or characterized by a more sophisticated organization, e.g., hierarchical, to cope with the increased volume and size of the messages and to allow mailbox sharing among users.

Figure 2 shows the flow and the components of the Internet mail model involved in transporting a message from a sender to a receiver. As can be seen, Mail User Agents are the source and the target of email. At the source an MUA collects messages to be transmitted from a user. At the destination, an MUA accesses and manipulates on behalf of a user the messages in the mailstore. A message originates from the sender's MUA and is transmitted to the MTA of the sender's mail server. This MTA determines the IP address of the receiver's mail server and sends the message to the receiver's MTA that hands it off to the MDA. The MDA deposits the message in the mailstore where it can be retrieved by means of the receiver's MUA. Note that the figure shows a simplified path as a message could travel through multiple mail servers.

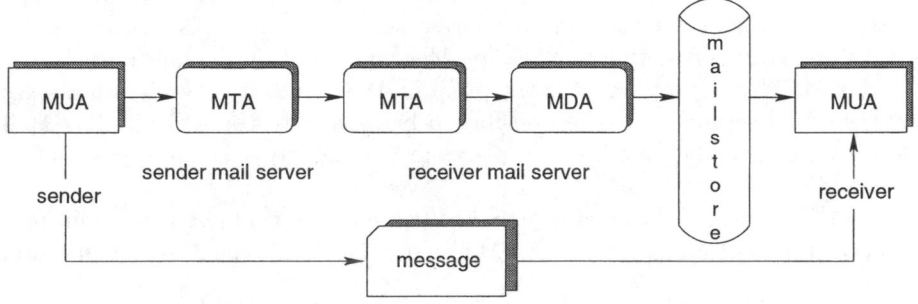

Fig. 2. Flow of a message in the Internet mail model.

Various standards and protocols are associated with the flow shown in Fig. 2. A message, before being transmitted, has to be formatted and encoded according to the Internet Message Format Standard [8] and to its MIME (Multipurpose Internet Mail Extensions) extensions [10], [11].
SMTP is the Internet standard protocol used to transport and deliver messages, whereas POP3 and IMAP4 are the standard mail access protocols. All these protocols are implemented within the corresponding agents. In particular, the MUAs implement the POP3 and IMAP4 protocols, whereas SMTP is implemented within the MTAs and MDAs.
To fully understand the performance and the workloads of mail systems, it is important to understand the characteristics and the behavior of the underlying transport and access protocols. Next sections present a brief survey of these protocols.

2.1 SMTP

The Simple Mail Transfer Protocol (SMTP) [17] is an Internet standard protocol whose objective is to transfer mail reliably and efficiently. The protocol defines the rules for exchanging messages between a client and a mail server and between mail servers. As already pointed out, SMTP is typically implemented by the Mail Transport Agent and Mail Delivery Agent running on the mail servers. SMTP is a push protocol, that is, the sender's mail server pushes the message to the receiver's mail server. The SMTP protocol consists of two components: the SMTP client that executes on the sender's mail server and the SMTP server that executes on the receiver's mail server. Both components reside on each mail server. When an SMTP client has a message to transmit, it establishes a connection to an SMTP server. An SMTP session is initiated when a client opens a connection to a server and the server responds with an opening message. The connection uses any available transport service, e.g., TCP/IP. Once the connection is established, the client and the server perform some handshaking before the actual transmission of the message.

SMTP is a stateful protocol, that is, both the client and the server know about the state of the connection. The communication between the sender (i.e., client) and the receiver (i.e., server) is an alternating dialogue controlled by the sender. The sender issues a command and the receiver responds with a reply. Each mail transaction involves a sequence of commands and replies. A session is terminated when the client sends a QUIT command.

The SMTP client is responsible to guarantee the reliability of the mail transport and delivery. The client either properly delivers the message or reports a failure. For this purpose, the protocol supports queueing capabilities. A message is put in a mail queue whenever a retryable error is encountered by the SMTP client, as in the case of receiver's mail server or network temporarily unreachable.

Sendmail [2], [6] is the most popular program that implements the SMTP protocol.

2.2   POP3

The Post Office Protocol – Version 3 (POP3) [22] is an Internet standard mail access protocol whose objective is to allow users to access their mailboxes located in the mail server. Like the SMTP protocol, POP3 is a stateful protocol consisting of two components: a POP3 client running on the user clients as part of their Mail User Agents and a POP3 server running on the mail server. The protocol is very simple and because of its simplicity it provides limited functionalities. A POP3 session starts when the POP3 client opens a connection to the mail server. The MUA running on the user client issues the commands to the mail server. The mail server responds to each command with a reply. Users can submit commands to retrieve information about their mailbox and to retrieve or delete messages. POP3 supports the off–line model, that is, once the messages have been stored in the client mailbox, the connection with the mail server is closed and the messages are manipulated on the user client. Moreover, messages will be deleted from the server mailbox depending on the option, i.e., download-and-delete, download-and-keep, specified by the POP3 client.

The POP3 protocol neither keeps any state information across sessions nor provides any mechanism to maintain consistency between client and server mailboxes. Thereby, it is the ideal protocol for users accessing their mailbox from a single client machine.

Today POP3 is the protocol of choice by ISPs as it consumes very little resources on the mail servers and puts the entire load on user clients.

2.3   IMAP4

The Internet Mail Access Protocol – Version 4 (IMAP4) [7] is an Internet standard protocol for mail access designed to overcome the POP3 limitations. IMAP4 is a very flexible and powerful protocol that provides complex messaging capabilities [15], [21]. Users can access and manipulate their mailbox directly on the mail server as it was a local resource. The IMAP4 protocol supports off–line, on–line and disconnected models. As for POP3, the off–line model allows users to download and manipulate messages on their local machine. With the on–line model, users access their messages directly on the mail server. The disconnected model complements the off–line model in that it allows users to download messages on their client machine, manipulate them and then upload the changes to synchronize the server mailbox such that it is always consistent. Hence, users can access their mailboxes from everywhere using different client machines without the need to replicate their mailboxes.

An IMAP4 session consists of commands sent by the client and responses of the mail server. A session goes through four possible states. Commands and responses are valid only in certain states. The protocol supports a large variety of commands that provide users with many features to manipulate their mailboxes. Users can maintain multiple mailboxes on the server with a hierarchical organization. They can share their mailboxes, search them on the server and selectively choose which messages to download based, for example, on their headers. The IMAP4 protocol puts a significant load on the mail server. It heavily exercises mail server resources for what concerns processing power and disk space. Moreover, demand of bandwidth is high as many IMAP4 sessions can be simultaneously open on the mail server and the duration of these sessions can be quite long.

Because of its flexibility and powerful characteristics, today IMAP4 is the mail access protocol of choice by enterprises and universities willing to offer advanced mail services.

3   Workload Characterization

The workload of mail systems consists of the commands issued by users towards a mail server and of the mail server responses. Moreover, the interaction between clients and server generates packets on the interconnecting network. The objective of workload characterization is to study the behavior of this composite workload, identify typical usage patterns and predict its future behavior.

Workload characterization can be addressed from different perspectives. The typical approach is experimental, that is, based on the analysis of measurements collected on the system under study [5]. Thereby, to characterize the workload of mail systems measurements of user clients, mail server and network have to be collected and analyzed. Mail servers typically provide facilities to log all the requests associated with the Internet standard protocols. A large variety of tools is available to collect measurements on the network. The hardest part of the analysis is represented by the clients. Clients are usually "dispersed" and it is difficult to install and run logging facilities on each individual machine. Moreover, users are somehow reluctant to have their activities monitored even for statistical purposes. As a consequence, measurements on the client side are seldom available. The behavior of the users is typically inferred from the analysis of the load of the mail server. Note that mail servers are the main target of the analysis as they are the most critical resources of mail systems, whereas performance of user clients is seldom addressed as an issue.

The level of details and the granularity of the measurements collected vary and depend on the protocol and on the measurement tools. For example, the measurements collected for the commands processed by a mail server include time stamps and other information that is a function of the command type, e.g., size of the message being fetched, number of messages in the mailbox, IP address of the sender and of the receiver of the message. Let us remark that before any processing, log files have to be "sanitized" to remove all sensitive information they might contain.

The analysis of the measurements is based on a typical workload characterization methodology consisting of several steps, namely:

1. identification of the parameters describing the characteristics and the properties of the workload
2. exploratory analysis of the identified parameters
3. application of statistical, probabilistic and numerical analysis techniques to build workload models
4. validation of the workload models.

The parameters have to provide a qualitative and quantitative description of the workload. The choice of the number and type of parameters is driven by the objective of the study. Once the parameters have been identified, the exploratory analysis provides some preliminary insights into the properties and behavior of the workload. Techniques, such as, clustering, factor analysis, numerical fitting, time series, are applied to obtain models able to reproduce the static and dynamic characteristics of the workload. The validity of the identified models is assessed using the representativeness criterion.

Moreover, to represent the hierarchical nature of the workload, it can be decomposed in a hierarchy of layers, where the workload of each layer is transformed into the workload of next lower layer. At the top layer of the hierarchy, we identify the *users* who are the source of the load of the mail server. Each user opens one or more sessions. Hence, the next lower layer is represented by the *sessions*. Each session consists of multiple *commands*. Each command involves a flow of *packets* on the network, whose number and type depend on the specific command that in turn depends on the mail protocol.

The workload characterization methodology has to focus on each layer of the hierarchy and take into account the properties and the characteristics of each layer together with the transformations between layers.

Note that due to the peculiarities of different mail protocols, workload characterization has to be performed on a per protocol basis, that is, on a per workload type. Next sections address the characterization of the three types of mail system workloads, namely, SMTP, POP3 and IMAP4 workloads.

3.1   SMTP Workload

The SMTP workload refers to the requests, i.e., commands, processed by a mail server to send and receive messages. The parameters used to describe the SMTP workload deal with its intensity and the characteristics of the messages being sent or received.

Workload intensity can be studied in terms of arrival rate, that is, number of requests arrived at the mail server per time unit, and exchange rate, that is, number of bytes sent/received by the mail server per time unit. Figure 3 shows the arrival rate of the SMTP requests over a time period of 24 hours. The measurements refer to a mail server of a small ISP. As can be seen, the

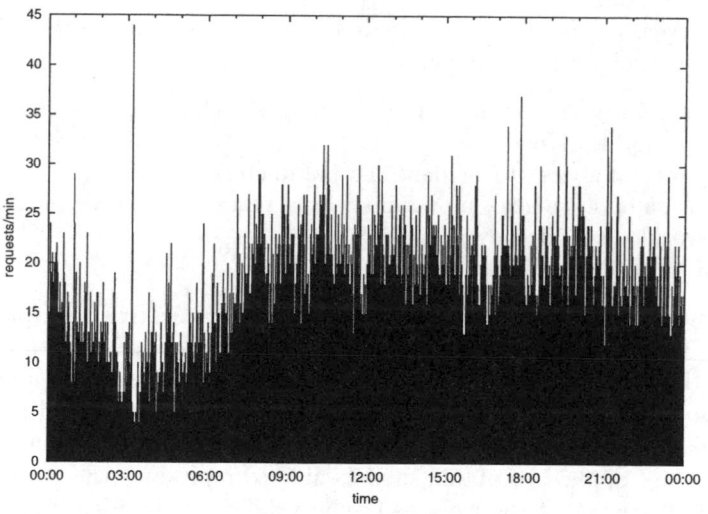

Fig. 3. Arrival rate over a 24-hours period.

arrivals exhibit large fluctuations and burstiness. The load is light during the early hours of the day, and then it slowly increases. The average arrival rate is 16.58 requests/min.

Messages can be characterized by quantitative parameters, such as, size, number of recipients, IP address of the sender and of the receiver. Figure 4 shows

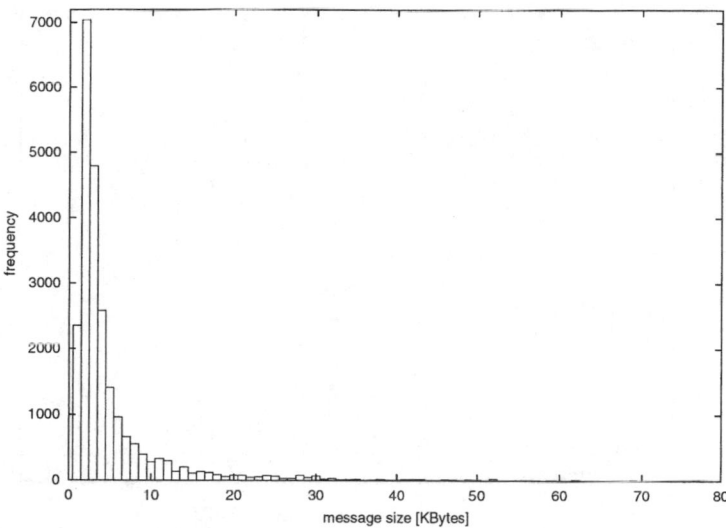

Fig. 4. Message size distribution.

an example of a message size distribution. The diagram plots a zoom of the distribution in the range 0-80Kbytes. As can be seen, the distribution is highly positively skewed. The average message size is equal to approximately 7.7Kbytes with a standard deviation more than six times larger than the corresponding mean. Most of the messages are short, even though there are messages as large as 5Mbytes.

Another important parameter that describes the SMTP workload is the interarrival time, that is, the time elapsed between two consecutive messages sent or received by the mail server. Figure 5 shows the interarrival time distribution obtained from measurements collected over a 12 hours period. The figure also shows the corresponding models obtained applying numerical fitting techniques. Two distributions, namely a Weibull distribution and a Pareto distribution, have been identified to represent the empirical data. A Weibull distribution, whose analytic expression is given by:

$$f(x) = \frac{b}{a}\left(\frac{x}{a}\right)^{b-1} e^{-\left(\frac{x}{a}^b\right)}$$

with shape parameter a equal to 3.027 and scale parameter b equal to 1.124, best fits the body of the empirical distribution. A Pareto distribution, whose analytic expression is given by:

$$f(x) = \alpha k x^{-(\alpha + 1)}$$

with shape parameter α equal to 2.459 and location parameter k equal to 3.958, best fits the tail. The dashed vertical line shown in the figure denotes the threshold, identified by right censoring techniques, between the body and the tail of the empirical distribution. This value is equal to approximately 7 seconds.

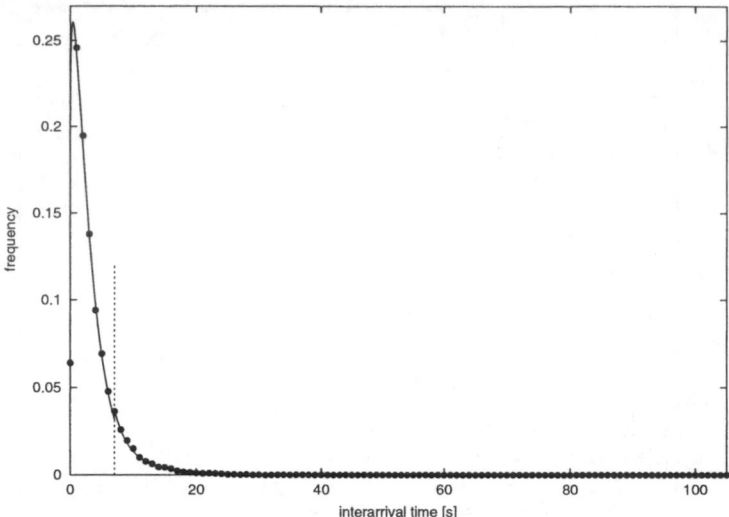

Fig. 5. Empirical interarrival time distribution (dotted curve) and fitting models (solid curves).

3.2 POP3 Workload

The analysis of the POP3 workload follows the hierarchical approach outlined in Sect 3. POP3 users are characterized by the number of sessions they open towards the mail server and by the size and number of messages in their mailbox. Each session is described by its length, that is, number of commands issued within the session, its duration, that is, the time elapsed between opening and closing the connection, and the time between two consecutive sessions. The POP3 protocol supports very few command types. Commands within a session are typically described in terms of their operations on the server mailbox, that is, number and size of the messages retrieved from the server mailbox and number and size of the messages deleted from the server mailbox. Moreover, to derive typical usage patterns, it is important to identify within each session commands issued by human users and commands automatically issued by mail readers to check, for example, for new messages or to keep the connection between client and mail server open.

As for the SMTP workload, models of the POP3 workload are obtained by applying statistical, probabilistic and numerical techniques. In [3], an application of probabilistic graphs to describe the behavior of POP3 users is presented.

3.3 IMAP4 Workload

The description of the IMAP4 workload has to reflect all the advanced features provided by the protocol. IMAP4 users are described in terms of number of sessions they open towards the mail server, number of different mailboxes, size and number of messages per mailbox. As for the POP3 workload, a session is described by its duration, length and time between two consecutive sessions. Figure 6 shows the distribution of the session duration as measured over a 24 hours time period for a population of about 80 users. The diagram plots a zoom

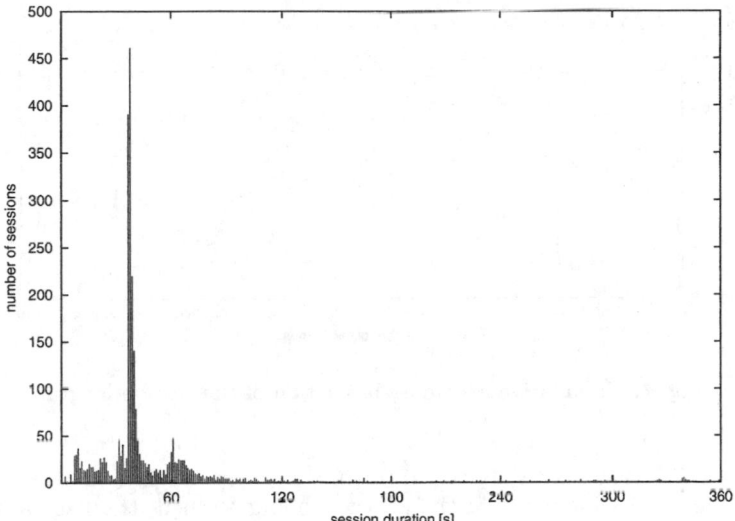

Fig. 6. Distribution of the session duration.

of the distribution for durations in the range 0-360 seconds. Most of the sessions are short, even though there are sessions spanning the whole 24 hours. The duration of approximately 10% of the sessions is between 32 and 37 seconds. About 92% of the sessions are shorter than 2 minutes and 94% end within 6 minutes.

Figure 7 plots the cumulative distribution of the session length. The distribution refers to lengths in the range 0-400 commands. The diagram shows the tendency of the users to issue few commands per session, even though we have observed a session with as many as 39000 commands. The 75-th percentile of the distribution corresponds to sessions with 25 commands at most. The 98-th percentile of the distribution corresponds to sessions with less than 400 commands. We notice that about 20% of the sessions consists of 3 commands only. These sessions correspond to the automatic checks for new messages performed by mail readers and ending up with no new messages in the mailbox.

The IMAP4 protocol supports a large variety of commands. The set of commands implemented varies as a function of the mail reader. Because of the heterogeneity of the commands, their characterizing parameters are specific of each type. For example, a SELECT command is described by its time stamp and by the identifier of the mailbox being selected. A FETCH command is described by

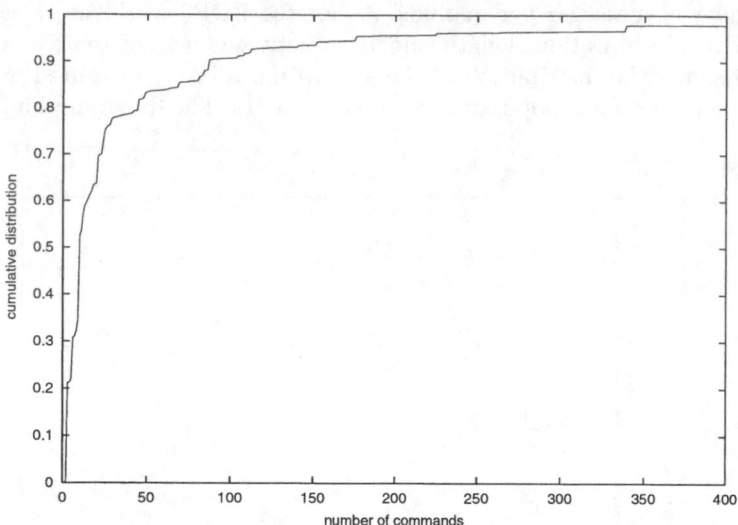

Fig. 7. Cumulative distribution function of the session length.

its time stamp, the identifier of the message being fetched, the flags associated with the message and the message size. The quantitative characterization of the commands has to take into account these parameters as well as the description of the traffic generated on the network. Indeed, each command generates multiple packets on the network, whose number, type and size depend on the command type and the data to the transferred.

The functional characterization of the IMAP4 workload takes into account the sequences of commands issued by the users within a session. The models of these sequences are based on probabilistic graphs. The nodes of these graphs correspond to the different command types. The arcs, with their associated probabilities, represent the transitions between commands. Figure 8 shows an example of a graph that describes the behavior of an IMAP4 user. As can be seen, the user issues nine different command types. The first command of the session is a CAPABILITY command. The last command of the session is a LOGOUT command. Once the mailbox has been selected, the user fetches the messages from the mailbox, then issues various commands to search, store or expunge messages. Note that the graph contains an extra arc, with probability 1, between the LOGOUT and CAPABILITY commands. This arc represents the "transition" between sessions, that is, at the end of a session a new session eventually starts.

The behavior of each user is then described by one graph. To obtain more compact and manageable models, clustering techniques are applied to the graphs described by their transition probability matrices. As a result, classes of graphs that identify users characterized by similar behavior are obtained.

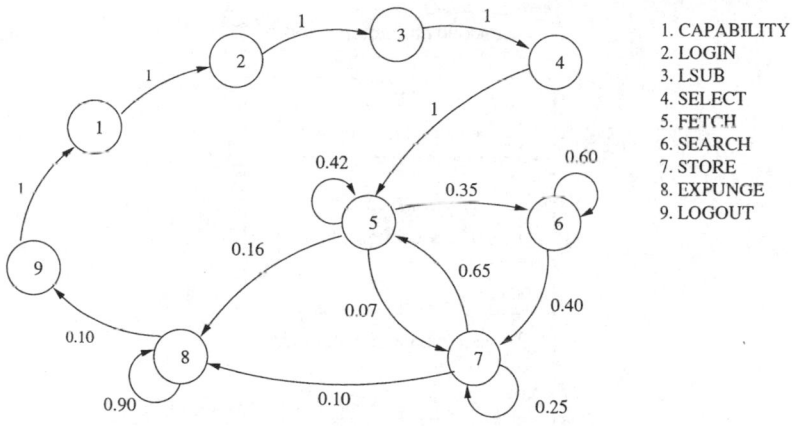

Fig. 8. Probabilistic graph of an IMAP4 session.

4   Benchmarking

Benchmarking is a technique extensively used to evaluate and compare the performance of different systems [9]. A benchmark of a mail system has to exercise its client/server architecture and reproduce the static and dynamic characteristics of the workload. All the components of the system under test (SUT), i.e., user clients, mail server and network, are involved in a benchmark experiment. Clients play a key role in that they have to load the SUT and emulate the behavior of the users. The accuracy of a benchmark experiment is heavily influenced by the representativeness of the workload being executed. Figure 9 shows the typical phases of a benchmark experiment. As can be seen, the characterization of the workload is a preliminary step towards the construction of an experiment. The workload is specified in terms of several parameters described by their corresponding models. The number and type of users to be emulated, the user profile, that is, the number and type of commands each user will issue towards the mail server and the user think time, that is, the time between two consecutive commands, are examples of parameters describing the workload of a mail system. As a result of the workload description, the timed sequences of commands that will be executed by the clients are generated. The setup of the SUT deals with the initialization of the working environment, that is, the installation and configuration on the SUT of the software modules required by a benchmark experiment and the distribution of the workload to the individual clients. Moreover, to avoid a "cold start" of the experiment, the mailstore has to be initialized by populating the user mailboxes according to their initial message distribution. After the setup phase, the actual execution of the experiment takes place. The emulated users send their commands to the mail server that processes them and responds with its replies. Note that the execution of a benchmark experiment requires the full availability of the SUT in a dedicated mode. During

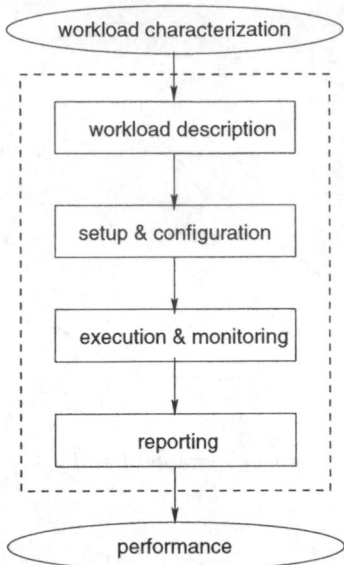

Fig. 9. Phases of a benchmark experiment.

the execution of the experiment, measurements of the workload and traffic of the network are collected. The reporting phase deals with the analysis of the measurements. As a result of this phase, indices that describe the performance achieved by the SUT are obtained.

Benchmarking tools have to implement all the phases outlined in Fig. 9 and guarantee the repeatability and the reproducibility of the benchmark experiments. SPECmail2001 [25], a standard benchmark that focuses on SMTP and POP3 workloads, describes the user profile in terms of parameters, such as, number of messages sent per user, number of recipients per message, percentage of messages to local destination, message size, number of mail checks per user. Running and reporting rules are defined as part of the benchmark specifications to ensure that the generated results are meaningful, comparable to other generated results and repeatable.

The research prototype presented in [4] focuses on SMTP, POP3 and IMAP4 workloads. Several types of statistical distributions, e.g., uniform, exponential, are provided to describe the workload parameters. Moreover, models based on probabilistic graphs can be used to specify the behavior of the IMAP4 users. To guarantee the repeatability of the experiments, the tool is provided with an archive that acts as a repository of the workload specifications, log files and reports of the benchmark experiments.

The metrics used to compare the performance of different mail systems typically refer to the throughput of the mail server, i.e., number of commands processed by the mail server per time unit, and response time, that is, time elapsed between the start and the end of a command. A composite metrics, known as "messages-per-minute", is used by SPECmail2001 to indicate the load the server can sustain with a reasonable quality of service.

Figure 10 shows an example of an output of a benchmark experiment. The

Benchmark Results

Experiment duration 0h 1min 49s 180ms
Throughput   18.78 commands/s

Number of SMTP users: 0
Number of POP3 users: 0
Number of IMAP4 users: 10

Sessions duration [s]

Protocol	Min	Max	Average	Standard Deviation	Coefficient of Variation	
IMAP4	107.54	108.99	108.20	0.47	0.00	

IMAP4 workload: response time [ms]

Command	Min	Max	Average	Standard Deviation	Coefficient of Variation	Number of commands
LOGIN	5.05	5.80	5.49	0.28	0.05	10
CAPABILITY	0.20	23.55	4.56	6.52	1.43	10
SELECT	9.30	638.38	41.05	69.32	1.69	385
GET	0.11	244.56	9.34	32.15	3.44	465
FETCH	0.21	448.20	68.34	80.41	1.18	283
SEARCH	0.40	162.74	11.15	19.85	1.78	251
STORE	0.21	160.76	3.97	15.40	3.88	302
EXPUNGE	0.16	560.48	16.32	46.09	2.82	334
LOGOUT	0.50	17.82	4.19	6.71	1.60	10

Fig. 10. Output of a benchmark experiment.

experiment refers to a run with 10 IMAP4 users, where each user opens one session. The report summarizes the performance of the mail server in terms of duration of the experiment and throughput of the mail server, that is, commands processed by the mail server per minute. Statistics of session duration and response time are also provided. Response times are reported on a per command basis.

5   Conclusions

Performance and reliability of mail services are important requirements for the modern society to work properly. Even excluding spam messages, the number of messages being exchanged world–wide is rapidly increasing. Similarly, the number of new users approaching mail services is increasing. Mail servers have to meet user demands and guarantee a good QoS. In this framework, performance evaluation plays a key role. This paper focuses on performance evaluation of mail systems that rely on the Internet standard protocols SMTP, POP3 and IMAP4.

The analysis of the workload of these systems shows the properties of each type of workload and the behavior of the users. Benchmarking is a powerful technique to assess and compare the performance of different mail systems. An accurate benchmark experiment has to exercise all the components of the systems under a realistic workload.

It would be interesting to consider QoS for mail systems in the more general context of QoS as discussed in this volume in [13] and [18], and also to view the specific situation of mail in the context of wireless systems as discussed in [1] and [26]. Moreover, open issues in the characterization of workload of mail systems deal with the analysis of the behavior of "spammers" and the identification of atypical usage patterns that might indicate the presence of a security attack against mail servers. Often security threats, e.g., virus propagation, denial of service attacks, come via email and are source of severe performance degradation as they lead to unexpected peaks of very heavy load on the mail servers. It is then important to detect these patterns and design mechanisms for the mail servers to quickly react and adapt to sudden changes in their workload.

References

1. K. Al-Begain. Performance Models for 2.5/3G Mobile Systems and Networks. In M. Calzarossa and E. Gelenbe, editors, *Performance Tools and Applications to Networked Systems*, volume 2965 of *Lecture Notes in Computer Science*. Springer, 2004.
2. F.M. Avolio and P.A. Vixie. *Sendmail: Theory and Practice*. Digital Press, 1995.
3. L. Bertolotti and M. Calzarossa. Models of Mail Server Workloads. *Performance Evaluation - An International Journal*, 46(2/3):65–76, 2001.
4. M. Calzarossa. A Tool for Mail Servers Benchmarking. In G. Kotsis, editor, *Performance Evaluation - Stories and Perspectives*, OCG Schriftenreihe, pages 231–240. Austrian Computer Society, 2003.
5. M. Calzarossa, L. Massari, and D. Tessera. Workload Characterization - Issues and Methodologies. In G. Haring, C. Lindemann, and M. Reiser, editors, *Performance Evaluation – Origins and Directions*, volume 1769 of *Lecture Notes in Computer Science*, pages 459–484. Springer, 2000.
6. B. Costales and E. Allman. *Sendmail - 3rd Edition*. O'Reilly, 2002.
7. M. Crispin. Internet Message Access Protocol - Version 4 rev1. RFC 3501, March 2003. http://www.ietf.org/rfc/rfc3501.txt.
8. D.H. Crocker. Standard for ARPA Internet Text Messages. RFC 822, August 1982. http://www.ietf.org/rfc/rfc822.txt.
9. R. Eigenmann, editor. *Performance Evaluation and Benchmarking with Realistic Applications*. MIT Press, 2001.
10. N. Freed and N. Borenstein. Multipurpose Internet Mail Extensions (MIME) Part One: Format of Internet Message Bodies. RFC 2045, November 1996. http://www.ietf.org/rfc/rfc2045.txt.
11. N. Freed and N. Borenstein. Multipurpose Internet Mail Extensions (MIME) Part Two: Media Types. RFC 2046, November 1996. http://www.ietf.org/rfc/rfc2046.txt.

12. E. Gelenbe, K. Hussain, and V. Kaptan. Enabling Simulation with Augmented Reality. In M. Calzarossa and E. Gelenbe, editors, *Performance Tools and Applications to Networked Systems*, volume 2965 of *Lecture Notes in Computer Science*. Springer, 2004.

13. E. Gelenbe, R. Lent, M. Gellman, P. Liu, and P. Su. CPN and QoS Driven Smart Routing in Wired and Wireless Networks. In M. Calzarossa and E. Gelenbe, editors, *Performance Tools and Applications to Networked Systems*, volume 2965 of *Lecture Notes in Computer Science*. Springer, 2004.

14. P. Hoffman. Putting it together: Designs on Internet mail. *netWorker*, 2(1):19–23, 1998.

15. The IMAP Connection. http://www.imap.org.

16. W. Jackson, R. Dawson, and D. Wilson. Understanding Email Interactions Increases Organizational Productivity. *Comm. of the ACM*, 46(8):80–84, 2003.

17. J. Klensin. Simple Mail Transfer Protocol. RFC 2821, April 2001. http://www.ietf.org/rfc/rfc2821.txt.

18. G. Kotsis. Performance Management in Dynamic Computing Environments. In M. Calzarossa and E. Gelenbe, editors, *Performance Tools and Applications to Networked Systems*, volume 2965 of *Lecture Notes in Computer Science*. Springer, 2004.

19. P. Loshin. *Essential Email Standards: RFCs and Protocols Made Practical*. Wiley, 2000.

20. W. Miles. A High–Availability High–Performance E–Mail Cluster. In *Proc. of the ACM SIGUCC Conf.*, pages 84–88, 2002.

21. D. Mullet and K. Mullet. *Managing IMAP*. O'Reilly, 2000.

22. J. Myers and M. Rose. Post Office Protocol - Version 3. RFC 1939, May 1996. http://www.ietf.org/rfc/1939.txt.

23. B. Pope. Characterizing Lotus Notes Email Clients. In *Proc. IEEE Third Int. Workshop on Systems Management*, pages 128–132, 1998.

24. M.T. Rose and D. Strom. *Internet Messaging: From the Desktop to the Enterprise*. Prentice Hall, 1998.

25. SPECmail2001. Standard Performance Evaluation Corporation. http://www.spec.org/mail2001.

26. C. Williamson. Wireless Internet: Protocols and Performance. In M. Calzarossa and E. Gelenbe, editors, *Performance Tools and Applications to Networked Systems*, volume 2965 of *Lecture Notes in Computer Science*. Springer, 2004.

27. E.V. Wilson. Email Winners and Losers. *Comm. of the ACM*, 45(10):121–126, 2002.

CPN and QoS Driven Smart Routing in Wired and Wireless Networks[*]

Erol Gelenbe[1], Ricardo Lent[2], Michael Gellman[1], Peixiang Liu[2], and Pu Su[2]

[1] Imperial College, London SW7 2BT, UK
{e.gelenbe,m.gellman}@imperial.ac.uk,
[2] University of Central Florida, Orlando, Florida 32816, USA
{rlent,pliu,psu}@cs.ucf.edu

Abstract. There exists an increasing need for dynamic mechanisms that take into account quality of service provisions in the establishment of routes in communication networks. Recently, we introduced a quality of service (QoS) driven routing algorithm called "Cognitive Packet Network" (CPN), which dynamically selects paths through a store-and-forward packet network so as to offer best effort QoS to an end-to-end traffic. This paper discusses a number of extensions to the algorithm: the incorporation of selective broadcasts to support the operation of an ad hoc network, the use of delay, loss, and energy information as metrics for routing, and the use of genetic algorithms to generate and maintain paths from previously discovered information by matching their "fitness" with respect to the desired QoS. We discuss implementation considerations as well as simulation and experimental results on a network testbed.

1 Introduction

Broadly speaking, quality of service (QoS) is the level of performance that a specific user (or a class of users) expects from the network. QoS becomes important when we consider network-sensitive applications such as real-time voice and video, which tend to have stringent requirements for the amount of loss, delay, or jitter that they can tolerate [5]. There is a great body of work dedicated to QoS including: routing schemes which try to achieve QoS objectives [11,12,13, 14,15], QoS performance of Internet protocols [17,18,16], and control techniques which offer QoS guarantees to traffic generated by specific applications [19]. A thorough survey of all of these subjects is beyond the scope of this paper and to do justice to the literature on QoS we would have to cite at least several hundred authors. However several other papers in this volume deal directly with QoS, including [8,9].

The Cognitive Packet Network (CPN) [1,2,3] is a flexible paradigm for quality-of-service driven routing. It is a store-and-forward network in which routes are chosen dynamically by "smart packets" (SPs) which belong to a connection, according to user specified QoS Goals. CPN routing stores information

---

[*] This work has been supported by a grant from the **UK Engineering and Science Research Council (EPSRC)** on **Self Aware Networks and Quality of Service** to Imperial College.

gathered by SPs and acknowledgement packets (ACKs) in mailboxes (MBs) at intermediate nodes and at the source node of a connection. Mailboxes store data about the delay and loss on paths in the network that subsequent SPs will use to select better paths.

This paper describes two distinct and important networking environments under which we have implemented CPN: wired and wireless, mobile, ad hoc networking. These two networking environments each have their own distinct requirements. While the wired domain is the most traditional networking environment, wireless ad hoc networking has a great many more challenges that must be overcome due to its dynamic nature. We briefly discuss these challenges and our solutions to them.

In addition, we describe an extension to CPN that uses genetic algorithms to generate new paths. Our approach generates new paths by combining paths which have been previously discovered. The QoS of these new paths can be estimated from the observed QoS of known paths. This QoS estimate use then employed to discern good from bad routes. If the estimated QoS is better than that of the paths which are already known, the newly discovered path is added to the "gene pool" and may be used in the future either for conveying traffic or for generating new paths in the same way. Thus the estimated QoS acts as the "fitness function" of the GA.

The remainder of this paper is organized as follows. Section 2 recalls the principles that have presided over the design of CPN routing. In Section 3 we discuss and show results of CPN in the wired domain. We comment on the use of delay and loss and routing metrics as well as the use of GA to discover new paths. Section 4 presents concepts and simulation results of CPN in the wireless ad hoc domain. Conclusions about this work are drawn in the final section, where we also discuss future research directions.

2   CPN Routing

CPN includes three different types of packets which play different roles:

- Smart or cognitive packets (SP) are used to discover routes for connections; they are routed using a reinforcement learning (RL) algorithm [22] based on a QoS "goal". We use the term goal to indicate that there are no QoS guarantees; rather CPN provides best effort service to optimize the desired QoS metrics. The role of SPs is to find routes and collect measurements; they do not carry payload. CPN's RL algorithm uses the observed outcome of a decision to "reward" or "punish" the routing algorithm, so that its future decisions are more likely to meet the desired QoS Goal. The goal is the metric which characterizes the success of the outcome, such as packet delay, loss, jitter and so on. In CPN, the specific Goal used by a SP will depend on the user's QoS requirements.
- When a smart packet arrives to its destination, an acknowledgment (ACK) packet is generated and the ACK stores the route followed by the original

packet and the measurement data it collected. An ACK being returned as a result of a SP will be source-routed along the "reverse route" of the SP. The reverse route is established by taking a SP's route, examining it from right (destination) to left (source), and removing any sequences of nodes which begin and end in the same node. That is, the path $< a, b, c, d, a, f, g, h, c, l, m >$ will result in the reverse route $< m, l, c, b, a >$. Note that the reverse route is not necessarily the shortest reverse route, nor the one resulting in the best QoS. ACKs deposit QoS information in the mailboxes (MBs) of the nodes they visit.

- Dumb packets carry payload and use source routing. Dumb packets also collect measurements at nodes. The route brought back to a source node by an ACK of a SP is used as a source route by subsequent dumb packets of the same QoS class having the same destination, until a new route is brought back by another ACK.

MBs in nodes are used to store QoS information. MB entries in a given node are identified by QoS class and Destination. Each MB is organized as a "least-recently-used" stack: old information can be discarded from fast memory when the MB is full, and new information is rapidly accessible from the top.

For each SP in the node's input buffer, the node will run the CPN routing code. Then the packet is placed in an output buffer which is selected by the CPN routing algorithm. If a DP or ACK enters a node, and the node number does not correspond to the node it should now be visiting, then the dumb packet is discarded.

As an example of how QoS information is processed, consider how delay is treated. When an ACK for a packet which was going from S to D and was of class K enters some node N from node M, the following operation will be carried out: the difference between the local time-stamp and the time-stamp stored in the ACK for this particular node is computed and divided by two. The resulting time is stored in the mailbox as the value $W(K, D, M)$ – it is an estimate of the forward delay for a packet of QoS class K going from node N to D and which exited node N via the port leading to M. Note that the identity of the local node N is obvious and need not be stored. The source node S is also not relevant since the $W(K, D, M)$ refers to the time to go from N to D using the next node M. The QoS class K is needed since the decision at each node, and the resulting observed delay, will depend on the requirements expressed by the QoS class K. The quantity $W(K, D, M)$ is inserted in the Goal function (see equation (3) of the RL learning algorithm for the delay value W). Delay is also used to compute jitter (variance or standard deviation of delay) which is a very important QoS metric [5] for "voice type" connections.

2.1   Routing Using Reinforcement Learning

Different approaches to learning could in principle be used to discover good routes in a network, including Hebbian learning, back-propagation [20], and reinforcement learning (RL) [22,7]. Hebbian learning is notoriously slow and was excluded from our consideration. Simulation experiments conducted on a 100 node network showed that RL is the most effective approach, and that it provides significantly better QoS than shortest-path routing. RL uses a *one step* update of weights so that it is very fast. It uses the most recent information to modify a neural network's internal state by rewarding successful outcomes and penalizing unsuccessful ones, so that new decisions are significantly impacted by the most recent outcome. In order to guarantee convergence of the RL algorithm to a single decision (i.e., selecting an output link for a given smart packet), CPN uses the random neural network (RNN) [4,6] which has an unique solution to its internal state for any set of "weights" and input variables. In general, at each node we will have a separate RNN for each QoS class and destination. In the RNN, the state q_i of the $i-th$ neuron denotes the probability that it is excited. The state is used to make the decision of a SP to select the output link i at the node: the largest q_i designates the output link i which is selected.

As an example, the QoS Goal G that SPs pursue may be formulated as minimizing Transit Delay W, Loss Probability L, Jitter, or some weighted combination captured in the numerical Goal function G and the reward $R = 1/G$. Successive values of R, denoted by R_l, $l = 1, 2, ..$, are computed from the measured delay and loss data, and are used to update the neural network weights.

The CPN RL algorithm first updates a threshold value:

$$T_l - aT_{l-1} + (1-a)R_l, \tag{1}$$

where a is some constant $0 < a < 1$, typically close to 1 and R_l is the most recently measured value of the reward. T_l is a running value that is used by the RL algorithm to keep track of the historical value of the reward and is is used to determine whether a recent reward value is better or worse than the historical value.

Suppose that the $l-th$ decision selected output link k, and that we received feedback from the network which measured the $l-th$ reward R_l. We first determine whether R_l is larger than, or equal to, the threshold T_{l-1}. If that is the case, then we conclude that the previous decision worked well; we increase the excitatory weights to the neuron k that was the previous winner (in order to reward it for its success), and make a small increase of the inhibitory weights leading to other neurons. On the other hand, if R_l is less than the threshold, we conclude that the previous decision was not so good and moderately increase **all** excitatory weights leading to **other** and increase significantly the inhibitory weights leading to the previously selected neuron k in order to punish it for being unsuccessful:

- If $T_{l-1} \leq R_l$
    - $w^+(i, k) \leftarrow w^+(i, k) + R_l$,
    - $w^-(i, j) \leftarrow w^-(i, j) + \frac{R_l}{n-2}$, *for all* $j \neq k$.
- Else
    - $w^+(i, j) \leftarrow w^+(i, j) + \frac{R_l}{n-2}$, *for all* $j \neq k$,
    - $w^-(i, k) \leftarrow w^-(i, k) + R_l$.

Finally the node with the largest q_i is identified, and the smart packet is placed in the corresponding output buffer i.

3 Wired CPN

CPN was first implemented in the wired domain. Two important QoS metrics for this domain are loss, and delay. In this section, we describe the way in which these two QoS metrics are treated in CPN.

3.1 Reward Based on Loss

In order to minimize the user's loss, we must first be able to measure the loss along a given path. In order to do this, we take advantage of our ACK packets. If we have sent N dumb packets, and A dumb ACK packets, it can easily be shown that the round-trip loss rate is given by $1 - \frac{A}{N}$. It is less trivial to derive the forward loss rate L_f. However, if we assume that the forward loss rate is equal to the backwards loss rate, we can derive that $L_f = 1 - \sqrt{\frac{A}{N}}$. In addition, we keep a smoothed average $\overline{L_f}$ of the most recent observed forward loss rates to measure *short term* changes in the network.

In order to formulate a reward equation for loss, we want to capture the relationship that as the loss increases, the reward should decrease. Therefore, we use a reward R of the form:

$$R = \frac{\alpha}{\overline{L_f} + \epsilon},\tag{2}$$

where ϵ is a constant representing some minimal value for the loss. In the experiments we report, we have used $\epsilon = 10^{-5}$ and $\alpha = 0.5$.

3.2 Reward Based on Delay

In section 2 we discussed how we measure delay. Based on the same principles as we used for loss, we propose a reward formula as follows:

$$R = \frac{1}{\beta \overline{W}},\tag{3}$$

where β is set to 0.75 in our experiments.

3.3 Measurements

The measurements that we report in this section were performed under a variety of conditions on the testbed consisting of 26 nodes shown in Figure 1. All tests were conducted using a flow of UDP packets entering the CPN network with constant bit rate (CBR) traffic and a packet size of 1024 Kb. All CPN links used 10Mbps point-to-point Ethernet. UDP data was sent from Node 10 to Node 7. The CPN protocol was run in the usual manner with paths being discovered by SPs, while DPs were used to carry and source route the payload UDP packets.

As shown in the top figure of Figure 2, using the goal function based only on the loss results in the lowest observed loss rates. The curves in the bottom figure show that the lowest delay is obtained by using only delay in the goal function.

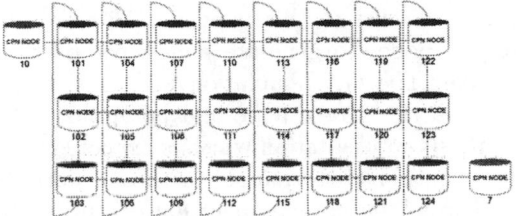

Fig. 1. CPN testbed

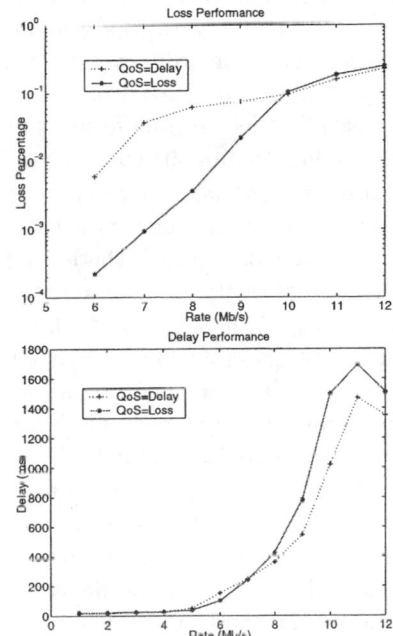

Fig. 2. Loss (top) and delay (bottom) in the large testbed

3.4   Genetic Algorithms and CPN

A GA is a learning algorithm which operates by simulating evolution [21]. Key features that distinguish a GA from other search methods include:

- A *population* of *individuals* where each individual represents a potential solution to the problem to be solved.
- A *fitness function* which evaluates the utility of each individual as a solution.
- A *selection function* which selects individuals for reproduction based on their fitness.
- The *genetic operators* which we use to alter selected individuals to create new individuals are crossover and mutation.

In the approach we discuss in this paper, a GA will run as a background process at each source node, to generate and select paths for dumb packets based on the QoS goal. The GA population will consist of individuals which represent paths between the source node and potential destination nodes. We will use a variable length representation which is expected to allow the GA more flexibility to evolve in response to changes in the network [23]. The fitness of a path is determined from the measurement data returned by an ACK that is received in response to sending a dumb packet along that path. New paths are constructed by genetic operators, e.g. mutation constructs new paths via small modifications to existing paths while crossover constructs new paths from two existing paths that share a common intermediate node. The GA also receives input from the CPN as new paths are discovered by SPs. Exploitation and preservation of existing good paths (or partial paths) is accomplished through fitness-based selection. A path is selected if its fitness function is better than, or within a range of, the fitness function of the current population.

We define a *routing word* or *word*, w, as a variable length sequence of nodes which begin with the source node S and end with the destination D. For any nodes a and b, ab is a subsequence in w only if there is a link (i.e. path of length 1) going from a to b. Thus w represents any viable path from S to D.

Each routing word w has a goal value $G(w)$ which is determined by the user of the network. G is similar to the goal function of CPN, and describes how effective the path described by the word w is. Thus, a smaller value of $G(w)$ means that w is more desirable. A simple example is when $G(w)$ is merely the number of letters, or nodes, on the path from S to D.

$G(w)$ can also be the delay or loss over the path w.

We shall say that a goal value is *additive*, if for any word $w = \alpha\beta$ which may be expressed as the concatenation of two words α and β, we have $G(w) = G(\alpha)+G(\beta)$. Note that goal value such as packet delay is additive. Note also that the CPN algorithm for measuring the forward delay of smart or dumb packets based on the information brought back by an ACK packet allows us to collect at the source not just the source-to-destination delay but also the delay from any intermediate node to the destination, and therefore also to deduce the forward delay from the source to any intermediate node.

We will assume that new words are generated in two different ways:

- Smart packets in CPN discover routes, and ACK packets bring back valid routes to the source; thus CPN already provides a way of generating new words w using both random generation of smart packets, and reinforcement learning based search for routes. The source will keep the words which have been brought back by the ACKs in a list sorted in the order of increasing $G(w)$ values which we call the *Stack*.
- At the source we generate additional words using the following *path crossover* operation: Suppose two words w_1 and w_2 share some intermediate node, so that for some node a, we have $w_1 = u_1 a v_1$ and $w_2 = u_2 a v_2$. The crossover operation will then generate the strings $w_3 = u_2 a v_1$ and $w_4 = u_1 a v_2$.

If any of these strings is not already in the *Stack*, then it is placed in the *Stack* with the corresponding goal values $G(w_3) = G(u_2a) + G(v_1)$ and $G(w_4) = G(u_1a) + G(v_2)$.

In this way, the *Stack* is enriched both with new paths obtained by crossover, and by paths which are obtained via the CPN smart packet search process. Whenever a dumb packet needs to be forwarded to the destination, the word at the top of the *Stack* (i.e. the one with the smallest goal value) will be used as the source route.

Every complete route discovered by SPs and brought back by ACKs naturally becomes an individual in the GA population for enhancement of CPN routing. New routes will be evolved as offsprings. Collectively, the individuals with the same source S and destination D form a GA population repository $P(S, D)$ which is organized as an LRU stack with some predefined maximum size.

Implementation. The GA algorithm operates by doing "crossovers" between pairs of paths and periodically discarding the paths which have the lowest fitness, or equivalently the worst QoS. In order for this selection to reflect the current state of the network, these calculations must be based on the latest data available. For example let us consider two paths A and B which share some common hops. When we evaluate the fitness of paths A and B in order to compare them, for this comparison to have a meaning, the underlying measurements on the hops (delay, loss) must reflect *the same network conditions*. As these paths were very likely brought back at different times, it is necessary to update measurements concerning hops as new information is brought back from the network, so that the QoS evaluation of paths A and B is constructed from data taken at roughly the same time. Since duplicating data is not desirable, the GA should store a pool of available *hop* measurements, and paths should just be a collection of references to hops in this pool.

The data structure storing the pair of nodes which constitute the hop and the corresponding measurement information is referred to as a *hop* in the rest of this document. Hops also have a timestamp field which represents the age of the hops. As newer measurements are sent over by the CPN module, hops are updated in place to reflect the new network conditions, and their timestamp is updated as well. The timestamp allows us to periodically rid the hop pool of obsolete hops, so that the GA engine does not make decisions based on obsolete measurements. A path is a collection of hops and is referred to as an *individual*. An individual does not store its fitness as the delays associated with the hops that built up the path will change over time. The individuals are arranged in subpopulations, one for each destination, and within these subpopulations they are periodically ordered by fitness. The data structures are shown in Fig.3.

The size of the data structures that the GA uses make it unlikely that a kernel-level implementation can be used. Also, as stated before, the GA processing should be done in the background. Thus the GA algorithm has been implemented as a system daemon and will be referred to henceforth as "the GA daemon".

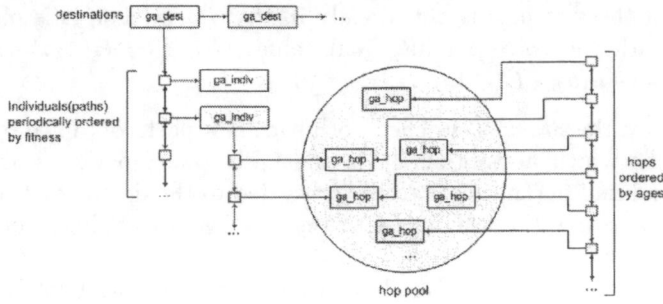

Fig. 3. Principal data structures of the GA daemon

The GA-enabled module works in a similar fashion to the regular CPN module, the main changes concern the handling of the ACK when it reaches the source. However, a modification to the way CPN generates and sends ACKs back was needed for GA to function properly. While CPN deals with entire paths, GA operates at a finer level and needs accurate measurements on individual hops and this means that caution needs to be taken when stripping a route of loops. Loops are no longer removed at the destination but instead as the ACK travels back, the nodes on the path check if they were present twice or more in the original packet's path and if that is the case, they remove the loop and adjust the timestamp of nodes between itself and the source to compensate for the extra delay introduced by the loop.

When the ACK reaches the source, we have a complete path to the destination and the measurements for each hop. This data is put into a FIFO buffer for the GA daemon to process whenever it becomes available, which is shown in Fig.4. Unlike the regular CPN module, the GA-enabled module does not update the dumb packet route repository unless there is no route for the current destination or the GA daemon is dead (defined as "if the GA daemon has not talked to the module for the past second").

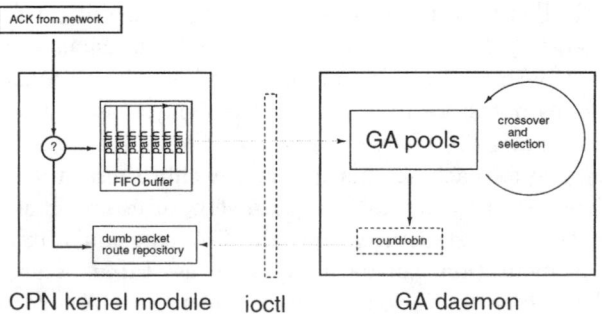

Fig. 4. Interaction between module and daemon

The GA daemon is started when the CPN module is loaded with no initial knowledge of the network. It consists of a main loop which polls the kernel for new paths, checking its internal data structures for size and consistency, selecting individuals for crossover and doing the actual crossover and periodically updating the CPN module's dumb packet route repository.

The role of the path selection process is to return a pair of individuals which are suitable for crossover. Depending on the situation, the GA daemon either requests a "match" for a specific individual or just asks for a pair of individuals to crossover.

The data exchange operations between the CPN kernel module and the GA daemon is bidirectional as the module passes measurements to the daemon and in return the daemon periodically updates the module's route repository with the paths who have the best fitness at that time. The GA daemon is always the initiator of the data exchange so that the kernel module does not need to keep track of the presence or absence of the daemon to route CPN traffic. This way we also eliminate lockups which would occur if the kernel were to probe the daemon while it is sleeping.

In a first implementation, the GA daemon would always send the kernel the best path for a given destination, but after some testing this was changed. The two problems encountered were route saturation and a depletion of the individual pool.

In order to solve both these problems, instead of systematically returning the best route, the daemon does a round-robin on the routes whose delay does not exceed the best delay by more than 5%.

Measurements. To evaluate the performance of CPN with and without the GA, measurements were conducted on the testbed shown in Figure 5. Initially, all the machines in the testbed start with an empty dumb packet repository and empty GA pools. For this particular set of experiments, the source node is the node at the left edge of the testbed as shown in Figure 5. The CPN path finding algorithm is started at the source node using SPs and ACKs. Once the first ACK comes back from the destination carrying the first path that has been discovered, DPs from source to destination are sent over the CPN network at a constant rate. Note that the GA daemon only runs at the source node.

For the experiments in which the GA algorithm is *not* in operation, we simply disabled the GA daemon. On the other hand if the GA daemon is enabled, when the first ACK arrives at the source, the GA daemon is automatically started up and will generate paths as described in the previous section. If a node sends packets to several destinations, the GA daemon divides the paths into subpopulations based on the destination and crossover is done for each given destination. However, the hop pool is common to the all sub-populations, so that when a source sends packets to several destinations, the GA hop pool will contain more data as well as potentially more *recent* data about hops, than if only one destination is used. However in this case the GA still creates independent paths for each destination from the hop pool.

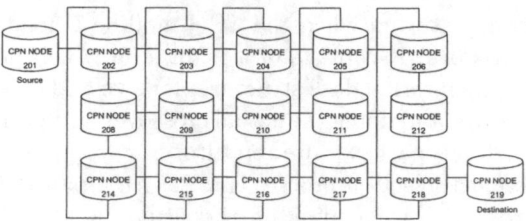

Fig. 5. The CPN-GA testbed topology

Throughout the experiments, the QoS Goal that is used for selecting paths is to minimize source to destination delay. However we measured the QoS both in terms of observed average delay and observed average packet loss rate for *DPs*, i.e. for the payload packets. both when the GA was disabled and when it was enabled. Note that SPs' routing is not affected by the GA in either case. We obtain the delay and loss measurements for DPs as follows. At the source, we count the number of ACKs received for DPs, say N_{AD} and the number of DPs transmitted, N_D. We can evaluate the DP loss rate as $L_{DP} = 1 - N_{AD}/N_D$. Note that this value is pessimistic because some ACKs will be lost and we are in fact measuring the total loss of DPs plus ACKs. On the other hand, we estimate the average forward delay(in milliseconds) by taking the half of the time at which the ACK comes to the source minus the time at which the corresponding DP is sent forward. Again, this is obviously a rough estimate.

Experiments were conducted with different levels of background traffic, where the background traffic is defined as additional traffic which is added locally at some fixed rate on *each link* in the network. Background traffic is composed of fixed size packets of 1024 bytes travelling from one end of a link (between two adjacent nodes) to the other.

We first ran experiments without any background traffic, meaning that the whole testbed only contained the DPs, SPs and ACKs for a single source to destination connection. Experiments were run without the GA daemon being enabled, and then with the GA daemon. The size of all DPs at the source was fixed at 1024 bytes. We varied the DP rate for a connection between 100 packets/sec to 800 packets/sec. for each traffic rate we ran 10 experiments and in each experiment the source node sent out 10,000 DPs to the destination. The average forward delay and loss rate were computed as an average for each DP transmission rate over all the 10 experiments, and the results (without background traffic) are shown on Fig.6. We observe that in the absence of background traffic, CPN with the GA outperforms the "regular" CPN. However the improvement is not considerable. When the connection's DP rate is less than 600 packets/sec, the delay with the GA enabled is only 80% of that without GA and the loss rate are both small enough to be considered to be zero. When the DP rate is great than 600 packets/sec, some packet losses are observed and the average delay increases dramatically in both cases. If we only consider the delay, the performance was improved by about 20% due to the introduction of GA. The experiments

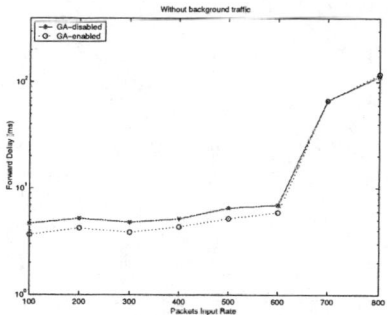 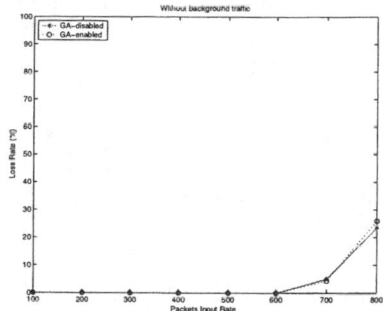

Fig. 6. QoS comparison of the system with and without the GA, without background traffic

were then conducted with approximately 0.8 Mbps background traffic on each link and the results are almost the same as shown in Figure 6. One likely explanation for this improvement is that in CPN without the GA, whenever a new route is brought back by an ACK it is immediately used for subsequent DPs, regardless of the performance of this new route as compared to the performance of the previous route. Thus paths are replaced by whatever path is discovered next. On the other hand, if the GA is enabled, if paths with shorter delay are discovered they tend to be used, while if a new path has higher delay then it will not be used because from the GA's perspective it is less fit than an existing path, and rightly so.

When we increased the background traffic level to 2.4Mbps and to 4Mbps on each link, the results were somewhat disappointing as shown in Figures 7 and 8. The performance of the system with the GA was again better that that of regular CPN when the DP traffic rate was low. When the DP traffic rate exceeded a certain value (e.g. 300 packets/sec in the 4Mbps excess link traffic scenario), the performance of regular CPN was better than that of CPN with the GA. We may explain this as follows. When the load of the network becomes heavier, since the DP traffic has an additive effect on the existing link traffic, the DP traffic will significantly increase the observed delay. SPs will try other routes which do not have DP traffic, and the corresponding ACK will bring back information about paths which are momentarily less loaded. These paths will be immediately used by CPN when the GA is not enabled, leading to path switching and the distribution of the traffic on multiple paths. On the other hand, CPN with the GA will be operating with information which is always "old" because round-trip delays are high, and will therefore recommend paths which may have worse QoS by the time the decision is taken. We then increased the background traffic level on each link even further to 5.6 Mbps, and ran the same experiments. The results are shown in Figure 9. We can draw the same conclusions as previously, except that the DP traffic rate beyond which the system with the GA is clearly worse than that without the GA is even smaller, i.e. 200 packets/sec.

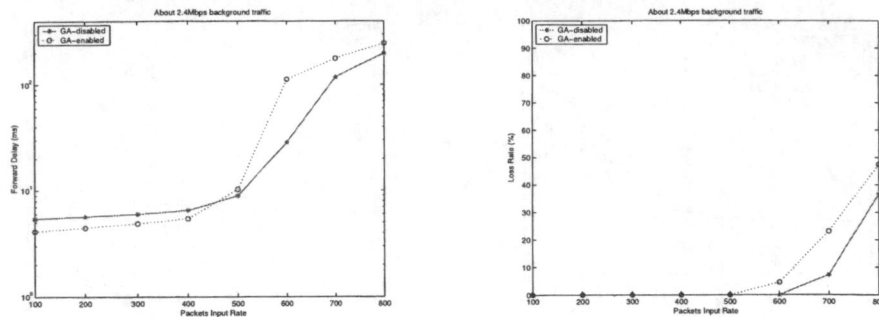

Fig. 7. Delay and loss rate for DPs with 2.4 Mbps background traffic on links

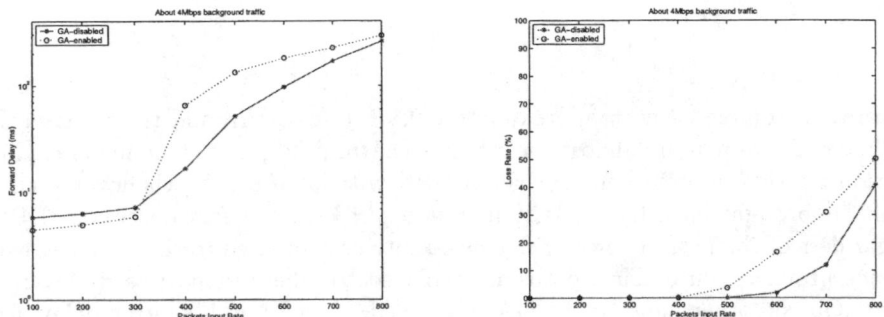

Fig. 8. delay and loss rate with about 4 Mbps background traffic on links

When we increase the background link traffic to 8 Mbps, as shown in Figure 10 we notice that the the network is saturated so that there is little difference that can be made by using or not using the GA: we always have high delay no matter which route the DPs use.

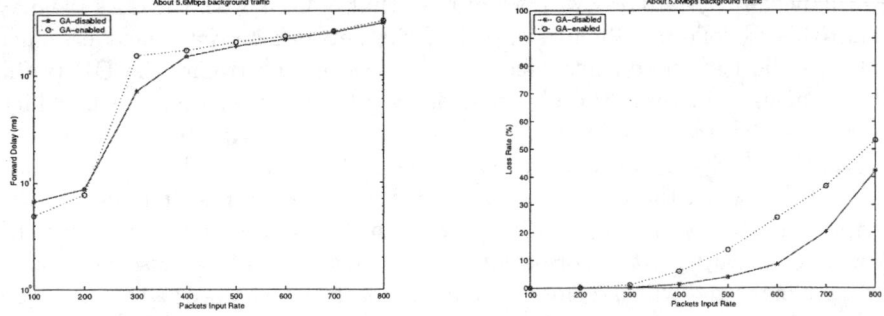

Fig. 9. Delay and loss with 5.6 Mbps background traffic on links

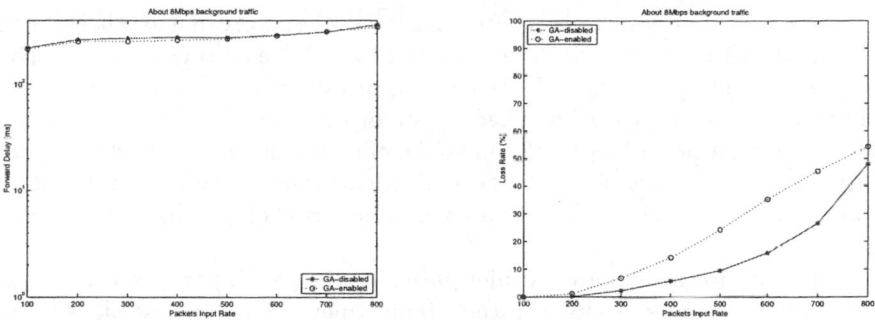

Fig. 10. Delay and loss with 8 Mbps background traffic on links

3.5 Comparison of QoS with and without the GA

When the network load is light and the packets input rate is low, the Genetic Algorithm helps to reduce the delay of dumb packets by about 20%. When the background traffic becomes heavier and at high packets input rate, the delay of dumb packets of regular CPN is a little lower than that of GA. When the packets input rate is higher than some threshold value, both suffer from the high loss rate. If the whole network is saturated, there is almost no difference on delay between GA and regular CPN.

With GA enabled, a GA daemon must be run in background at the source node. Of course it will occupy some system resources. When the network is idle, The GA daemon uses a total of 376kB of memory, 316kB of which is in fact shared memory used by standard libraries. As for processor usage, it is difficult to give an exact figure. After running for an hour with an input rate of 800 packets/sec, the cumulated processor time used by the GA daemon was below a second. Therefore, running a GA daemon at the source node requires for little system resources.

4 Wireless Ad Hoc CPN

In ad hoc cognitive packet networks (AHCPN), route discovery is triggered whenever a node needs to send data packets to an unknown destination. While this discovery process takes place, data packets await in a local buffer until a new path is discovered. As in CPN, smart packets depart from the source node to find routes and at each node they use a RNN/RL algorithm to select the next hop (which is reached via unicast). In addition to unicast transmissions, smart packets may use broadcasts as an alternative propagation mechanism that is employed whenever sufficient information is not available to make a regular RNN/RL decision. The introduction of broadcasts helps to overcome the problems associated with topological changes. Broadcasts are mainly used as a last-resource alternative. Whenever possible, unicasts are preferred to maintain a low routing overhead.

Because smart packets may create a partial or total broadcast query and await-reply cycle, they are tagged with a packet identifier to help them identify previously visited nodes. For this, after an arrival, they store in a database within the node their packet identifier and source address. When a smart packet arrives at a node, it only proceeds with the routing process if no previous information (source address and packet identifier) exists in the database. This process prevents the formation of loops and limits the effective number of packet replicas on the network. Furthermore, via the same identification mechanism, the destination node replies only to the first arriving smart packet and discards all future replicas.

Nodes acquire neighborhood information by inspecting the immediate originator of the packets awaiting in their input queue. This process allows them to learn the nodes that are actively sending packets within their area of coverage. Note that a pure broadcast query and await-reply cycle forces each node to transmit one packet and therefore allows them discover neighbors. Also, as nodes continuously listen to their neighbors' transmissions (regardless of the packet destination), the neighborhood information is often available before a smart packet needs such knowledge.

Smart packets in CPN replace a small fraction of RNN/RL based decisions with random decisions. This random exploration of paths increases the probability of information collection about links not being used by any connection. Random decisions are replaced by broadcasts in AHCPN, which have the advantage of a more invasive exploration.

After a route has been established, the transmission of data packets proceeds immediately by removing them from the waiting buffer and sending them into dumb packets. A copy of the original data packet remains at the source node until an acknowledgment is received. If no acknowledgment arrives within a time frame, the packet is retransmitted. Packets may be retransmitted up to a finite number of times and they are discarded thereafter. Retransmissions at the AHCPN layer are implemented to improve end-to-end performance.

4.1 Power-Aware Routing Goal

We define as *path availability*, the probability to find all nodes and links available for routing on a path. Formally, suppose that a smart packet takes the path: $(n_1, n_2, \ldots, n_d)$, where n_i represents the i-th node on the path and (n_i, n_{i+1}) represents the link between nodes n_i and n_{i+1}.

Assume that the node n_i is available for routing with probability $P_n(n_i)$ and that the link (n_i, n_{i+1}) is available with probability $P_l(n_i, n_{i+1})$.

We define the path availability $P_p(n_i, n_d)$ as:

$$P_p(n_i, n_d) = \prod_{j=i}^{d-1} P_n(n_{i+1}) P_l(n_i, n_{i+1}) \tag{4}$$

We formulate a recursive, combined routing goal function G_i as follows which consist of round-trip delay and path availability:

$$G_i = P_p(n_i, n_d) D(n_i, n_d) + [1 - P_p(n_i, n_d)](T_o + G_i) \tag{5}$$

where $D(n_i, n_d)$ is the round-trip delay from n_i to n_d and T_o is a predefined timeout interval.

In real systems, many factors contribute to the values of P_n and P_l. For the purposes of this paper, we assume that $P_l = 1$. Because batteries are the major source of energy in mobile nodes, they represent one major factor in determining P_n. To provide greater portability, batteries need to be small and lightweight, which unfortunately restricts the total energy that they can carry. Once the batteries exhaust their energy, they need to be replaced or recharged, which typically reduces the independence of the mobile node to a few hours of operation.

The energy consumption, in communication related tasks, depends on the communication mode of the node. A node may be transmitting, receiving, or in idle mode. Transmission naturally consumes more energy than the other two modes. From the routing perspective, our interest is in selecting routes in such a way that the transmission and reception of packets is distributed on the network so as to maximize the overall average battery lifetime of the nodes. Therefore, we are interested in getting smart packets to select—with greater frequently—the nodes with the longest remaining battery lifetime.

Let us set $P_l(n_i, n_{i+1}) = 1$ in Equation 4, for all $i \in [1, d-1]$, to simplify our discussion. Therefore:

$$P_p(n_i, n_d) = \prod_{j=i}^{d-1} P_n(n_{i+1}) \tag{6}$$

Assume B_i is the remaining battery lifetime of node n_i. We formulate $P_n(n_i)$ in terms of B_i as:

$$P_n(n_i) = \frac{B_i}{B_m} \tag{7}$$

where B_m is the lifetime of a fully charged battery. From Equation 6 and Equation 7 we obtain the desired path availability to destination n_d from any node on the path:

$$P_p(n_i, n_d) = \prod_{j=i}^{d-1} \frac{B_{j+1}}{B_m}$$

4.2 Simulation

We developed and integrated a simulation model of AHCPN into *Network Simulator 2* (NS-2). The model was used to perform various studies to observe the effectiveness of smart packets in learning how to establish reliable paths. The simulations were conducted on the the network depicted in Figure 11 and by employing a test traffic between nodes 0 and 19. The batteries of all nodes

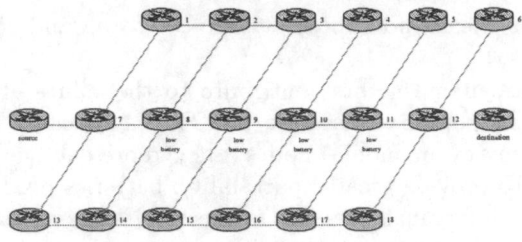

Fig. 11. Wireless network topology for routing simulation

were fully charged except for nodes 8–11, which were 25% charged. These low-charged nodes were intentionally located on the shortest path (in the number of hops sense) between the two end nodes.

Smart packets were sent at a rate of 1, 10, and 100 packets per second to discover and maintain a route between the aforementioned nodes. Dumb packets were suppressed during this experiment. The packets employed random neural networks with reinforcement learning to make routing decisions except for a fraction of them which employed broadcasts (neighborcasts). The number of effective broadcasts was controlled by a pre-defined *broadcast selection ratio* value. When using RNN/RL, smart packets moved on the network by deciding at each node the next hop that best matched their routing goal. After arriving at their destination, acknowledgments departed on the reverse path towards the source node, to inform the discovered route and compute the path availability metric (Equation 5) for that route.

Figure 12 illustrates the path availability as measured by successive smart packets at the source node. As expected, smart packets were able to choose paths with higher availability when the broadcast selection ratio was small. As shown in Figure 12(a) and Figure 12(b), initial smart packets encountered low path availability because they attempted to move through the shortest path. However, subsequent packets managed to increase this value with the selection of alternative paths as they became more aware of the status of the network. After reaching a maximum level, path availability decreased as a consequence of energy consumption in the nodes involved.

5   Conclusions

We have discussed the design, implementation, and performance of CPN under two paradigms: wired networking and wireless ad hoc networking. We presented the use of delay, loss, and energy information in the routing goal of smart packets to discover and maintain routes.

In addition, we considered the use of GA to create new routes out of the information discovered by smart packets. For this, a GA process at the source node, composes new routes from existing routes and selects those that offer a good "fitness" with respect to the desired QoS for the path. We found that a

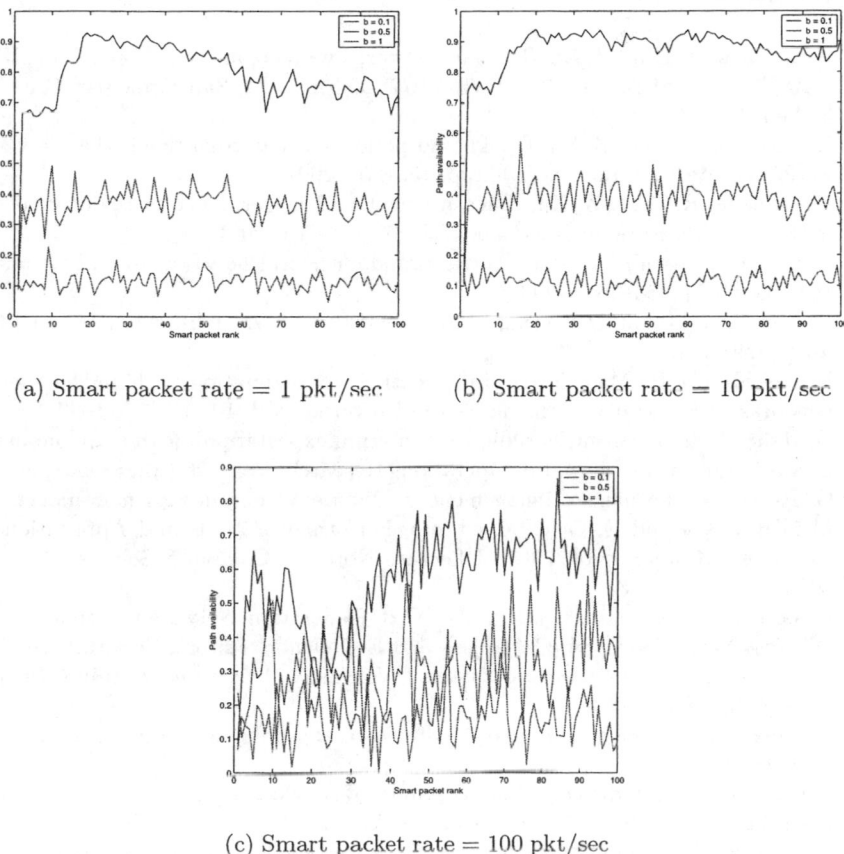

(a) Smart packet rate = 1 pkt/sec   (b) Smart packet rate = 10 pkt/sec

(c) Smart packet rate = 100 pkt/sec

Fig. 12. Path availability as measured by successive smart packets

simple round-robin policy on the discovered routes serves well to balance the system load. Our results when using GA are mixed. The Genetic Algorithm daemon significantly improves QoS under light network traffic but not under high traffic conditions. The explanation is in that our GA tends to delay the decision making of a route and select outcomes based on longer term trends.

To assist in the dynamic process of neighbor and route discovery, we introduced a broadcast mechanism to CPN to operate in ad hoc networks. In addition, we defined an specific routing goal that allow us take into account the availability of resources on the paths so as to better utilize network assets. Our simulation results showed that the CPN algorithm can improve the working lifetime of a mobile network by introducing energy information in the routing goal of packets.

Our ongoing research includes further experimentation with larger, geographically separated network testbeds, interoperation with existing IP routing algorithms, and the implementation of the algorithm in embedded architectures.

References

1. E. Gelenbe, R. Lent, Z. Xu, Towards networks with cognitive packets, Proc. IEEE MASCOTS Conference, ISBN 0-7695-0728-X, pp. 3-12, San Francisco, CA, Aug. 29-Sep. 1, 2000.
2. E. Gelenbe, R. Lent, Z. Xu, Design and performance of cognitive packet networks, Performance Evaluation, Vol. 46, pp. 155-176, 2001.
3. E. Gelenbe, R. Lent, Z. Xu, "Measurement and performance of cognitive packet networks", J. Computer Networks, Vol. 37, 691–701, 2001.
4. E. Gelenbe, Learning in the recurrent random neural network, Neural Computation, Vol. 5(1), 154–164, 1993.
5. D. Minoli, E. Minoli, Delivering Voice over IP Networks, John Wiley & Sons, New York, 1998.
6. E. Gelenbe, Z.-H. Mao, Y. Da-Li, Function approximation with spiked random networks, IEEE Transactions on Neural Networks, Vol. 10 (1), 3–9, 1999.
7. U. Halici, Reinforcement learning with internal expectation for the random neural network, European Journal of Operations Research, Vol. 126 (2),288–307, 2000.
8. G. Kotsis, Performance Management in Dynamic Computing Environments, in M. Calzarossa and E. Gelenbe, editors, Performance Tools and Applications to Networked Systems, Vol. 2965 of Lecture Notes in Computer Science, Springer, 2004.
9. P. Lorenz, IP–Oriented QoS in the Next Generation Networks: Application to Wireless Networks, in M. Calzarossa and E. Gelenbe, editors, Performance Tools and Applications to Networked Systems, Vol. 2965 of Lecture Notes in Computer Science, Springer, 2004.
10. S. Moon, Measurement and analysis of end-to-end delay and loss in the Internet, PHD thesis, 2000.
11. S. Chen, K. Nahrstedt, Distributed Quality-of-Service routing in ad-hoc networks, IEEE Selected Areas in Communications, Vol. 17, No. 8, 1–19, 1999.
12. F. Hao, E.W. Zegura, M.H. Ammar, QoS routing for anycast communications: motivation and an architecture for Diffserv networks, IEEE Communications Magazine, vol. 46, No. 2, 48–56, June 2002.
13. Y-D. Lin, N-B. Hsu, R-H. Hwang, QoS routing granularity in MPLS networks, IEEE Communications Magazine, vol. 46, No. 2, 58–65, June 2002.
14. S. Nelakuditi, Z-L. Zhang, A localized adaptive proportioning approach to QoS routing granularity, IEEE Communications Magazine, vol. 46, No. 2, 66–71, June 2002.
15. M. Kodialam, T.V. Lakshman, Restorable quality of service routing, IEEE Communications Magazine, vol. 46, No. 2, 72–81, June 2002.
16. A. Chaintreau, F. Baccelli, C. Diot, Impact of TCP-like congestion control on the throughput of multicast groups, IEEE/ACM Transactions on Networking, vol. 10, No. 4, Aug. 2002.
17. M. May, J-C. Bolot, C. Diot, A. Jean-Marie On Internet QoS performance evaluation, INRIA Tech. Report, 1997.
18. E. Altman, K. Avrachenko, C. Barakat, P. Dube, TCP over a multi-state Markovian path, in K. Goto, T. Hasegawa, H. Takagi and Y. Takahashi (eds.), Performance and QoS of Next Generation Networking, 103–122, Springer Verlag, London, 2001.

19. J.C.S. Lui, X.Q. Wang, Providing QoS guarantee for individual video stream via stochastic admission control, in K. Goto, T. Hasegawa, H. Takagi and Y. Takahashi (eds.), Performance and QoS of Next Generation Networking, 263–279, Springer Verlag, London, 2001.
20. D.E. Rumelhart, J.L. McClelland Parallel distributed processing vols. I and II, Bradford Books and MIT Press, 1986.
21. J.H. Holland, Adaptation in Natural and Artificial Systems, University of Michigan Press, 1975.
22. R.S. Sutton, Learning to predict the methods of temporal difference, Machine Learning, vol. 3, 9–44, 1988.
23. D.S. Burke, K.A. De Jong, J.J. Grefenstette, C.L. Ramsey, A.S. Wu, Putting more genetics in genetic algorithms, Evolutionary Computation, vol. 6, No. 1, 387–410, 1998.

Modeling and Simulation of Mesh Networks with Path Protection and Restoration

Hassan Naser and Hussein T. Mouftah

School of Information Technology and Engineering (SITE)
University of Ottawa, Ottawa, Ontario, Canada K1N 6N5
{hnaser, mouftah}@site.uottawa.ca

Abstract. This article presents a modeling and simulation methodology to analyze the availability performance of demands when carried over mesh-based transport networks with path (end-to-end) protection and restoration schemes. Three main schemes are examined and formulations for diverse routing and capacity allocation are given. These schemes are: (a) shared path restoration, (b) dedicated path protection, and (c) unprotected path. The impact of several network parameters including link failures and repair times on the achievable performance is investigated. Techniques of object-oriented modeling and notations in the Unified Modeling Language (UML), including class diagrams and state diagrams, are used to characterize the structure of states and behaviors of prevalent network components.

1 Introduction

Survivable transport networks are defined as networks that continue functioning correctly in the presence of non-functioning (failed) network components. The architecture of transport networks consists of two basic components: (1) nodes and (2) links that interconnect these nodes. The focus of survivable networks is to ensure the connectivity of the network in case of a single link or node failure. An important category of survivable networks is the category of "mesh restorable" networks. In these networks, the restoration mechanism is able to identify the failed network component and reroute quickly the traffic traversing it. This is done by exploiting mesh topology and through highly diverse and efficient routing mechanism.

There are two survivability paradigms in mesh networks [1] [2] [5]: (1) link protection/restoration and (2) path protection/restoration. In link protection/restoration, when a link fails, the two nodes that terminate the link reroute the traffic traversing the link over a set of replacement (backup) paths between these nodes. The source and destination nodes of the traffic traversing the failed link are oblivious to the link failure. In path protection/restoration, the source and destination nodes of the traffic traversing the failed link (or failed node) reroute the traffic over a set of replacement paths between these nodes. In this sense, link protection/restoration mechanisms are considered "local", whereas path protection/restoration mechanisms are "end-to-end".

Protection and restoration mechanisms are generally different in the following manner. In protection mechanisms, the backup paths have been identified and spare (backup) capacities allocated statically in advance, whereas in restoration mechanisms the backup paths are established dynamically after the failure occurs. Protection

M.C. Calzarossa and E. Gelenbe (Eds.): MASCOTS 2003, LNCS 2965, pp. 88–117, 2004.
© Springer-Verlag Berlin Heidelberg 2004

mechanisms are *proactive*, whereas the restoration mechanisms vary from *totally-reactive* to *semi-proactive*. In totally-reactive restoration mechanisms, the backup paths are identified and resources configured after the failure occurs. In semi-proactive restoration mechanisms, the backup paths are identified and signaled prior to failure, but the resources along these paths are configured only after the failure occurs.

There are two approaches to backup resource allocation in survivable networks: (1) *dedicated-resource* protection and (2) *shared-resource* restoration. In dedicated-resource protection, the backup resources are exclusively reserved for each failed connection (in the path protection) or failed link (in the link protection). In shared-resource restoration, the backup resources are shared among different connections (in the path restoration) or links (in the link restoration). The connections (or links) sharing the backup resources are to be chosen in such a way that any single failure cannot put out of service more than one connection (or link) at one time. Double failures leading to the loss of more than one connection (or link) at one time may be tolerated, provided the probability of such failures is low.

In this article, we describe a simulation framework for modeling protection and restoration mechanisms in transport mesh networks. The focus will be on the end-to-end path protection/restoration mechanisms, including, in particular, dedicated path protection and shared mesh restoration. One of the main characteristics of the simulation framework is its generality: it can be simply extended to mesh networks with "local" link protection/restoration mechanisms or to non-mesh topologies (such as rings).

The article is organized as follows. Section 2 introduces the network model and outlines protection and restoration schemes implemented in the network. An overview of the diversity routing mechanisms to be employed by each scheme is given. Section 3 describes the simulation apparatus and the functional model of the relevant network components. Network objects, classes, and diagrams are demonstrated by using a number of commonly used notations in the Unified Modeling Language (UML) with minor adaptations in syntax for the sake of consistency with C/C++. Section 4 presents algorithms for constrain-based path computation and protocols for bandwidth reservation in the simulated network. Section 5 gives the state transition diagram of the simulated network. Section 6 presents the core of the simulation program implemented in C++ code. Section 7 characterizes the performance of the implemented protection and restoration mechanisms by showing the results from the simulation.

2   Network Model

A network illustrated in Fig. 1 is modeled as nodes interconnected by bidirectional links. An agent called Network Agent collects information about the nodes, links, and demands, and ensures that requests received via demands for connection setup are satisfied. The agent handles events occurring in the network. It includes an entity called "Path Computation Entity (PCE)", which computes network paths using a selection of constrain-based routing algorithms. The basic functions of the network agent are described later.

A connection is defined by a working (service) path, and if requested, by a secondary (backup) path. In order to avoid a single point of failure, the working and backup paths of a connection must be failure-disjoint. In this paper, we ignore node failures, and assume that network links are the only components that can and will fail. We assume also that failures of every pair of links are mutually disjoint. Therefore, two paths are said to be failure-disjoint if and only if they are link-disjoint.

A connection is computed when a demand arrives at the network requesting some level of protection services. The network supports three levels of protection services:

1. Unprotected service: a request for this service is satisfied by establishing a working path from the source node of the connection to the destination. Fig. 2 (left) illustrates bi-directional working paths (solid arcs) computed for demands A-E and A-F. Each demand has requested 1 unit of bandwidth, which is reserved on every link along the corresponding path. The reserved bandwidth on each link is enclosed within brackets.

2. Dedicated protection service: a request for this service is satisfied by establishing a pair of working and backup paths from the source node of the connection to the destination. The resources (such as bandwidth) on links along the backup path are exclusively reserved for the connection. Fig. 2 (middle) illustrates bi-directional working paths and their corresponding backup paths (dashed arcs) for demands A-E and A-F. The backup path of demand A-E traverses links AB and BE, whereas the backup path of demand A-F traverses links AB, BE, and EF. One unit of bandwidth is exclusively reserved for each demand on links along their backup paths.

3. Shared restoration service: a request for this service is satisfied by computing a pair of working and backup paths from the source node of the connection to the destination. The resources along the working and backup paths are reserved immediately after the paths are computed. The working path is established when the resource reservation is completed. However, the backup path is established only after a failure occurs along the working path. Fig. 2 (right) illustrates bi-directional working paths and their corresponding backup paths for demands A-E and A-F. Both backup paths traverse links AB and BE. The working paths are link-disjoint, and, the backup paths may share reserved resources in their mutual links. If a link along one of the working paths fails the reserved bandwidth on the mutual links is allotted to the failed demand. However, if two or more links along the working paths fail, the reserved bandwidth is allotted on a first-come first-serve basis and one demand is blocked. From the standpoint of a "tagged" demand receiving this service, we denote by *interfering* demand a demand whose backup path shares reserved resources with the tagged demand on at least one mutual link.

The network is said to be "down" if it is unavailable (failed) to provide the service. From the standpoint of a demand receiving unprotected service, the network is down for as long as a link along the working path of that demand has failed. The longer that it takes to repair the failed link the longer will be the downtime experienced by the demand. For a demand receiving dedicated protection service, the network is down if the working and backup paths of the demand have failed. Since these paths were bound to be link-disjoint, they can both fail if at least one link along each path fails.

For a demand receiving shared restoration service, the network is down if (a) the working and backup paths of the demand have failed, or, (b) if the working path of this demand and the working path of an interfering demand have failed and the reserved bandwidth on the mutual backup link(s) has already been allotted to the interfering demand.

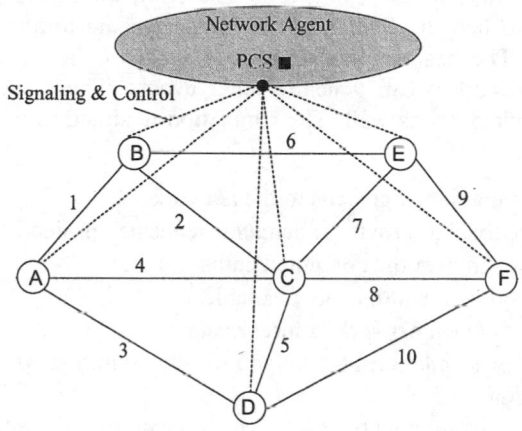

Fig. 1. Network Model

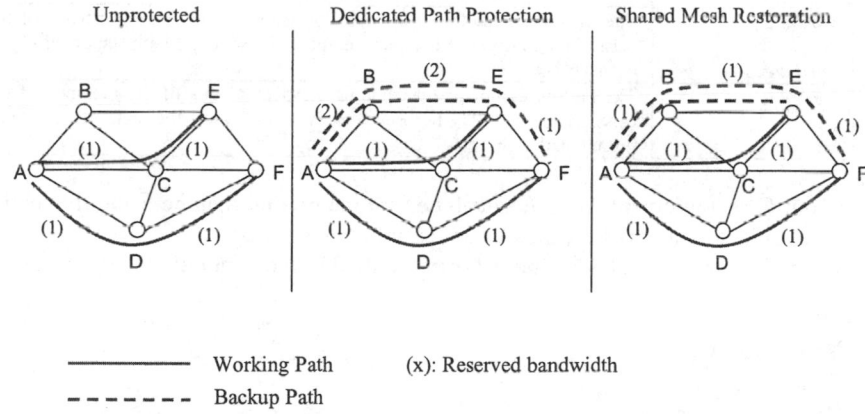

—————— Working Path   (x): Reserved bandwidth
- - - - - - - Backup Path

Fig. 2. Connection protection classes

3   Simulation Model

A discrete-event continuous-time simulation technique [14] is used to model the network and dynamical behavior of demands and failures. Objects in the simulation model are objects in the real network, and are programmed to behave as much closely as possible to the real objects. A linked list is used to store a representation of events that are waiting to happen. Events in the linked list are arranged in order of their occurrence time. The head of the list always includes the earliest event to occur. When an event occurs it can generate other events. These subsequent events are placed into the linked list as well. The simulation is aimed to perform the following tasks:

– Route demands upon their arrivals to the network
– Find alternative (backup) routes if demands requested protection
– Allocate bandwidth over the computed paths
– Block demands if bandwidth is not available
– Restore demands when network failures occur
– Repair link failures and return the failed link to a condition where it can perform its intended function
– Take various measurements such as connections downtime and allocated bandwidth on links.

We are now ready to define various classes (objects) implemented in the simulation. We employ the widely used UML notation to describe these classes, which provide the structure for the simulation activities. Reference [13] defines a UML notation for classes as a rectangular box with as many as three compartments:

Class Name	The top compartment shows the class name.
attribute₁ ... *attributeₙ*	The middle compartment includes the declarations of all attributes of the class. These attributes together compose the state of the instantiated object of this class.
method₁ ... *methodₙ*	The bottom compartment contains the declarations of the methods of the class. A method may be used to access or modify the state of the object instantiated in this class.

In our C++ implementation, the attributes are the private members and the methods are the public members of the classes unless otherwise stated.

The following basic classes (objects) are defined in the simulation:

– Event
– Network Agent
– Link
– Node
– Demand
– Random Number.

Each of these classes and their key attributes and methods will be discussed in a separate section in the following pages.

3.1 Events

There are four basic activities associated in the network: arrival of a demand, departure of a demand, failure of a network link, and repair of a network link. This is reflected in four types of events enumerated by the type-definition EventType in the following:

EventType = {DemandArrival, DemandDeparture, LinkFailure, LinkRepair}.

A base Event class is represented in UML, as follows:

Event
EventType type double time int id
double getTime() EventType getType() int getId()

Events are represented as derived classes of the base class. An Event has a type, identity (id), and the time that it occurs in the network. If the event type is DemandArrival or DemandDeparture, the event id will be the id of the demand that is arriving or departing. If the event type is LinkFailure or LinkRepair the event id will be the id of the link that is failed or repaired. Public methods in the Event diagram are used to access the Event attributes. In pure object oriented programming, the Event class would normally include a method to handle the event when it occurs. In this article, we choose a different approach to handle the events. We define four event-handler methods--- each method executing one of the event types above--- as public methods in the Network Agent class.

3.2 Network Agent

As outlined in Section 2, we have adopted a centralized network model where the routing and link state information is collected and managed in one central place. This model greatly simplifies the managing tasks and satisfies the objectives of the simulation. Based on this model, a Network Agent object is created to maintain various components and perform numerous tasks in the network. Most notable among these tasks are the following:

– Configure links and nodes
– Manage simulation clock
– Maintain a list of future incoming events
– Call appropriate event handler routines when events arrive
– Process demands upon their arrivals and compute routes
– Manage the capacity of network links
– Detect network failures, and if required, initiate restoration process
– Take various measurements including connection downtime and capacity usage.

Network Agent will be the brain of the simulated network. The Network Agent class is represented in UML as follows:

Network Agent
double simulationTime int numberOfNodes int numberOfLinks int numberOfDemands Node **nodeArray Link **linkArray Demand **demandArray List<Event> EventList int **phi
void readNetworkTopology() List<Event>::iterator nextEvent() void demandArrivalEventHandler(Event *) void demandDepartureEventHandler(Event *) void linkFailureEventHandler(Event *) void linkRepairEventHandler(Event *) void assignLinksCosts() bool dijkstraAlgorithm(int node1, int node2, int bandwidth) void allocateBandwidth(int *path, int bandwidth, Bwtype type) void run() void outputMeasurment() void configure()

The attribute simulationTime is provided to hold the current simulation time. This attribute is updated every time an event occurs in the network. Network Agent includes pointers to the arrays of all nodes, links, and demands in the network. It holds a list of pending events in the EventList list. When shared mesh restoration is in place, Network Agent records backup bandwidth reservations on links in a two-dimensional array **phi. The mechanism that is used to record the bandwidth is given in Section 4.2. Other less important attributes have been deleted from the UML representation for the sake of space.

Method readNetworkTopology in Network Agent is invoked to read (from an input file) the topology of the network being simulated. The method is normally invoked during the initial phase of the simulation. Upon execution of this method all network links and nodes are instantiated and corresponding objects are created.

Network Agent identifies the next incoming event by iterating through the event list. The iteration is performed by invoking the method nextEvent. As we noted before, the head of the event list always points to the next pending event. Once the event occurs, Network Agent examines the type of the event and executes an appropriate event handling method. Four methods are demonstrated in the UML representation, each method corresponds to one of the event types in the network. We describe each of these methods in the following sections.

3.2.1 Demand Arrival Handling Method

Method demandArrivalEventHandler is invoked if Network Agent determines that the pending event is of type DemandArrival. The following tasks are performed when this method is executed:

- The demand is instantiated by creating the corresponding object.
- Depending on the demand's requested protection service, one or two paths between the source node of the demand and the destination node are computed. Section 2 presented the protection services supported in the network and for each service outlined constrains for path computation. Section 4 will present the design of constrain-based path computation algorithms implemented in the simulation. The computation of paths encompasses several methods in Network Agent. Examples of these methods are `assignLinksCosts` and `dijkstraAlgorithm`. Whereas the former method assigns costs (weights) to network links based on the protection services requested, the latter method searches for an optimal path in the network with the assigned link costs.
- Bandwidths of links along the computed path(s) are updated via signaling the path(s) through the network. Method `allocateBandwidth` performs this task. The arguments to this method are: a pointer to the computed (working/backup) path, the bandwidth requested by the pending demand, and the type of bandwidth requested. The latter argument will be described in Section 3.3.
- A departure time is scheduled for the demand, based on some stochastic models with historical data. It is widely accepted that the holding times of demands (calls) in transport networks generally follow an exponential distribution. As a result, in the simulation, the exponential distribution is used to schedule a departure time for the demand. Once it is scheduled, an event object is created and placed in the event list. This object must have a type identifier `DemandDeparture`, an id equal to the demand's id, and occurrence time equal to the scheduled departure time.

3.2.2 Demand Departure Handling Method

Method `demandDepartureEventHandler` is called if Network Agent determines that the pending event has a type identifier `DemandDeparture`. The following tasks are performed when this method is executed:

- The demand object is deleted and the allocated bandwidth on network links is released.
- Measurements taken for this demand are saved. Notable examples of such measurements are: the fraction of time that the demand has been down (blocked) due to network failure(s); or the length of the working path, and if applicable, the length of the backup path.
- A new arriving time may be drawn for the demand, based on some widely accepted inter-arrival models. In the simulation, a new arrival time for the demand is arranged by using an exponential inter-arrival distribution. After the new arrival time is arranged, an event object is created and placed in the event list. The object must have a type identifier `DemandArrival`, an id equal to the demand's id, and occurrence time equal to the arranged arrival time.

3.2.3 Link Failure Event Handling Method

Method `linkFailureEventHandler` is called if Network Agent determines that the pending event has a type attribute `LinkFailure`. In this case, the id of the event is the id of the link that is failing. The following tasks are performed when this method is executed:

– Network Agent examines the impact of the link failure on every demand currently existed in the network. Depending on the protection service requested, one of the following situations can arise for every demand:

1. The demand will be blocked if it has requested an unprotected service and if the failed link is located along its working path.
2. The demand will be blocked if it has requested a dedicated protection service and if one of the following conditions is satisfied: (a) the failed link is located along the working path of the demand and a link along the backup path has already failed, or, (b) the failed link is located along the back path of the demand and a link along the working path has already failed.
3. The demand will be blocked if it has requested a shared restoration service and if condition (a) or (b) in the above, or condition (c) in the following is satisfied. (c) The failed link is located along the working path of the demand and the reserved bandwidth on a link along the backup path has already been allotted to an interfering demand.

– When a demand is blocked, an attribute "`downTimeStartTime`" in the demand object (see Section 3.5) is set to the current value of `simulationTime`. This is done to record the beginning of this downtime period.
– A repair is scheduled for the failed link. The distribution of the time to repair of links is described in Section 3.3 when we present the Link class. Nevertheless, after the repair time is scheduled, an event object with type identifier `LinkRepair` is created and placed in the event list. The object will have an id equal to the id of the failed link, and occurrence time equal to the time that the repair is completed (scheduled).

3.2.4 Link Repair Event Handling Method

Method `linkRepairEventHandler` is invoked if Network Agent determines that the pending event has a type attribute `LinkRepair`. The id of the event is the id of the link that has just been repaired. The following tasks are performed when this method is executed:

– The repaired link is placed back in service and its status is changed to "functional".
– Network Agent examines the impact that the repair of this link has on every blocked demand in the network. For a blocked demand to become functional it is sufficient that all links along its working path are functional. This is true regardless of the protection service requested by the demand. As a result,

Network Agent examines the working path of every blocked demand first. Network Agent will change the status of a blocked demand to functional if all links along its working path are functional. If the above condition is not met there is still a possibility that the blocked demand can become functional if and only if the demand has requested a protection service. Network Agent examines this possibility by performing the following additional task:

1. If the blocked demand has requested dedicated protection service and if all links along the backup path of that demand are functional, the status of the demand will be changed to functional.
2. If the blocked demand has requested shared restoration service, the status of the demand will be changed to functional if (a) all links along the backup path of the demand are functional, and, (b) the reserved bandwidth on every link along the backup path has not been allotted to an interfering demand.

- When a blocked demand becomes functional, Network Agent inquires from the demand object about the state (value) of its downTimeStartTime attribute. Network Agent then calculates the partial downtime for the demand by subtracting downTimeStartTime from the current simulationTime. The partial downtime is then added to the attribute downtime in the demand object (see Section 3.5).
- Once the repaired link is returned to service, Network Agent asks the link object to provide a number that represents the next time that the link will fail. The agent then creates an event object with the type identifier LinkFailure, the id equal to the link id, and occurrence time equal to the number returned by the link object. The agent places the event object in the event list.

3.3 Link

The Link class represents links in the network. The class contains all attributes and methods of a real network link that are relevant to this study. The following UML notation illustrates the most important declarations in the Link class:

Link
int id
int vertex[2]
int capacity
double workingBandwidthRatio
double backupBandwidthRatio
int *allocatedBandwidth
int length
double failureRate
double failureRatePerUnitLength
Exponential *failureDistribution
Deterministic *repairDistribution
double cost

```
StatusType status
LinkType type
Demand **backedupDemands
Demand **workingDemands
```
```
bool isBandwidthAvailable(BWtype type, int amount)
bool    acceptDemand(Demand    *,    BWtype    type,    int
bandwidth)
void upgradeCapacity(int amount)
```

A short description of each of the above attributes and methods is in order:

- id: an id of the link. An integer that can take a value from 0 to *J-1*, where *J* is the number of links in the network.
- vertex[2]: an array of two integers; each stores the id of one of the two nodes (vertices) that the link is incident upon.
- capacity: the capacity of the link measured in Gigabits per second. The capacity may be allocated wholly or partially for working traffic. In the former case the backup traffic cannot be accommodated on the link. In the latter case, a portion of the capacity--- up to a configurable maximum--- is used for the working traffic and the remaining portion is used for backup traffic. Attributes workingBandwidthRatio and backupBandwidthRatio record the configurable maximum portion of the capacity that can be allocated to working and backup traffic, respectively. For example, if a link is configured with workingBandwidthRatio=1 and backupBandwidthRatio=0 the link is usable only for the working traffic. If the link is configured with workingBandwidthRatio=0.5 and backupBandwidthRatio=0.5 only half of the capacity is used for the working traffic, and the remaining half is reserved for the backup traffic.
- *allocatedBandwidth: a pointer to an array whose dimension is equal to the number of values in the following enumeration:

BWtype={Working=0, DedicatedBackup=1, SharedBackup=2}.

Entry 0 of the array allocatedBandwidth keeps the total amount of working bandwidth reserved on this link. Similarly, entries 1 and 2 of the array store the total amount of dedicated backup bandwidth and the total amount of shared backup bandwidth reserved on the link, respectively. At any time, the total reserved working bandwidth on the link must be less than the portion of the link capacity configured for the working bandwidth. Also, the sum of the total reserved dedicated backup bandwidth and shared backup bandwidth on the link must be less than the portion of the link capacity configured for the backup bandwidth.

- length: the length of the link measured in miles.
- failureRate: there are typically two types of failures in transport networks links: *soft failures* and *hard failures*. Possible causes of soft failures are slow frequency drifts, excessive jitter and wander on the synchronization signals. Soft failures can yield erroneous bits, and more severely, occasional *Loss of Frame*

(LOF) in the received signal. Hard failures are those caused by equipment or cable outages. Outages are mostly caused by construction machines, which may damage (or cut) the cables during road works. We do not consider soft failures in this study, and thus, all link failures are regarded as link outages. It is customary to assume that the number of outages per unit time of a link (cable) is a linear function of the link length. Employing this assumption, the attribute `failurePerUnitLength` has been defined and is configured with the number of failures per mile per 10^9 hours (FITS/mile) of the link operation. The term FITS has been defined in telecommunications literature as the number of failures per 10^9 hours of operation. If one multiplies the value stored in `failurePerUnitLength` by the length of the link the total failure rate of the link is yielded. This parameter is stored in the attribute `failureRate`.

- `*failureDistribution`: it is a pointer to an object of type exponential distribution with which the link failure incidences are drawn. The mean of this distribution will be the inverse of the value stored in `failureRate`. Refer to Section 3.6 for more explanation of random variable classes.

- `*repairDistribution`: it is a pointer to an object of type deterministic distribution with which the duration of link repairs is obtained. The repair time of telecommunications systems has been shown to follow closely a deterministic distribution [9]. Refer to Section 3.6 for more explanation of deterministic distribution.

- `cost`: every link is associated with a cost (weight) that is used as a metric during the route computation and optimization. The link cost is dynamically calculated each time that a new route is computed. Refer to Section 4 for further explanation on the assignment of link costs.

- `status`: the status of the link that can be `Functional` or `Failed`.

- `type`: the type of the link, which is defined by `LinkType` enumeration. In its most generality, a link can be of one of the following types:

`LinkType={SonetRing, Local1+1, Local1:N, MeshLink}`.

A short description of these link types is in order. If a link is of type `SonetRing` it will be used solely in SONET self-healing rings [11] [15]. A link of type `Local1+1` or `Local1:N` has a local automatic protection switching mechanism. As illustrated in Fig. 3, for every link with `Local1+1` protection there is a dedicated backup link. At one end of the link (called near-end) the signal (traffic) is split and sent over both the working and backup links. The working and backup signals are identical. At the far-end of the link, both signals are monitored independently for failures. The far-end selects the best quality signal between the working and backup signals. In `Local1:N` protection illustrated in Fig.3, there is 1 backup link for N working links between the near-end and far-end nodes. The traffic is normally sent only over the working links with the backup link kept free until a working link fails. The traffic "reverts" to the working link as soon as the failure has been repaired. Finally, a link is said to be of type `MeshLink` if it does not provide any local or SONET protection mechanism. The traffic carried over a mesh link is considered unprotected. Protection may be provided end-to-end between the source and destination nodes of the traffic, using a mesh protection/restoration mechanism. A network may generally include links of different types discussed above. In such network, a demand may be routed over several links, some of which are SONET links, local 1:N links, local 1+1 links, or mesh links. A problem that might arise in this case is that a single failure can trigger multiple failure recovery mechanisms simultaneously. For instance, suppose that a demand has requested a dedicated path protection, and a link with local 1+1 protection capability existed along its working path. If this link fails it will likely trigger the local protection switching on the link, and the end-to-end mesh protection switching between the source and destination nodes of the demand. This situation can be avoided by "excluding" this link from the routes computed for the demand. As a result, in the simulation, if there exists any link in the network with link-type `SonetRing`, `Local1+1`, or `Local1:N` the link will be excluded from the path computation.

- `**workingDemands`: a pointer to an array of demands whose working paths contain this link.

- `**backupDemands`: a pointer to an array of demands whose backup paths contain this link.

- `isBandwidthAvailable(BWtype type, int amount)`: Network Agent invokes this method to investigate the availability of the bandwidth on the link when it is computing a path. The first argument passed to this method is the type of the bandwidth investigated. The second argument carries the amount of the bandwidth required. After the link object executes the method it will return a negative (false) response if the link does not have enough available bandwidth of the requested type. Network Agent will exclude (prune) the link from the path computation upon receiving the negative response.

- `acceptDemand(Demand *, BWtype type, int bandwidth)`: Network Agent invokes this interface during the path signaling (establishment) phase, in order for the link object to modify its bandwidth reservations and to insert the demand object into its list of routed demands. Three arguments are passed to the link object. The first argument is a pointer to the demand object being routed over this link. Second and third arguments are the type and the exact amount of bandwidth needed to be reserved on the link, respectively. The demand object contains an attribute `protection` that specifies the protection service requested by the demand. The link can access this information by calling `getProtection` interface of the demand object. Depending on the value returned by this interface, and the type of the bandwidth requested, the link will perform one of the following procedures:

1. If the demand has requested an unprotected service, the type of the bandwidth being passed to the link object cannot be anything other than `Working`. In this case, the link adds `bandwidth` to its total reserved working bandwidth (i.e., `allocatedBandwidth[0]`).

2. If the demand has requested a dedicated protection service, the type of the bandwidth being passed to the link object can be `Working` or `DedicatedBackup`. If the type is `Working` the link adds `bandwidth` to its total reserved working bandwidth, whereas if the

type is `DedicatedBackup` it adds `bandwidth` to its total reserved dedicated backup bandwidth (i.e., `allocatedBandwidth[1]`).

3.   If the demand has requested a share mesh restoration service, the type of the bandwidth being passed to the link object can be `Working` or `SharedBackup`. The link follows the same procedure as in item 2 if the type of the bandwidth being passed is `Working`. If, however, the type of the bandwidth is `SharedBackup` the link object will add `bandwidth` to its total reserved shared backup bandwidth (i.e., `allocatedBandwidth[2]`).

The link will add the demand object to the array of demands already accepted. If the link is going to be on the working path of the demand the demand object will be added to `workingDemands` array, otherwise, the demand object will be placed in `backupDemands` array.

—   `upgradeCapacity(int amount)`: when needed, Network Agent invokes this interface in order to add capacity to the link. The amount of additional capacity needed is passed as an argument.

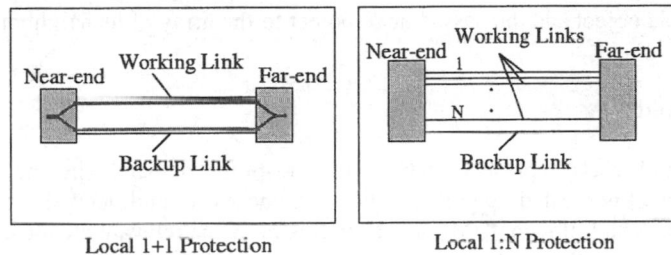

Fig. 3. Local protection switching mechanisms

3.4   Node

The Node class represents nodes in the network. It contains all attributes and methods of a network node that are relevant to this study. The following UML notation illustrates the most important declarations in the Node class:

Node
char *name int id double latitude double longitude int cost StatusType status Link **incidentLinks Node **neighborNodes
void addIncidentLink(Link *) void addNeighborNode(Node *)

- name: a pointer to a character string representing the name of the node. This can be the name of the city, corporate, or the local area network being modeled.
- id: an integer assigned to the node by Network Agent. It can take a value from 0 to *N-1*, where *N* is the number of nodes in the network.
- latitude/longitude: the geographical position of the node.
- cost: every node is associated with a cost (weight) that can be used as a metric during the path computation. This metric is not however used in the routing algorithms presented in this paper.
- status: the status of the node that can be Functional or Failed. In this paper, we assume that the network nodes are always functional.
- **incidentLinks: a pointer to an array of links that are incident upon this node.
- **neighborNodes: a pointer to an array of nodes that are directly connected to this node.
- addIncidentLink(Link *): Network Agent invokes this method to have the node object add the passed link object to the array of its incident links.
- addNeighborNode(Node *): Network Agent invokes this method to have the node object add the passed node object to the array of its neighbor nodes.

3.5 Demand

The Demand class represents active point-to-point demands in the network. A demand object is created upon its arrival to the network, and deleted when it departs the network. The following UML notation illustrates the relevant attribute and method declarations in the Demand class:

Demand
int   id int   bandwidth char *sourceName char *destinationName int   sourceId int   destinationId int *workingPath int *backupPath double downTimeStartTime double downTime StatusType status ProtectionService protection Exponential *interArrivalTimeDist double meanInterArrivalTime Exponential *holdingTimeDist double meanHoldingTime
int getBandwidth() ProtectionService getProtection() double getNextArrivalTime() double getNextDepartureTime()

- id: an integer assigned to the demand by Network Agent. It can take a value from 0 to D-1, where D is the total number of demands in the network. D is typically kept fixed during the simulation. However, not all demands are active at any time during the simulation.

- bandwidth: amount of bandwidth requested by the demand.

- *sourceName/*destinationName: a pointer to a character string representing the name of the demand's source/destination node.

- sourceId/destinationId: the id of the demand's source/destination node.

- *workingPath: a pointer to an integer array representing the id of links along the working path of this demand. Network Agent computes array workingPath during the path computation time.

- *backupPath: a pointer to an integer array representing the id of links along the backup path of this demand. Network Agent computes array backupPath during the path computation time. This array will be null (empty) if the demand has requested unprotected service.

- downtime: is used to accumulate the amount of time that the demand has been down (blocked) since its arrival to the network. It is generally comprised of multiple noncontiguous intervals during which the demand has been blocked.

- downtimeStartTime: records the start time of a downtime interval.

- status: the status of the demand that can be Functional or Failed (blocked).

- protection: represents the protection service requested by the demand. A declaration ProtectionService enumerates all possible values that this parameter can take. These are:

ProtectionService={Unprotected, SharedMesh, DedicatedMesh,
                Local1:N, Local1+1, Local1:1}.

In this paper, we only consider demands that request the first three protection services enumerated above. The remaining services are included for completeness, and have no significance here.

- *interArrivalTimeDist: a pointer to an object of the type Exponential distribution with which a new arrival time is drawn for the demand. The demand object has an attribute meanInterArrivalTime that is configured initially, and used as the mean of the above exponential distribution (see Section 3.6).

- *holdingTimeDist: a pointer to an object of the type Exponential distribution with which a new departure time is drawn for the demand. The demand object has an attribute meanHoldingTime that is configured initially and used as the mean of the above exponential distribution.

- The methods declared in the Demand UML notation are self describing. These methods are generally called by Network Agent in order to get the state of the demand's attributes. For example, Network Agent may call `getNextDepartureTime` method in order to get the next departure time scheduled for the underlying demand. The demand object will in turn have the `holdingTimeDist` member object draw a random number that will represent the next departure time for the demand.

3.6 Random Variables

Random variables are divided into two general categories: discrete and continuous. They can all be represented as derived classes of a base class, which we call *Random Variable*. The base class, as illustrated in the UML diagram below, simply performs the basic operations required by all random variables. One notable example of such operations would be to generate a pseudorandom floating-point number in the interval [0, 1]. Method `drawUniformRand` in the base class is designed to perform this operation. Another common operation would be to set a seed for the random number generator. This operation is performed by method `setSeed` in the base class.

Pure virtual methods `genDoubleNumber` and `genIntegerNumber` will be invoked to draw random numbers according to the concrete implementation of these methods in the derived classes. Continuous random variables will include a concrete implementation of `genDoubleNumber` according to their actual distribution of data, whereas discrete random variables will typically include a concrete implementation of `genIntegerNumber`.

Random Variable

`private virtual double drawUniformRand()`
`private virtual void setSeed()`
`public virtual double genDoubleNumber()=0`
`public virtual int genIntegerNumber()=0`

Exponential random variable is a continuous random variable that is derived from Random Variable base class. The UML notation of exponential random variable is illustrated below. It includes a concrete implementation of `genDoubleNumber`, and will invoke this method each time in order to draw a number. Exponential random variable will be completely specified by its mean, which must be configured when the object is instantiated.

Exponential
`double mean`
`public virtual double genDoubleNumber()`

For simplicity of programming, we consider Deterministic distribution as a derived class of Random Variable. The UML diagram of the Deterministic distribution class is shown below. Here, when `genDoubleNumber` method is invoked it will always return a deterministic fixed value stored in `fixedNumber`.

Deterministic
`double fixedNumber`
`public virtual double genDoubleNumber()`

4   Routing and Bandwidth Allocation

How does Network Agent compute a pair of link-disjoint paths for a demand requesting a dedicated protection or shared mesh restoration service? How does it compute a path for a demand requesting unprotected service? What metric is used to decide whether a link should be included in the path computation? What mechanisms are used to monitor and allocate the bandwidth on the network links?

The rest of this section is devoted to answer the above questions. In Section 4.1, we present algorithms that Network Agent will use in order to compute and allocate a working path for a demand. These algorithms are independent of the protection service requested by the demand. Unlike the working path, the computation of a backup path depends largely on the protection service requested. Section 4.2 is devoted to the computation of a backup path under the assumption that shared mesh restoration is requested. Section 4.3 presents the computation of a backup path when dedicated mesh protection is requested.

A 2-step heuristic algorithm is used to compute the optimal working and backup paths. In step 1 of the algorithm, as described in Section 4.1, a shortest working path is computed, whereas in Step 2, as described in Sections 4.2 and 4.3, an optimal (second shortest) backup path is computed. In both steps, Dijkstra algorithm [16] is used to find the shortest paths, with appropriate assignment of link costs.

For ease of formulation, we introduce a slight change in the notations in this section. We use short letters and Greek symbols to describe various network parameters instead of long ASCII words given in the previous sections. For each link with id j, we define the following notations:

- C_j: capacity of link j
- α_j: configured working portion of the capacity (`workingBandwidthRatio`) on link j
- β_j: configured backup portion of the capacity (`backupBandwidthRatio`) on link j
- W_j: total allocated working bandwidth (`allocatedBandwidth[0]`) on link j
- B_j: total allocated shared backup bandwidth (`allocatedBandwidth[2]`) on link j
- D_j: total allocated dedicated backup bandwidth (`allocatedBandwidth[1]`) on link j
- Q_j: Residual (available) working bandwidth on link j
- R_j: Residual backup bandwidth on link j

The residual working and backup bandwidths on link j are obtained as follows:

$$Q_j = (\alpha_j \times C_j) - W_j \tag{1}$$

$$R_j = (\beta_j \times C_j) - (B_j + D_j)$$

4.1 Working Path Computation

Upon the arrival of a new demand, Network Agent will invoke `demandArrivalEventHandler` method in order to service the demand. Let us denote by r the id of the newly arrived demand, and by b the bandwidth that it requires. The method follows the following procedure to compute a working path for this demand, regardless of the protection service that it requested.

Let ξ_i denote the cost (weight) of choosing link i to be on the working path of demand r. ξ_i is defined as follows:

$$\zeta_i = \begin{cases} 1 & if & b \le Q_i \\ \\ \infty & otherwise \end{cases}, \qquad \forall i \in [0,...,J-1] \tag{2}$$

The cost of link i is set to 1 if the link has enough residual working capacity to accommodate the bandwidth requested by r. Otherwise, the cost is set to an arbitrary large number, denoted by ∞ here. The above cost assignment is used by Dijkstra algorithm (implemented in method `dijkstraAlgorithm`) to find a shortest working path for demand r. If such a path cannot be found, the demand will be rejected.

Once the path is identified the total reserved working bandwidth on links along the working path must be updated. The procedure for updating the bandwidth on these links is straightforward. For every link i along the computed working path, bandwidth b is simply added to W_i.

If demand r has requested a protection service (dedicated or shared mesh) the above procedure for bandwidth-update along the working path is postponed until after the backup path is also computed. This is done to make sure that the demand will not be rejected due to the unavailability of a backup path.

4.2 Shared Mesh Restoration

The shared mesh restoration process ensures that the amount of backup bandwidth required on a link to restore the failed traffic resulting from any single link failure will not be more than the total amount of share backup bandwidth reserved on that link. Network Agent records the share backup bandwidth reserved on links in the following matrix [6] [7] [8] [10]:

$$\Phi = \begin{bmatrix} 0 & \varphi_{01} & \varphi_{02} & \cdots & \varphi_{0J-1} \\ \varphi_{10} & 0 & \varphi_{12} & \cdots & \varphi_{1J-1} \\ \varphi_{20} & \varphi_{21} & 0 & \cdots & \varphi_{2J-1} \\ \cdots & & & & \\ \varphi_{J-10} & \varphi_{J-11} & \varphi_{J-12} & \cdots & 0 \end{bmatrix} \tag{3}$$

Element φ_{ij} is the amount of share backup bandwidth needed on link j if link i fails ($0 \le i, j \le J-1$). Network Agent records the elements of matrix Φ in its two-dimensional ASCII equivalent array ** phi introduced in Section 3.2.

The total amount of share backup bandwidth needed on link j is indeed the maximum of all elements in column j of matrix Φ, that is:

$$B_j = \max_{\forall i < J}[\varphi_{ij}] \tag{4}$$

Assuming that the working path has already been found for the arrived demand r (see Section 4.1), method demandArrivalEventHandler will follow the procedure below to compute an optimum backup path between the demand's source and destination nodes. Let us define the following additional notations:

- S_W: set of links along the working path of demand r. This set is already known.
- S_B: set of links along the backup path of demand r, to be found in this section.

Let T_j denote the maximum amount of backup bandwidth required on link j if a link in S_W fails. It follows that T_j will simply be:

$$T_j = b + \max_{\forall i \in S_W}[\varphi_{ij}]. \tag{5}$$

Let θ_j denote the cost (weight) of choosing link j to be on the backup path of demand r. θ_j can be defined as follows [3]:

$$\theta_j = \begin{cases} \infty & \text{If} & j \in S_W \\ \varepsilon & \text{elseif} & T_j \le B_j \\ T_j - B_j & \text{elseif} & T_j - B_j \le R_j \\ \infty & \text{otherwise.} \end{cases} \tag{6}$$

The cost of link j is set to ∞ if j is along the working path of demand r. The cost is set to a very small number (ε) if j is not in S_W and if T_j is less than B_j. In this case, demand r can be restored on link j without need to reserve any additional backup bandwidth on this link. If neither of the above conditions is satisfied the cost is set to $T_j - B_j$ if this quantity is less than the residual backup capacity on this link. In this case, $T_j - B_j$ will be the amount of additional backup bandwidth required on link j in order to restore demand r. If the residual backup capacity on link j is not adequate to accommodate this additional required bandwidth, the cost of link j is set to ∞ in the forth term above.

The optimal backup path for demand r is computed using Dijkstra algorithm in which the cost of every link j is set to θ_j. The computed path, if any, will be an optimal shared backup path that is also link-disjoint from the working path of demand r computed earlier. Demand r will be rejected immediately if such a backup path cannot be found.

Once the backup path is computed the total reserved share backup bandwidth on links along the backup path must be updated. This is performed via updating the elements of matrix Φ: for every link i along the computed working path and every link j along the computed backup path, bandwidth b will be added to element φ_{ij}. That is:

$$\forall i \in S_W, \forall j \in S_B : \varphi_{ij} \leftarrow \varphi_{ij} + b. \tag{7}$$

After the elements in matrix Φ are updated, the new total reserved share backup bandwidth on the backup links in S_B can be obtained from (4). On some of these links, the new value of this quantity can be the same as the old value (i.e., before demand r arrived). If link j is such a link, T_j must have been less than or equal to B_j when θ_j was computed.

4.3 Dedicated Mesh Protection

In dedicated mesh protection, the cost of each link (θ_j) is set to the following when a backup path is computed for demand r:

$$\theta_j = \begin{cases} \infty & \text{if} & j \in S_W \\ 1 & \text{elseif} & b \leq R_j \ . \\ \infty & \textit{otherwise.} \end{cases} \tag{8}$$

Here the cost of link j is set to unity if this link is not along the working path and if it has enough residual backup capacity to accommodate the requested bandwidth, b. Otherwise, it is set to infinity. A shortest backup path is obtained by executing Dijkstra algorithm with the link cost assignment above. The computed path, if any, will be link-disjoint from the working path computed earlier for demand r. If a backup path is not found, demand r will be rejected immediately.

Once the backup path is computed the total reserved dedicated backup bandwidth on links along the backup path must be updated. For every link j along the computed backup path, b is simply added to D_j.

5   State Diagram

A state diagram for the network is given in Fig. 4 below. The network passes through a series of discrete states during operation. The diagram consists of the following states:

- Initial: a pseudo state denoted by a solid circle. It shows the starting point or first activity of the network.
- Configuration: represents the state in which the network links and nodes are configured and various objects are instantiated. During this state, network simulation parameters are configured and counters are reset. Method configure of Network Agent is invoked when the network is in this state.
- Wait: represents the state in which the network waits for an event to occur. When a qualified event takes place the network transits to one of the following four states based on the type of the event.
- Path Setup: this state is reached when the event-type is DemandArrival. In this state, Network Agent invokes its method demandArrivalEventHandler to handle the event and compute routes for the arrived demand. Upon the completion of this state the network returns to wait state.
- Path Teardown: this state is reached when the event-type is DemandDeparture. In this state, Network Agent invokes its method demandDepartureEventHandler to handle the event by tearing down the path(s) of the departing demand. Upon the completion of this state the network returns to wait state.
- Restoration: this state is reached when the event-type is LinkFailure. In this state, Network Agent invokes its method linkFailureEventHandler to handle the event. The method initiates repair process and restores demands affected by the failure. Upon the completion of this state the network returns to wait state.
- Recovery: this state is reached when the event-type is LinkRepair. In this state, Network Agent invokes its method linkRepairEventHandler to handle the event.   The method performs several tasks including reverting demands with shared/dedicated protection service to their normal working route, and examining the impact of the repaired link on blocked demands. Upon the completion of this state the network returns to wait state.
- Output: a transition occurs to this state when the network simulation clock (simulationTime) reaches a preset configured value. When this time is reached the simulation is stopped and various measurements are outputted.

6   Implementation

The heart of the simulation is the method run that is implemented in C++ code below. The method defines a while loop that controls the simulation time, and makes use of a switch structure that selects a proper event handling method. Upon the occurrence of an event the simulation time is updated with the time that the event occurred. Before the while loop begins, the method `configure` is invoked to initialize simulation parameters and configure the network topology. After the while loop ends, the method `outputMeasurment` is invoked to print the simulation measurement.

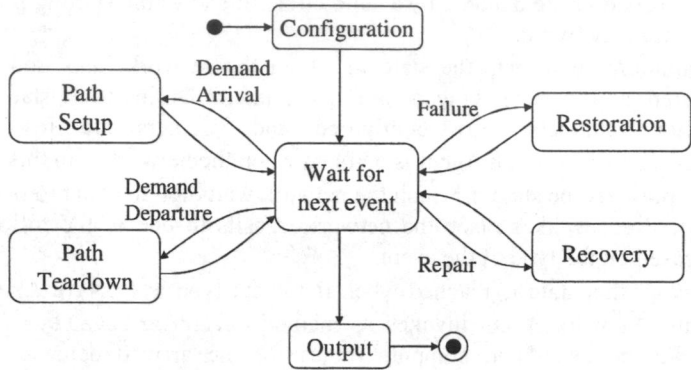

Fig. 4. Network state diagram

```
void NetworkAgent::run()
{
    double simulationTime = 0;
    configure();
    while( simulationTime <= PresetTime )
    {
        List<Event>::iterator event = nextEvent();
        simulationTime = (*event).getTime();
        EventType type = (*event).getType();
        switch( type )
        {
            case LinkFailure:
                int linkId = (*event).getId();
                linkFailureEventHandler(linkId,
            simulationTime);
                break;
            case LinkRepair:
                int linkId = (*event).getId();
                linkRepairEventHandler( linkId,
            simulationTime);
                break;
            case DemandArrival:
                int demandId = (*event).getId();
                demandArrivalEventHandler(demandId,
            simulationTime);
                break;
```

```
      case DemandDeparture:
          int demandId = (*event).getId();
          demandDepartureEventHandler(demandId,simulationTime);
          break;
   } //end of switch
 } //end of while loop
 outputMeasurment(simulationTime);
} //end of run method
```

7   Performance Evaluation

In this section we will compare the performances of the following mesh protection and restoration services: dedicated path protection, shared path restoration, and unprotected path. Our criteria for performance comparison are: (a) the amount of downtime experienced by demands receiving these services, and (b) the usage of bandwidth on network links. A demand's downtime is measured as the number of minutes per year that the demand is blocked due to link failures. We use a backbone network described in the next section to test the formulations and programs presented in this paper.

7.1   Test Network

We simulated a network topology with 16 nodes and 25 links as illustrated in Fig. 5. This is the NSF (National Science Foundation) network, which is a representative of North American backbone networks. Nodes are representative of popular North American cities. Distances between nodes are shown in miles. They represent the lengths of optical fiber cables terminated at these nodes.

We generated a traffic demand matrix called *distributed traffic* between nodes in the NSF network. Point-to-point demands between nodes were obtained in proportion to the population of these nodes. We assumed the following model to relate the traffic generated by a node with the population of the node. We let P_i denote the population of node i, and G_i denote the total traffic in Gigabits/second generated at this node. The following relation was used between P_i and G_i:

$$\forall i:\ G_i = \frac{2 \times P_i}{10^6} \quad \text{Gigabits/second.} \tag{9}$$

We do not intend to prove the correctness of the above model. The model has been used in the telecommunications industry though.

The portion of G_i being destined for any arbitrary node j was then obtained based on the population of node j and overall population of all nodes in the network.

Accordingly, if we denote G_{ij} the amount of traffic generated at i and terminated at j, G_{ij} is obtained as:

$$\forall j \ (j \neq i): \ G_{ij} = G_i \frac{P_j}{\sum_{n=1}^{16} P_n} \qquad \text{Gigabits/second.} \tag{10}$$

From (9) and (10) one can simply prove that the traffic generated between any two nodes i and j is symmetric: that is $G_{ij}=G_{ji}$.

We did not use matrix $[G_{ij}]$ directly in the simulation. Instead, we normalized each element G_{ij} to the bandwidth required by an OC48 channel. An OC48 channel requires 2.488 Gigabits/second bandwidth. Thus, denoting D_{ij} the number of demands between nodes i and j such that each demand required one unit of OC48 bandwidth, D_{ij} was derived as:

$$D_{ij} = \left\lceil \frac{G_{ij}}{2.488} \right\rceil. \tag{11}$$

We used demand matrix $[D_{ij}]$ as the input to the simulation. We correspondingly measured all bandwidth quantities in units of OC48 channels. Demands between every pair of nodes were considered individually and routed independently from each other. There were in total 128 bi-directional OC48 demands between all pairs of nodes in Fig. 5.

7.2 Results

As a baseline, we chose the failure rate per mile of all links in the network to be $\lambda_0=500$ FITS/mile. We also chose the link repair times to be constant and equal to 3 hours. Link capacities were chosen large enough to accommodate all incoming demands. Demands were arrived at the network during the initial phase of the simulation, and never departed the network once they arrived. We ran the simulation for 2×10^8 hours, long enough to eliminate the initial transient behavior due to the sequential arrival of demands to the network. It also allowed the link-failure generators to generate enough failure events. Statistics were collected after all demands arrived at the network. Longer simulation times did not change significantly the statistics shown in the rest of this section.

A mix of protection services was not allowed in the simulation. All demands in the network were allowed to request either unprotected, dedicated protection, or shared restoration service, and not the mix. Fig. 6 illustrates the average downtime achieved when each of these protection services were used in the network. Fig. 7 illustrates the corresponding absolute maximum downtime achieved in the network. In both figures, the presented downtime is a function of the number of hops (links) along the working path. For every fixed value h_W along the horizontal axes, the average downtime was the sum of the downtimes experienced by all demands with working path sizes h_W, divided by the number of such demands. Similarly, the maximum downtime is the maximum downtime experienced by all these demands. Three curves are shown in both figures; each curve is properly labeled with the name of the protection service used.

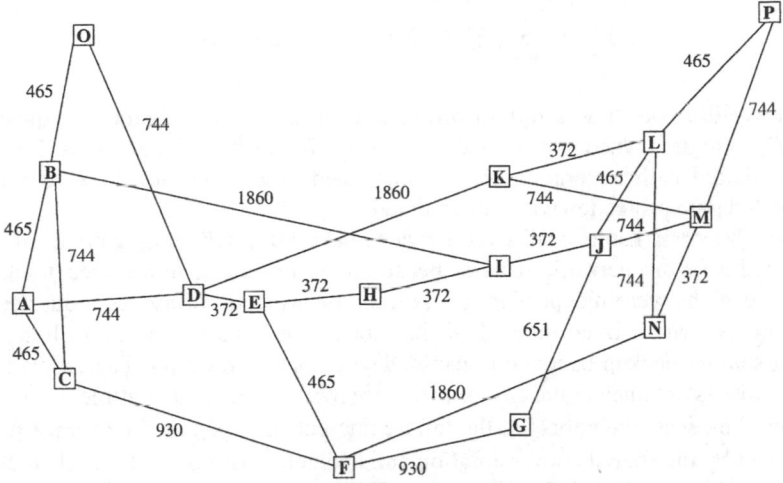

Fig. 5. Topology of NSF network

As it is seen in Figs. 6 and 7, the unprotected service has yielded an unacceptably high downtime over all range of horizontal values. Client applications of transport networks would typically require a much lower downtime, not beyond a few tens of minutes/year. One important class of these applications is the class of carrier grade voice applications. They require less than 5.256 minutes/year downtime. Obviously, the unprotected service cannot be used when these applications are concerned.

As evident from Figs. 6 and 7, dedicated protection and shared mesh restoration schemes have achieved much lower downtimes than the unprotected service. This shows the effectiveness of assigning alternate routes to demands that can be used when link failures occur along their primary routes.

Shared mesh restoration has yielded a relatively higher downtime than dedicated mesh protection, particularly when h_W is large. The average downtime achieved by shared mesh restoration reaches 2.75 times higher than the average downtime achieved by dedicated mesh protection. The same is true in terms of the maximum achieved downtime. Shared mesh restoration has never achieved a downtime of less than 5.256 minutes/year required by carrier grade services.

Table 1 illustrates the total reserved working and backup bandwidths in the network when each of the above protection services was used. The total reserved working bandwidth (W_{tot}) is defined as the sum of the working bandwidth reserved on all links in the network. Using the symbolic notations defined in Section 4, we can obtain W_{tot} as:

$$W_{tot} = \sum_{\forall i < J} W_i \quad \text{(OC48 units)}. \tag{12}$$

Similarly, the total reserved backup bandwidth (B_{tot}) is defined as the sum of the backup bandwidth reserved on all links in the network. That is:

$$B_{tot} = \sum_{\forall i < J} (B_i + D_i) \qquad \text{(OC48 units)}. \qquad (13)$$

Since we did not allow a mix of protection services in the simulation, quantities B_i and D_i were zero when unprotected service was implemented; quantities D_i were zero when shared path restoration was implemented; and quantities B_i were zero when dedicated path protection was implemented.

As shown in Table 1, all three services have allocated equal amount of working bandwidth in the network. This is because, the criteria that we used to determine working paths were independent of the type of protection services requested. When backup bandwidth is considered in the table, shared path restoration has allocated much smaller backup bandwidth than dedicated path protection. The reduction in the bandwidth is obtained at the cost of mediocre increase in the downtime.

Fig. 8 presents the impact of the failure rate per mile (λ_0) on the average downtime achieved by the shared path restoration and dedicated path protection schemes. λ_0 has been varied from 100 FITS/mile to 900 FITS/mile in steps of 200 FITS/mile. The downtime tends to vary exponentially with λ_0. With shared path restoration, the downtime of less than 10 minutes/year is achieved only when λ_0 does not exceed 300 FITS/mile.

Fig. 9 studies the impact of repair time on the average downtime achieved by the shared path restoration and dedicated path protection schemes. The repair time has been varied from 3 to 24 hours. The downtime tends to vary exponentially with the repair time as well. With the shared path restoration scheme, the downtime exceeds 120 minutes/year when the repair time is longer than 6 hours.

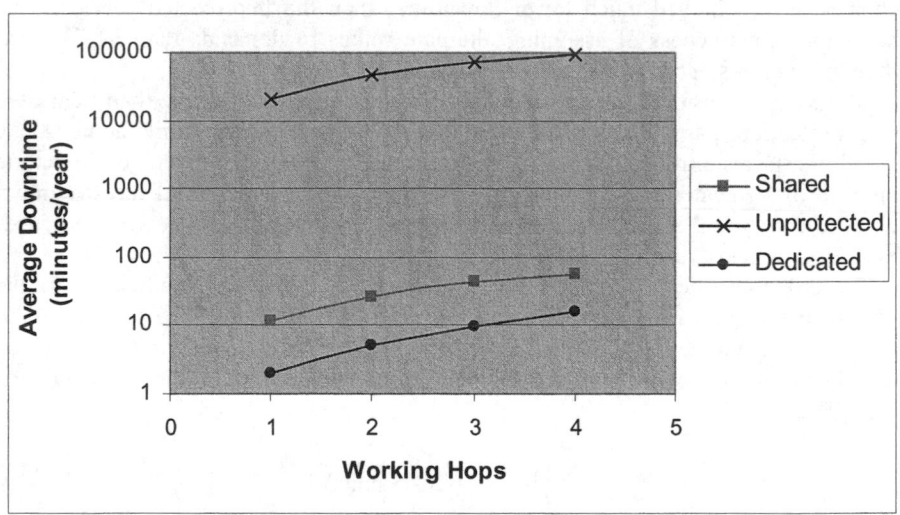

Fig. 6. Average downtime as a function of number of hops along working path

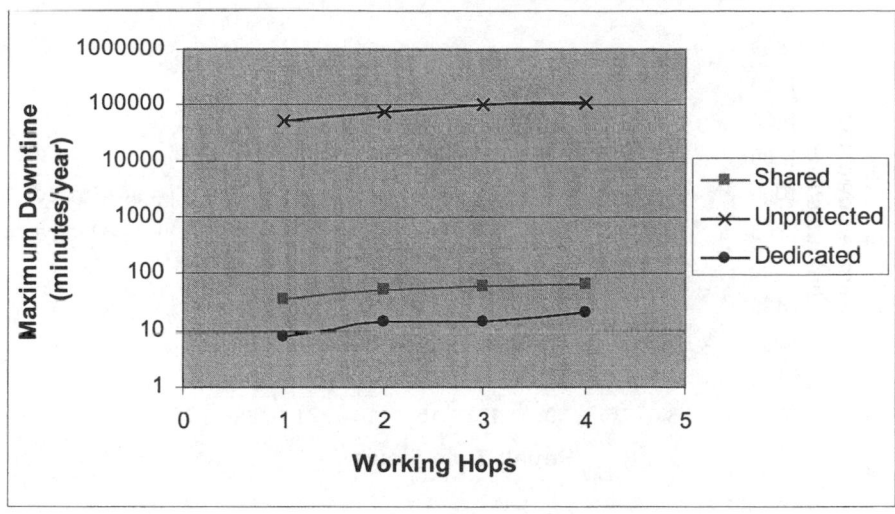

Fig. 7. Maximum downtime as a function of number of hops along working path

Table 1. Total measured working and backup bandwidths

Protection Scenario	Total Working Bandwidth (OC48 Units)	Total Backup Bandwidth (OC48 Units)
Unprotected	291	0
Shared Path Restoration	291	160
Dedicated Path Protection	291	467

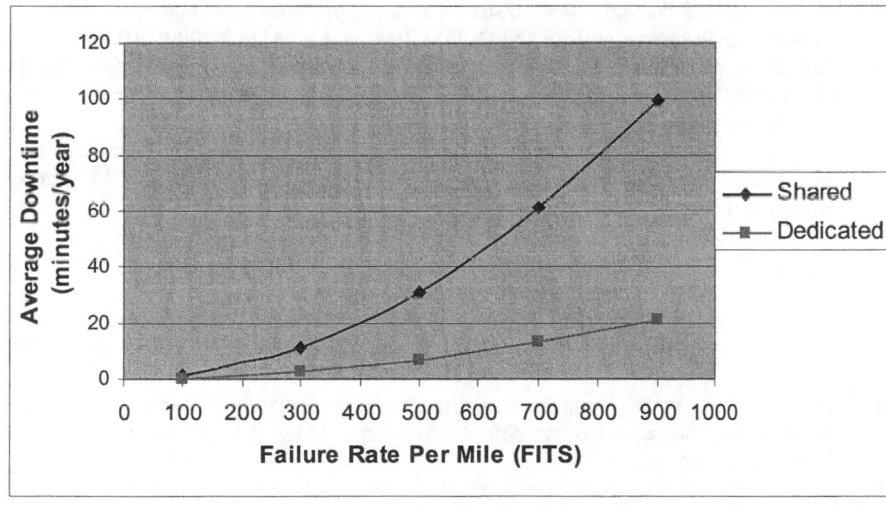

Fig. 8. Average Downtime as a function of failure rate per mile

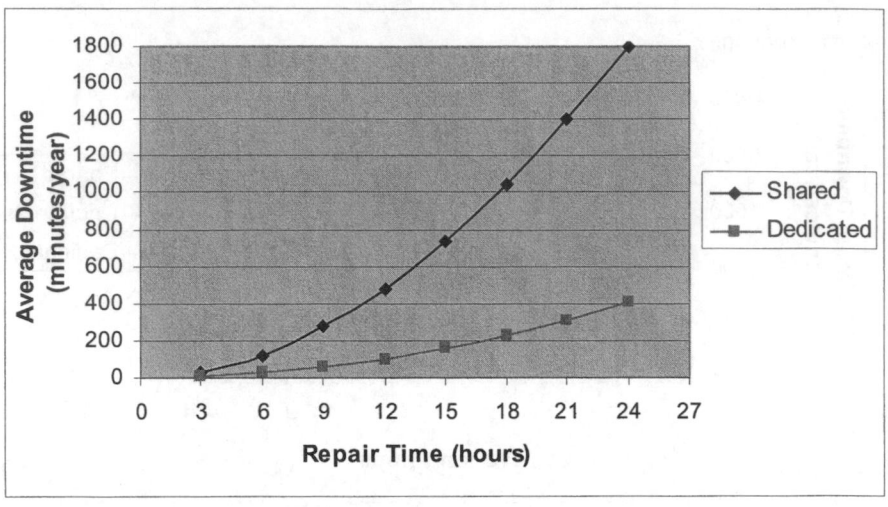

Fig. 9. Average downtime as a function of link repair time

References

1. Ramamurthy, S., Mukherjee, B.: Survivable WDM Mesh Networks, Part I – Protection. Proceedings IEEE INFOCOM'99, vol. 2. New York (March 1999) 744–751
2. Ramamurthy, S., Mukherjee, B.: Survivable WDM Mesh Networks, Part II – Restoration. Proceedings IEEE ICC'99, (June 1999) 2023–2030
3. Kodialam, M., Lakshman, T.V.: Dynamic Routing of Bandwidth Guaranteed Tunnels with Restoration. Proceedings IEEE INFOCOM'2000, vol. 2. (March 2000) 902–911
4. Kodialam, M., Lakshman, T.V.: Dynamic Routing of Locally Restorable Bandwidth Guaranteed Tunnels using Aggregated Link Usage Information. Proceedings IEEE INFOCOM'2001, vol. 1. (March 2001) 376–385
5. Iraschko, R.R., MacGregor, M.H., Grover, W.D.: Optimal Capacity Placement for Path Restoration in STM or ATM Mesh-Survivable Networks. IEEE/ACM Trans. On Networking, vol. 6, No. 3. (June 1998) 325–336
6. Naser, H., Mouftah, H.T.: A Multi-Layer Differentiated Protection Services Architecture. Submitted to IEEE Journal of Selected Area in Communications, Special Issue on Advances in Metropolitan Optical Networks
7. Li, G., Wang, D., Kalmanek, C., Doverspike, R.: Efficient Distributed Path Selection for Shared Restoration Connections. Proceedings IEEE INFOCOM'2002, vol. 1. (March 2002) 140–149
8. Liu, Y., Tipper, D., Siripongwutikorn, P.: Approximating Optimal Spare Capacity Allocation by Successive Survivable Routing. Proceedings IEEE INFOCOM 2001, vol. 2. (2001) 699–708
9. Bell Communications Research: Methods & Procedures for System Reliability Analysis. Special Report SR-TSY-001171, Issue 1. (January 1989)
10. Datta, S., Sengupta, S., Biswas, S., Datta S.: Efficient Channel Reservation for Backup Paths in Optical Mesh Networks. Proceedings IEEE GLOBECOM 2001, vol. 4. San Antonio, TX. (November 2001) 2104–2108

11. Stern, T.E., Bala, K.: Multiwavelength Optical Networks: A Layered Approach. Prentice Hall (1999)
12. Wen, W., Ben Yoo, S.J., Mukherjee, B.: Quality-of-Service Based Protection in MPLS Control WDM Mesh Networks. Journal of Photonic Network Communications, vol. 4, issue ¾. (2002) 297–320
13. Jia, X.: Object-Oriented Software Development Using Java: Principles, Patterns, and Frameworks. 2nd edition, Addison Wesley, Boston. (2003)
14. Law, A.M., Kelton, W.D.: Simulation modeling and analysis. McGraw-Hill, New York. (1982)
15. Bell Communications Research: SONET Bidirectional Line-Switched Ring Equipment–Generic Requirements. GR-1230-CORE, Issue 4. (December 1998)
16. Bhandari, R.: Survivable Networks: Algorithms for Diverse Routing. Norwell, MA: Kluwer Academic. (November 1998)

Wireless Internet: Protocols and Performance

Carey Williamson

Department of Computer Science, University of Calgary, Calgary, AB, Canada

Abstract. This tutorial article describes the IEEE 802.11b Wireless Local Area Network (WLAN) standard, which is commonly referred to as "WiFi". This standard offers up to 11 Mbps of transmission capacity at the physical layer of the protocol stack, and is one of the key enabling technologies for wireless Internet, mobile computing, and ad hoc networking applications. After introducing the standard and its features, the latter part of the article discusses protocol interactions that occur when popular Internet applications, such as multimedia streaming and the World Wide Web, operate over IEEE 802.11b WLANs. These interactions can lead to performance problems in the TCP/IP Internet protocol stack.

1 Introduction

Two of the most exciting and fastest-growing Internet technologies in recent years are the World Wide Web and wireless networks. The Web has made the Internet available to the masses, through its TCP/IP protocol stack and the principle of layering: Web users do not need to know the details of the underlying communication protocols in order to use network applications. Wireless technologies have revolutionized the way people think about networks, by offering users freedom from the constraints of physical wires. These technologies are available today, in laptop or handheld form, at relatively modest cost. Mobile users are interested in exploiting the full functionality of the technology at their fingertips, as wireless networks bring closer the "anything, anytime, anywhere" promise of mobile networking.

One of the primary challenges in this new networking context is "performance transparency": providing an end-user Internet experience that is hopefully no worse than that in the traditional wired-Internet desktop environment. Significant advances are taking place in both wired and wireless networking environments that substantially increase the raw bit rate available at the physical layer. However, these advances are of little value if the extra bandwidth cannot be delivered all the way up to the application layer. In some cases, performance problems occur at intermediate layers of the protocol stack.

This tutorial paper focuses on one particular wireless networking technology, namely the IEEE 802.11b Wireless Local Area Network (WLAN) standard [ANSI 1999], and the protocol performance issues that arise in that environment. The first part of this tutorial provides an overview of IEEE 802.11b WLAN protocols, as well as TCP/IP protocols, and some popular Internet applications used on wireless LANs. The last part of the tutorial focuses on protocol

M.C. Calzarossa and E. Gelenbe (Eds.): MASCOTS 2003, LNCS 2965, pp. 118–142, 2004.
© Springer-Verlag Berlin Heidelberg 2004

performance issues for wireless Internet applications. To illustrate the issues, practical examples are used. These include wireless TCP performance, multimedia streaming, TCP performance in multi-hop ad hoc networks, and Web performance in wireless ad hoc networks.

Let us point out that all of these issues are also directly or indirectly discussed in several other papers in the present volume. Wireless networks and related end-to-end protocols are examined in [AB 2004,BN 2004,GLGL 2004]. Other papers which discuss adhoc networks in this volume include [KLW 2004,MTG 2004].

2   Background

Figure 1 provides an illustration of the Internet protocol stack [Tan 2003]. A protocol stack provides a modular architecture and a conceptual framework for discussing communication protocols and their functionality. Note that this diagram shows only a 5-layer protocol stack, compared to the 7-layer protocol stack in the classic OSI network reference model [Tan 2003].

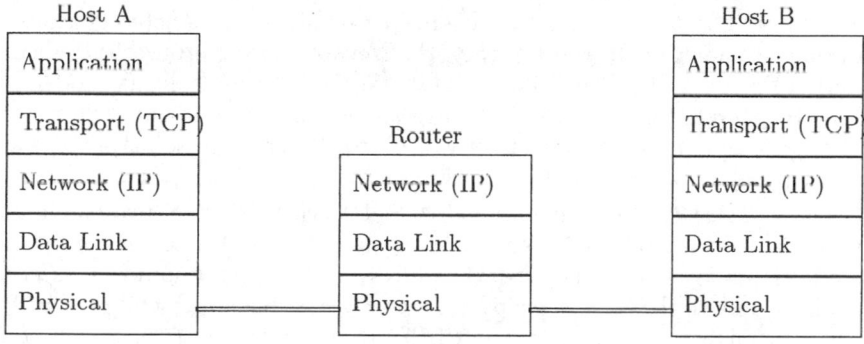

Fig. 1. Illustration of the Internet TCP/IP Protocol Stack

The lowest layer of the protocol stack is the *Physical Layer*. The physical layer deals with the raw transmission of bits between two communicating devices. Many different transmission media are possible at the physical layer, including wired (guided) media such as twisted pair (copper), coaxial cable, or optical fiber, and wireless (unguided) media such as microwave, satellite, IR (Infra-Red), or RF (Radio Frequency) transmission. The physical layer performs the signalling and modulation required to encode information (e.g., binary 0's and 1's) on the channel, by varying physical characteristics of the signal (e.g., amplitude, frequency, phase). The coding techniques used are highly dependent upon the properties of the transmission medium chosen at the physical layer.

The next layer up the protocol stack is called the *Data Link Layer*, or the Link Layer for short. This layer deals with a larger logical unit called a *frame*. A frame typically carries several hundred or several thousand bits. Frames may be fixed-size or variable-size, depending on the specific networking technology being used. For example, Asynchronous Transfer Mode (ATM) networks use fixed-size frames called *ATM cells*, while Ethernet and IEEE 802.11b LANs allow variable-size frames, with upper and lower limits on the legal frame sizes permitted.

The Link Layer provides two main services. First, it regulates access to the channel amongst the contending stations. In a broadcast network, this *Medium Access Control* (MAC) mechanism is important, since at most one station can successfully transmit on the shared channel at a time. In a point-to-point network, the MAC protocol has a very minor role, since each link has only two endpoints. Second, the Link Layer provides framing, flow control, and error control services, to provide reliable hop-by-hop communication. Commonly-used mechanisms at this *Logical Link Control* (LLC) sublayer are checksums, sequence numbers, acknowledgements (ACKs), timeouts, and retransmissions.

The *Network Layer* of the protocol stack builds upon the Link Layer services, by adding addressing, routing, and internetworking functionality at the *packet* level. Addressing uniquely identifies any endpoint host on the network. Routing determines a path for reaching a destination. Internetworking support allows communication across different networks by defining how to translate packet formats and how to accommodate diverse packet sizes across heterogeneous networking technologies. In the Internet, the Network Layer protocol is called the Internet Protocol (IP). It provides a "best effort" datagram delivery model. Most of the IP packets that are sent will correctly arrive at the intended destination, but there is no guarantee that they will do so. Packets are sometimes delayed, lost, duplicated, or corrupted in transit.

The *Transport Layer* provides end-to-end services between two communicating entities on the Internet. While IP routing gets a packet to the correct host, an additional layer of transport-level addressing (e.g., port numbers) is needed to deliver data to the correct recipient (of many possible recipients) on that host. The Transmission Control Protocol (TCP) on the Internet is one example of a Transport Layer protocol. It provides end-to-end reliable data delivery. More details on TCP are provided in Section 4. Another example is the User Datagram Protocol (UDP), which is a minimal mechanism transport-layer protocol. It provides a connection-less service model similar to IP.

The topmost layer of the Internet protocol stack is the *Application Layer*. Many user-level network applications reside here: electronic mail, file transfer, network news, media streaming, peer-to-peer, and the World Wide Web. Each of these applications has a well-defined application-layer protocol, such as SMTP (Simple Mail Transfer Protocol), FTP (File Transfer Protocol), or HTTP (Hyper-Text Transfer Protocol). These protocols offer services to end users of the Internet.

This layered Internet protocol stack model provides a reference point for the discussion of IEEE 802.11b protocols in the next section, as well as the discussion of TCP/IP protocol performance issues later in the article.

3 The IEEE 802.11b WLAN Standard

One of the most popular solutions for wireless Internet access today is the IEEE 802.11b wireless local area network standard [ANSI 1999], commonly referred to as "WiFi" (Wireless Fidelity). This section provides information on the basic operation of IEEE 802.11b, as well as some of its features.

3.1 Overview

Wireless networking refers to the use of infrared (IR) or radio frequency (RF) signals to transmit information between devices, without requiring physical cabling between them. Commonplace examples are using remote control devices to change the channel on your television, open your garage door, or advance the slides on your laptop during a conference presentation. Simple wireless devices convey only control information (e.g., open or close your garage door), while more sophisticated ones allow the transmission of arbitrary data (e.g., using a wireless keyboard to interact with the TV/computer in your hotel room, or using your personal digital assistant (PDA) to 'beam' your business card and contact information to your colleague's PDA).

Wireless networking solutions typically require line-of-sight transmission, or at least close proximity to the devices being controlled. For example, your TV remote control does not work very well from the bathroom, and your garage door opener does not work when you are several blocks away from your house. One reason is the limited transmit power used by the source, which curtails the physical distance that an intelligible signal can propagate (e.g., the signal strength typically diminishes proportionally with the square of the distance travelled). The limit on transmit power is often regulated by the federal government to reduce the interference between devices operating in different jurisdictions, and to minimize health and safety concerns. A second reason is the operating frequency (in Hertz (Hz)) used in the electromagnetic spectrum. Some signals (e.g., broadcast radio, RF, X-rays) are able to pass through solid objects (e.g., walls, humans), while other signals (e.g., IR, visible light) are not. This is why a household baby monitor can be used to listen to an infant sleeping in the nursery upstairs, while your TV remote control cannot turn down the volume on the TV in the apartment next door to your own.

The IEEE 802.11b WLAN standard is an attractive and popular wireless networking solution based on these principles. IEEE 802.11b offers physical layer data rates of up to 11 Mbps, which makes it a cost-effective LAN solution similar to the classic 10 Mbps Ethernet LAN. This "WiFi" LAN solution is attractive in business, education, and research environments because it enables tetherless access to the Internet. With the wireless network card commonplace in laptops today, users can roam from office to office or lab to lab in their organization while still maintaining network connectivity for email, Web browsing, or other Internet-related activities. WiFi "hotspots" are widely deployed in many cities (e.g., at airports, hotels, coffee shops, and bookstores) for general Internet access.

The following sections provide greater detail on IEEE 802.11b.

3.2   Physical Layer

The IEEE 802.11 Working Group has developed an entire family of IEEE 802.11 protocols, which continues to grow and evolve today. The standards define both the Physical Layer and the Data Link Layer operation for IEEE 802.11 protocols.

The first version of the IEEE 802.11 standard supported a data rate of 1 Mbps, using a physical layer transmission technique called Frequency Hopping Spread Spectrum (FHSS). Later improvements doubled the data rate to 2 Mbps, while still maintaining backward compatibility with the 1 Mbps version. These standards were developed for both FHSS and DSSS (Direct-Sequence Spread Spectrum) transmission at the physical layer.

Within the DSSS portion of the IEEE 802.11 family, a high data rate extension called IEEE 802.11b was defined. This extension supports higher data rates of 5.5 Mbps and 11 Mbps, using more sophisticated modulation schemes for physical layer transmission. This standard also maintains backward compatibility with earlier devices, by supporting the 1 Mbps and 2 Mbps data rates in the original IEEE 802.11 standard.

The operating frequencies for IEEE 802.11b are in the Industrial, Scientific, and Medical (ISM) band of the electromagnetic spectrum, near 2.4 GHz. Specifically, the frequency band ranges from 2400 MHz to 2483.5 MHz. Many vendors have produced products that operate in this frequency range, including IEEE 802.11b, IEEE 802.11g, Bluetooth, baby monitors, microwave ovens, Home RF, and some cordless phones. Interference from other devices is a prevalent concern for IEEE 802.11b WLANs, since it can lead to unpredictable WLAN performance.

Within the assigned portion of the electromagnetic spectrum, IEEE 802.11b devices can choose from 14 "channels", each approximately 22 MHz wide. However, many of these channels overlap. Only channels 1, 6, and 11 are non-overlapping. The choice of channels can be manually configured, so that multiple WLANs in close proximity do not interfere with each other. An IEEE 802.11b device can automatically detect when it is in the range of more than one wireless Access Point (AP), and dynamically select the AP that provides the strongest signal. (See Section 3.5 for more details on the Infrastructure Mode of operation.)

The commonly used transmit power for IEEE 802.11b is 100 milliWatts, which typically provides about 100 meters of omni-directional coverage. Some users have been able to achieve much greater distance coverage (e.g., several kilometers) using higher transmit power and directional antennas.

A recent modification to IEEE 802.11b is the IEEE 802.11g standard. IEEE 802.11g offers data rates up to 54 Mbps, using the same basic technology as IEEE 802.11b, and the same busy portion of the electromagnetic spectrum (2.4 GHz). Many chip sets produced today support both IEEE 802.11b and 802.11g.

The follow-on to IEEE 802.11b is IEEE 802.11a, which offers data rates up to 54 Mbps. Two important differences exist between IEEE 802.11b and 802.11a. First, IEEE 802.11a operates in the 5 GHz band of the electromagnetic spectrum, which is distinct from that of IEEE 802.11b. This new frequency band is attractive because there is (so far) less interference in this portion of the spectrum. However, not all vendors have commodity chip sets designed for operation in this regime yet, so the product prices are significantly higher than

IEEE 802.11b. Second, IEEE 802.11a uses a different physical layer modulation technique called Orthogonal Frequency Division Multiplexing (OFDM). This technique provides greater error resilience than DSSS.

3.3   Channel Access Protocols

The IEEE 802.11b standard defines three channel access protocols that can be used at the Medium Access Control (MAC) sublayer. These protocols are called Distributed Coordination Function (DCF), Request-To-Send/Clear-To-Send (RTS/CTS), and Point Coordination Function (PCF).

Distributed Coordination Function (DCF). The most commonly used MAC protocol option, and the default in most IEEE 802.11b WLAN deployments, is DCF. This protocol defines a distributed algorithm that allows multiple stations to compete for the use of a single broadcast channel in the wireless coverage area, which is called a *cell*. All wireless devices in the cell must share use of the same channel.

The fundamental rule that must be obeyed in the WLAN cell is that at most one successful frame transmission can be in progress at a time on the network. If no stations transmit, the channel is idle. If exactly one station transmits, there is a very high probability that the receiver will receive the frame successfully. If two or more stations transmit, the result is a *collision* on the channel, which results in unintelligible data for the receivers. Such colliding frames waste network resources, since they require retransmission at a later time for successful delivery.

The purpose of the MAC protocol is to determine which station is allowed to transmit, particularly when multiple stations have frames ready for transmission. Desired properties for the MAC protocol include: a low channel access delay for acquiring the channel; a low collision rate on the network, so that few retransmissions are required; high efficiency under high load, so that the maximal network throughput can be achieved; and fairness, so that each station is equally likely to acquire the channel when it is available.

The DCF protocol used is called Carrier Sense Multiple Access with Collision Avoidance (CSMA/CA). The "Multiple Access" part of this name refers to the coordination problem defined previously: multiple stations competing for use of a single shared transmission channel. The "Carrier Sense" part of this name identifies an important aspect of the protocol: the stations can listen to the channel to see if it is available or not prior to transmitting a frame. In particular, a ready station with a frame to transmit must first listen to the channel. If the channel is idle, the station can then begin transmitting its frame. If the channel is busy, the station must defer (i.e., refrain from sending), because transmitting would surely cause a collision on the network (hence the term "Collision Avoidance" in the name of the protocol).

Randomization is also an important part of the channel access protocol. A variable called the Contention Window (CW) is used for this purpose. A station wishing to transmit a frame chooses a random BackOff (BO) time between 0 and CW, and this extra BO delay must elapse before the actual frame transmission can begin. If the channel is busy and collisions are observed, stations dynamically increase (e.g., double) CW. If frame transmissions are routinely successful, CW can be reset to its default value CW_{min}.

The CSMA/CA protocol reduces the number of collisions on the channel, but does not eliminate collisions entirely. For example, if two stations become ready at exactly the same time, and both sense the channel idle, then both could start transmission at the same time, and collide with each other.

Handling this type of collision problem is tricky. In an Ethernet LAN environment, a transmitting station can use its transceiver (transmitter/receiver) to listen to the channel during its own outgoing frame transmissions. This property allows a station to detect discrepancies between what it was trying to send and what was actually observed on the wire. Discrepancies between the two indicate a collision on the network (i.e., more than one station transmitting at the same time). A station detecting such a collision aborts the transmission of its frame, and generates a noise burst on the wire for all stations to hear. This protocol is called CSMA/CD, with the CD standing for Collision Detection. It is used in Ethernet wired LANs.

Unfortunately, the Collison Detection (CD) mechanism is not applicable for IEEE 802.11b WLANs. Physically, stations can either transmit or receive using their antenna, but they cannot do both at the same time. Even if they could, the transmit power would be so dominant that it would be almost impossible to detect a received signal. Further complicating matters are the noisy characteristics of the wireless propagation environment: not all stations may hear the collision if there was one, and some stations may hear a collision even if there wasn't one.

The IEEE 802.11b DCF protocol solves this problem with a combination of mechanisms: acknowledgements, timeouts, and retransmissions. Upon the successful arrival of a frame at the intended recipient, the receiver sends back a control frame with a positive acknowledgement to the sender. This acknowledgement tells the sender that its frame transmission was successful. In the absence of the acknowledgement, the sender will retransmit another copy of the same frame after a randomly chosen short timeout interval (e.g., up to 1 millisecond). This mechanism handles collision-related losses just the same as corruption-related losses due to wireless channel errors. It also recovers from the loss of either the data frame or its acknowledgement. In both cases, a frame retransmission is required. If repeated retransmissions are required for the same data frame, the timeout interval is repeatedly doubled, up to a maximum limit CW_{max}. If the maximum number of retries (e.g., default 8 in most implementations) is reached, then the frame transmission is aborted. Further error recovery is left to higher-layers of the protocol stack.

To ensure that the recipient of a successful frame can acquire the channel to send an acknowledgement, the IEEE 802.11b standard defines two separate time intervals. The Short Inter-Frame Space (SIFS) is the amount of time that the recipient waits before sending its acknowledgement. This time is 28 microseconds (μsec). The Distributed Inter-Frame Space (DIFS) is the amount of time that a station must observe a quiescent channel before concluding that it is idle, and proceeding with its random BackOff (BO) and frame transmission. This DIFS time is 128 μsec. With these settings, the recipient of a successful frame is the first station entitled to use the shared channel to send back an acknowledgement. This Positive Acknowledgement with Retransmission (PAR) protocol is used only for unicast (one-to-one) frames on the WLAN. It is not used for multicast or broadcast frames that are addressed to many recipients.

Figure 2 summarizes the CSMA/CA DCF protocol. The diagram illustrates an example frame transmission, say from station A to station B. After sensing the channel idle for the DIFS period, and waiting for its random BO period, station A grabs the channel and transmits its data frame. After receiving the frame successfully, station B waits for the SIFS period and then sends its positive ACK to A.

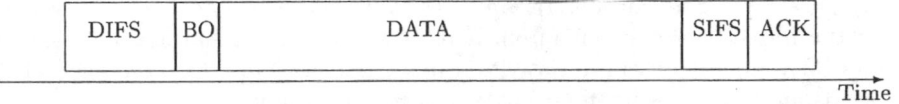

Fig. 2. Illustration of CSMA/CA DCF in IEEE 802.11b WLAN

Request-To-Send/Clear-To-Send (RTS/CTS). The RTS/CTS protocol is designed to handle the *hidden node* problem that can arise in wireless networks. Consider three stations called A, B, and C, who happen to be geographically in alphabetical order in the network. Depending on the relative distances between the nodes, it is possible that A and B can hear each other, that B and C can hear each other, but A and C cannot hear each other. In such a scenario, A might transmit a frame to B at the same time that C transmits a frame to B. The result is a collision (and unintelligible garbage) at B.

Neither A nor C is aware of this collision problem, since they are not within range of each other. Their only clue is the lack of a positive ACK from B, which will soon trigger retransmissions of the colliding frames. The larger the frame size, the greater the proportion of time wasted on collisions, and the greater the probability that the retransmitted frames will also collide. From A's viewpoint, C is a hidden node, and from C's viewpoint, A is a hidden node. The hidden node problem can dramatically degrade wireless channel performance.

The RTS/CTS protocol resolves this problem by having the intended receiver node (B) manage the shared channel amongst its neighbours (A and C, in this example). In particular, the neighbours must make advance reservations of the channel for their transmissions. For example, node A sends to node B a short control frame called the Request-To-Send (RTS) frame. The RTS indicates the size of the frame that A wants to send, and the intended recipient B. If this is the only RTS request that node B receives, then node B can send a short Clear-To-Send (CTS) control frame to A to grant its request. The broadcast nature of this CTS transmission tells node A to go ahead with its frame transmission, while also telling node C (since C can hear B) to refrain from sending. The information contained in the CTS frame conveys a Network Allocation Vector (NAV) that expresses the amount of channel time required for A and B to complete the exchange of their data frame and acknowledgement. Station C simply defers from accessing the channel for this NAV interval, and thus avoids colliding with A's frame.

Similarly, node C can initiate a frame transmission to B using the RTS/CTS exchange. If node B receives multiple RTS requests, it can choose which one to grant, as long as it has at most one CTS outstanding at a time. Because RTS control frames are very short, it is unlikely for them to collide. Nevertheless, if they do collide, the random timeout and retransmission mechanisms described previously resolve this. A station that does not receive a CTS response to its RTS request after a maximum number of RTS retries will abort its attempted transmission of the current frame.

Figure 3 provides an illustration of the RTS/CTS protocol. Again, assume that the frame transmission is from A to B. Once A receives the CTS response to its RTS request, it can initiate its frame transmission. The receiving station B sends an ACK to A upon receipt of a successful frame.

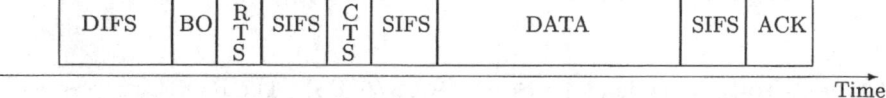

Fig. 3. Illustration of RTS/CTS in IEEE 802.11b WLAN

Some implementations of IEEE 802.11b trigger the use of RTS/CTS automatically when the average number of collisions in the DCF protocol exceeds a threshold. Other implementations require manual selection of either the DCF or RTS/CTS protocol.

Point Coordination Function (PCF). The two foregoing MAC protocols for IEEE 802.11b WLANs are not suitable for real-time applications, since there is no bound on the maximum latency for channel access and frame transmission. That is, there is no guarantee of when (or even if) a data frame will be successfully transmitted across the network. These protocols should not be used for hard real-time control systems applications (e.g., blast furnace operation, nuclear power plant, braking system on your car), but they may be satisfactory for some soft real-time applications (e.g., wireless video streaming, home security monitoring, network intrusion detection).

The IEEE 802.11b standard defines an additional MAC protocol that is better suited to real-time applications. This protocol is called Point Coordination Function (PCF). It is intended for WLANs that are operating in infrastructure mode (see Section 3.5), with an Access Point (AP) to coordinate usage of the shared wireless channel amongst multiple wireless devices. This PCF mode of operation is similar to the Master/Slave mode of operation in Bluetooth scatternets and piconets. Few vendors of IEEE 802.11b products support the PCF mode of operation.

PCF is a polling protocol. The AP has an explicit list of all wireless devices operating on the network, and polls (asks) each device in turn if it has any information to send. By bounding the number of devices in the network, and the maximum frame size used, the AP establishes a schedule of service with an upper bound on the channel access latency for each device. Typically, all devices in the network are equivalent in priority, so one slot of service is provided to each device in turn in the service cycle. In general, however, multiple slots can be assigned to some devices within each service cycle, to accommodate devices with different priorities or bandwidth requirements.

While the PCF protocol conceptually bounds the channel access delay for each device (by eliminating collisions from the MAC protocol), there is still the possibility of wireless channel errors (e.g., due to noise or external interference on the wireless LAN). The timeout and retransmission events in the PAR protocol can still occur, which in turn implies that there is no deterministic guarantee on when (or if) a given data frame is successfully transmitted on the WLAN. In other words, IEEE 802.11b WLANs are not a perfect solution for hard real-time applications.

3.4 Link-Layer Frame Formats

The purpose of the MAC sublayer protocol in the previous section is to ensure that at most one station is transmitting a frame on the WLAN at a time. The purpose of the Logical Link Control (LLC) sublayer above the MAC is to ensure that the frame is sent reliably.

The IEEE 802.11b link layer protocol uses variable length data frames, with the frame format illustrated in Figure 4. Two different versions of this frame format are supported, called the Long Preamble and the Short Preamble. The choice between these formats is usually software-settable in the configuration parameters of an IEEE 802.11b product.

Preamble (128 bits)	S O F	S p e c d	S e r v	L e n	C R C	Payload (0-2312 bytes)

Fig. 4. Illustration of Link-Layer Frame Format in IEEE 802.11b

The default frame format in IEEE 802.11b is the Long Preamble format. The transmission of a frame begins with 128 bits of preamble: an alternating bit pattern of 1's and 0's that is used to establish signal clocking and synchronization between sender and receiver. The preamble is followed by a 16-bit Start Of Frame

(SOF) delimiter. This bit pattern is "10101010 10101011". The last two bits of this field tell the receiver that the important control header of the frame is about to begin.

The Physical Layer Convergence Protocol (PLCP) header at the LLC layer is 48 bits long, and has four fields. The first 8-bit field indicates the data rate (Signal Speed) that will be used for the payload portion of the frame transmission. The valid choices are 1 Mbps, 2 Mbps, 5.5 Mbps, and 11 Mbps. The second 8-bit field specifies an optional Service Type, which is currently unused in IEEE 802.11b. The third field (16 bits) indicates the size in bytes of the payload portion of the frame. The fourth field is a 16-bit Cyclic Redundancy Check (CRC) for the frame.

The preamble and the PLCP header of the LLC frame are transmitted at 1 Mbps. This feature makes the IEEE 802.11b protocol backwards-compatible with older IEEE 802.11 devices that might be part of the network. While these devices would be unable to receive frames at any of the higher data rates, they can at least listen to the network and correctly determine when frames begin and end. The payload portion of an LLC frame is transmitted using the data rate indicated in the PLCP header of the frame.

The Short Preamble version differs from the foregoing format in two ways. First, the length of the preamble is reduced to 56 bits instead of 128 bits. This change makes the entire preamble (including the SOF delimiter) 72 bits instead of 144 bits, reducing the amount of channel time consumed by the preamble. Second, the 48-bit PLCP header is transmitted at 2 Mbps instead of at 1 Mbps. Again, this slightly reduces the time consumed on the network by each frame. The payload portion of an LLC frame is transmitted using the data rate (1, 2, 5.5, or 11 Mbps) that was indicated in the PLCP header of the frame.

Regardless of the frame format used, the payload portion of the frame contains additional control information (e.g., 48-bit MAC address of the sender, 48-bit MAC address of the intended receiver) plus the useful data, if any (e.g., a TCP/IP packet carrying user-level data). The maximum payload size is 2312 bytes. Many implementations use a Maximum Transmission Unit (MTU) size of 1500 bytes at the network layer, to be compatible with Ethernet LANs.

The combined overheads of the link-layer frame format and the channel access protocol limit the effective throughput that can be achieved over IEEE 802.11b WLANs. A general rule of thumb is that about 60% of the stated physical layer data rate can be achieved as user-level throughput at the application layer. For IEEE 802.11b, this means that about 6.5 Mbps of throughput is possible for TCP/IP. For IEEE 802.11a, the corresponding value is about 32 Mbps. A paper by Jun et al. [JPS 2003] provides a careful analysis of the maximum throughput that is theoretically possible for IEEE 802.11 networks.

3.5   Other Features

There are two different ways to use IEEE 802.11b WLANs: *infrastructure mode*, and *ad hoc mode*.

Infrastructure Mode. Infrastructure mode requires an Access Point (AP) that functions as a gateway or bridge between the wireless access network and the general wired Internet. Figure 5 shows a WLAN operating in infrastructure mode. The AP has two network interfaces: one for transmitting and receiving information on the WLAN, and one for transmitting and receiving information on the wired network, such as an Ethernet LAN.

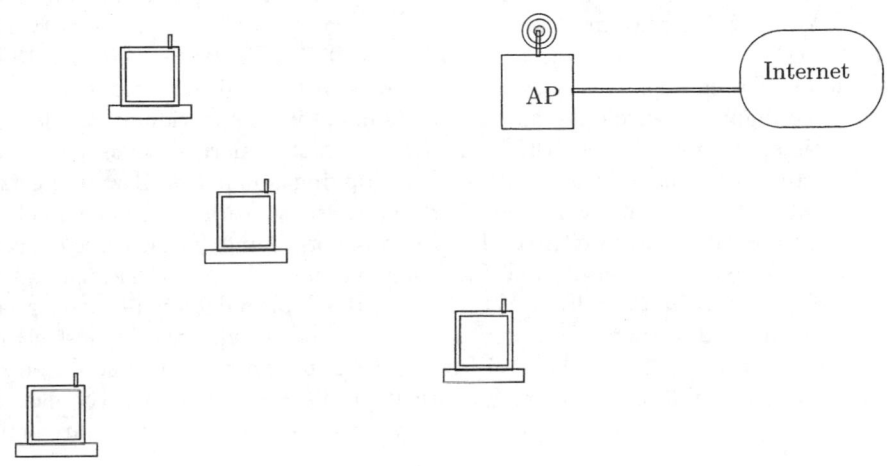

Fig. 5. Illustration of an IEEE 802.11b WLAN in Infrastructure Mode

In infrastructure mode, all communication to and from a mobile wireless device must traverse the AP. For example, if the mobile device wishes to access content from the general Internet, then the request is first transmitted to the AP, which forwards the request to the Internet on behalf of the mobile device. When the response returns to the AP from the wired Internet, the AP transmits the response on the WLAN as wireless data frames addressed to the client that initiated the request. Similarly, if mobile device A in the WLAN wants to communicate with mobile device B also in the WLAN, the request must be relayed via the AP. That is, A sends a frame to the AP, which acknowledges its successful delivery. Then the AP transmits the frame to B, which acknowledges its successful delivery. In this communication scenario, the frames between A and B are transmitted twice on the WLAN: once to the AP, and once by the AP. This can compromise the efficiency of the WLAN. (In ad hoc mode described below, nodes A and B can communicate directly with each other.)

The AP plays a central role in an infrastructure-based WLAN. An AP advertises its presence on the WLAN by broadcasting *beacon frames*, typically every 100 milliseconds (i.e., 10 times per second). These management (control) frames identify the AP, its MAC address, its Service Set Identifier (SSID), as well as whether it is using WEP (Wired Equivalent Privacy) encryption or not. These frames are broadcast omni-directionally by the AP, so that all wireless devices within the coverage area (cell) of the AP can detect its presence. Wireless devices in the cell, such as client laptops with IEEE 802.11b WLAN cards, can detect this signal and associate with the AP if desired. The nominal coverage area for an AP has a radius of about 100 meters, but the quality of coverage may vary depending on building construction materials (e.g., reinforced concrete, tinted windows, metallic coatings on glass). This range can be significantly extended with directional antennas, if desired.

Typically, the deployment of IEEE 802.11b WLANs requires multiple APs (e.g., for a university residence network, a campus-wide wireless network, or a WLAN in a business organization) [BB 1997,KE 2002,SB 2004,TB 1999, TB 2000]. To reduce interference, adjacent APs in the logical network topology can be configured to use different channel numbers in the portion of the electromagnetic spectrum used by IEEE 802.11b. In North America, there are eleven such channels to choose from, with each occupying about 22 MHz of spectral bandwidth. However, many of these channels partially overlap. Only channels 1, 6, and 11 are sufficiently well-spaced to be non-overlapping. These three channel numbers are commonly used in WLAN deployments. A mobile user can roam through a composite of multiple WLAN cells. If a laptop detects beacon signals from multiple APs, then the laptop can select the AP with the highest signal strength. Conceptually, the handoff from one AP to the next AP should be seamless and invisible to the user. Smooth handoffs are not always the case in practice, particularly if APs from multiple vendors or if IP subnetting are being used [BVBP 2002,KE 2002,SB 2004].

Ad Hoc Mode. The second mode of operation supported by IEEE 802.11b is *ad hoc mode*. In this mode, a collection of wireless devices can operate together as a standalone wireless network, completely independent of the Internet. That is, there is no AP required for access to the general Internet. This standalone mode of operation is often used for special purposes, such as military network applications, sensor networks, peer-to-peer networks, or multi-player wireless gaming. An example of an ad hoc network is shown in Figure 6.

In ad hoc mode, every wireless device is an equal peer with other wireless devices. Channel access is regulated using the DCF protocol most of the time, with the use of RTS/CTS to resolve hidden node problems, if they occur.

Ad hoc networks are especially interesting when they are *multi-hop* wireless ad hoc networks [FZLG 2003,SS 2002,WTLS 2002]. Suppose that there are hundreds of wireless devices in the WLAN, and that the geographic coverage area of the entire WLAN is far greater than the transmission range of any single wireless device (perhaps because of transmit power limitations to conserve energy and extend battery lifetime). In such a scenario, a device A at one location of the network (e.g., the left edge) may not be able to communicate directly with another device Z elsewhere in the network (e.g., the right edge).

Communication between A and Z can only be realized by having other nodes (e.g., B, C, D...) function as intermediate routers for forwarding frames on behalf of other nodes. This is the principle upon which multi-hop ad hoc networks are based. The routing protocols used in such networks are complicated, since they

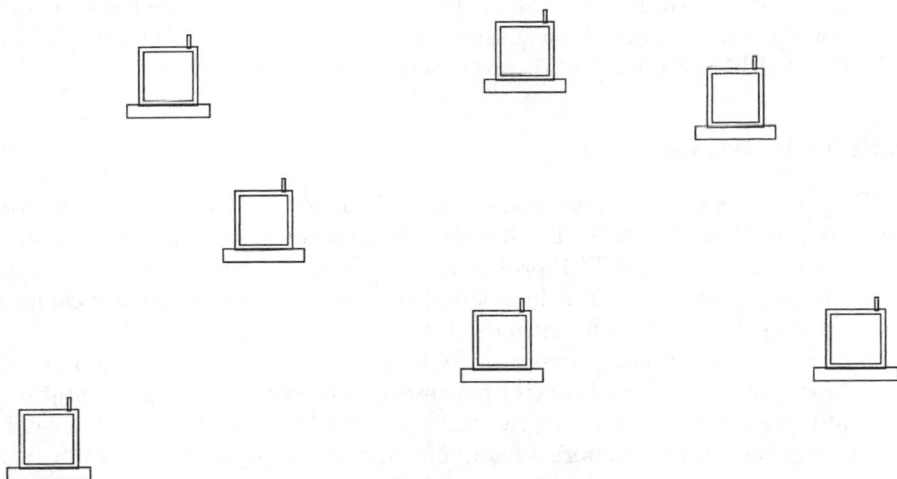

Fig. 6. Illustration of an IEEE 802.11b WLAN in Ad Hoc Mode

must dynamically determine valid routing paths to the intended destination. Nodes can move at any time, changing the topology of the network, and thus changing the quality of the routes used. Some routes may break, and other routes (better or worse) may become available at any time.

In addition to the routing problem, multi-hop ad hoc networks induce other performance problems for wireless Internet protocols. For example, the end-to-end performance of TCP degrades significantly over multi-hop ad hoc networks, because of contention and collision problems. Problems such as these are discussed in more detail in Section 5, following the TCP review in the next section.

4   The Web and TCP/IP

This section provides background on the Web and TCP/IP. An understanding of these protocols is required prior to the discussion of the wireless Internet protocol performance problems in Section 5.

4.1   The Web

The Web relies primarily on three communication protocols: IP, TCP, and HTTP. The Internet Protocol (IP) is a connection-less network-layer protocol that provides global addressing and routing on the Internet. The Transmission Control Protocol (TCP) is a connection-oriented transport-layer protocol that provides end-to-end data delivery across the Internet. Among its many functions, TCP has flow control, congestion control, and error recovery mechanisms to provide reliable data transmission between sources and destinations. The robustness of TCP allows it to operate in many network environments. Finally, the Hyper-Text Transfer Protocol (HTTP) is a request-response application-layer protocol layered on top of TCP. HTTP is used to transfer Web documents between Web servers and Web clients (browsers). Currently, HTTP/1.0 and HTTP/1.1 [RFC 1996,RFC 1997] are widely used on the Internet.

4.2   TCP Overview

TCP is a connection-oriented, end-to-end reliable-byte-stream transport-layer protocol [Ste 1994,Tan 2003]. The basic mechanisms of this general-purpose protocol are quite robust: the TCP protocol has undergone relatively minor changes in its 30-year existence, over a time period that has witnessed dramatic changes in computing and networking technologies.

The basic unit of data transfer in TCP is a *byte* (i.e., for sequence numbering, flow control, and error control purposes). However, TCP implementations generally work with a larger logical unit size called a *segment* when transmitting packets across an IP internetwork. The Maximum Segment Size (MSS) is a settable parameter for TCP. The choice of the MSS depends on the Maximum Transmission Unit (MTU) size supported by the underlying network layer. In most cases, each TCP segment is carried in one IP packet; hence the terms segment and packet are often used interchangeably. The task of TCP is to divide the application-layer data into one or more segments, transmit them across the network, and deliver them reliably (and in order) to the receiving TCP. Each segment carries an explicit sequence number, for the purposes of ordering and reliability.

There are several mechanisms in TCP to ensure reliable packet delivery. For example, when a sender transmits a segment, it sets a timer. If this segment is received successfully, then the receiver sends back an acknowledgement (ACK). TCP ACKs are cumulative, and always indicate the next expected TCP sequence number. The sender uses the ACK for flow control and error control purposes, as well as to estimate the round-trip time (RTT) to the destination. If the timer expires before an ACK is received, then the sender retransmits the outstanding segment. Another commonly used strategy is *Fast Retransmit* [Flo 2001], which uses duplicate ACKs to trigger the retransmission of a missing segment, often well before the retransmission timer expires. This approach works well in recovering from single packet losses [FF 1996].

TCP uses sliding window flow control to limit the maximum number of bytes outstanding (i.e., not yet acknowledged) between a sender and a receiver at any time. A sender is allowed to transmit the segments in a window as quickly as it wishes, providing that data is available to transmit. As ACKs are received, the flow control window advances, and new segments are transmitted.

A congestion control mechanism was added to TCP in 1988, based on algorithms proposed by Jacobson [Jac 1988]. These algorithms use adaptive window-based flow control to achieve congestion control, since the IP network layer in the Internet does not provide congestion control.

In TCP congestion control, the flow control window size is adjusted dynamically based on two TCP state variables: the congestion window (*cwnd*), and the slow-start threshold (*ssthresh*). The initial value of *cwnd* is one segment, and *cwnd* is increased as successful ACKs are received. The increase is exponential in the slow-start phase (i.e., doubling *cwnd* every RTT, until *ssthresh* is reached), and linear in the congestion avoidance phase (i.e., increasing *cwnd* by one segment for every complete window's worth of data exchanged) [Jac 1988].

TCP uses packet loss (due to buffer overflow at a router) as an implicit signal of network congestion. Each time a packet loss is detected, TCP updates its estimate of the slow-start threshold (e.g., $ssthresh = cwnd/2$), reduces its congestion window size (e.g., $cwnd = MSS$), and re-enters the slow-start phase. The *Fast Recovery* [Flo 2001] mechanism reduces the congestion window size by half (e.g., $cwnd = cwnd/2$) following a Fast Retransmit, rather than reducing it to one segment.

The foregoing algorithms are part of most TCP implementations, including Reno TCP and New Reno TCP that are widely used on the Internet today [Flo 2001].

5   Protocol Performance Issues

Many interesting protocol interactions occur when TCP/IP Internet applications are carried over wireless access networks such as IEEE 802.11b WLANs. This section discusses four examples of these Wireless Internet protocol performance problems.

5.1   Wireless TCP Performance

A well-documented TCP performance problem occurs if packets are lost in wireless networks [BB 1995,BK 1998,BPSR 1997,FPAY 2003,LLL 2002,Pent 2000]. Most TCP implementations use packet loss as an implicit signal of network congestion, and use backoff mechanisms to reduce the packet load offered to the network. This design assumption is clearly articulated in the TCP slow start congestion control mechanism [Jac 1988]. This approach works well for the wired Internet, since losses of packets due to congestion dominate packet losses due to transmission errors. In wireless networks, however, the situation is reversed: losses due to transmission errors dominate congestion losses. Upon losing a packet due to a transmission error, the desired behaviour in a WLAN is to retransmit right away, but a conventional TCP implementation instead experiences timeout and backoff. This behaviour can lead to very low TCP throughput on wireless networks [BK 1998].

One solution to this wireless TCP performance problem is to implement a "wireless-aware" version of TCP at the boundary between the wired backbone network and the wireless access network. This approach (referred to as Proxy-TCP, Indirect-TCP, or Snoop-TCP in the literature) logically splits the end-to-end TCP control loop into two smaller control loops, with one covering the wired segment of the network, and the other the wireless segment of the network. The parameters for these two control loops can be set and tuned separately, with the intermediate TCP "proxy" handling the buffering and forwarding of packets between the communicating TCP endpoints. While this approach technically violates the end-to-end semantics of TCP acknowledgements, it is effective in improving TCP performance over wireless network environments [BPSR 1997].

Our own network protocol performance research at the University of Calgary has identified three additional TCP-related performance problems in IEEE 802.11b wireless LANs [KXW 2003]. These problems are:

- low throughput from improperly configured wireless network cards;
- network thrashing from dynamic MAC-layer rate adaptation; and
- high collision rates between TCP data and acknowledgement packets.

In the following paragraphs, we briefly describe each of these protocol performance problems. Further details are provided in [KXW 2003].

The first anomaly that we observed in our IEEE 802.11b WLAN was low TCP throughput for bulk data transfers. The throughput was approximately 1.2 Mbps, compared to the expected value of approximately 6.0 Mbps. The primary cause for this problem was an improperly configured network card. In particular, the Linux device driver for the IEEE 802.11b card was statically setting the card's transmission rate to 2 Mbps rather than 11 Mbps. We diagnosed this problem with the help of a wireless network analyzer: the PLCP header of each transmitted frame showed a signal speed of 2 Mbps. Since the frame header is transmitted at 1 Mbps, and the payload at 2 Mbps, it is no surprise that our throughput barely exceeded 1 Mbps. Fixing the device driver setting increased throughput by about a factor of 4. A related, but much more subtle performance problem has to do with the Universal Serial Bus (USB) protocol used to transfer TCP/IP packets from the kernel to the network card or vice versa. Because the USB transfer size is much smaller than the TCP/IP packet size, the USB overhead can dominate. For the 8 different USB configurations considered in [KXW 2003], the TCP throughput varied by more than an order of magnitude, from 0.4 Mbps to 5.2 Mbps. For many of these configurations, the USB was the bottleneck. Only in a few of the configurations did the throughput approach the theoretical capacity of an IEEE 802.11b WLAN. These observations highlight the importance of proper parameter configuration at all layers of the protocol stack.

The second performance anomaly that we observed was related to the dynamic rate adaptation feature in IEEE 802.11b. That is, if the link-layer protocol detects an excessive number of retransmissions when using high data rate (11 Mbps) transmissions, it can revert to lower-rates (e.g., 5.5 Mbps, 2 Mbps, or 1 Mbps) on subsequent retransmissions, which may be more resilient to wireless channel errors. We tested this rate adaptation behaviour in a WLAN environment with poor signal quality and user mobility. We found that some implementations of dynamic rate adaptation produce a cyclic pattern of rate oscillation, which in turn results in frequent periodic TCP packet losses. Greater hysteresis is required in the channel quality estimation if dynamic rate adaptation is to perform well in a noisy WLAN environment.

The third and final performance anomaly that we noticed was an excessive number of collisions on the WLAN when the TCP protocol is used. For example, with UDP for wireless data transfers, collision rates are below 0.5% on the WLAN. With TCP, the collision rates are 4-7%. What is particularly surprising about these collision rates is that they occur with only a *single* wireless client (communicating with a server on a wired network), competing with the AP for use of the wireless channel. We have shown experimentally that this high collision rate is related to TCP: namely, the contention between TCP data packets sent by the client and TCP ACK packets being forwarded by the AP. The design of the IEEE 802.11b MAC protocol assumes that stations generate frames at random times. With TCP, this assumption is not true. A sending station often has a backlog of frames to send, as soon as the channel access protocol permits. Similarly, TCP ACK packets are traversing the reverse path in the network, contending for the channel at the wireless access link. The TCP-level ACKs induce correlated behaviours on the network, leading to the higher collision rates observed. Fortunately, the random backoff in the MAC-layer retransmission resolves most of these collisions without causing TCP packet loss, but the efficiency of the IEEE 802.11b WLAN still suffers.

5.2   Multi-hop Ad Hoc Networks

TCP performance can suffer greatly in multi-hop wireless ad hoc networks, where intermediate wireless nodes are used as routers for forwarding packets from a source to a destination. Performance problems arise from the physical characteristics of the wireless channel, the Medium Access Control (MAC) protocols, node mobility, and the dynamics of TCP.

Ignoring node mobility for the moment, even the topology of an ad hoc network and the wireless propagation characteristics can adversely affect TCP performance. For example, Fu *et al.* [FZLG 2003] use simulation and analysis to show that in an N-hop "chain" network topology, the end-to-end TCP throughput is much lower than the nominal throughput of the wireless channel. The problems occur because of the burstiness of TCP transmissions: multiple TCP data packets within the congestion window are competing for channel access on the forward path, multiple TCP ACK packets are competing for channel access on the reverse path, and interference effects at the wireless layer (e.g., hidden node, exposed node) preclude adjacent nodes on the routing path from forwarding packets at the same time. Fu *et al.* recommend a carefully chosen TCP window size that depends on the routing path length (e.g., a window size of N/4 packets for an N-hop chain topology gives the maximal TCP throughput [FZLG 2003]). Other authors propose rate-based flow control or multi-channel approaches to expedite TCP data transfer [KW 2003,WTLS 2002].

Node mobility in ad hoc networks can make the TCP performance problem even worse. In addition to network congestion and wireless channel errors, node mobility can produce a *third* type of packet loss: losses due to transient routing failures when IP routing paths are disrupted. The TCP protocol has no way to differentiate these types of packet losses. Some authors have proposed Explicit Loss Notification (ELN) [BK 1998], so that TCP packet transmissions are suspended while the IP route discovery process is re-initiated.

5.3   Wireless Media Streaming

The popularity of multimedia streaming on the Internet, combined with the growing deployment of wireless access networks, augurs the converging usage of these two technologies in the not-too-distant future. Experience with wireless multimedia streaming on today's networks can provide valuable insights into the design of future wireless multimedia networks and applications.

In a recent paper [KW 2002], we presented a measurement study of Real-Media streaming traffic on an indoor IEEE 802.11b wireless LAN. The traffic is analyzed hierarchically, from the application layer to the network layer to the data link layer. We focus on the traffic structure at each layer, and on interaction effects across layers.

Our main observation is that multimedia streaming quality is quite robust in all but the poorest channel conditions, despite the inherent burstiness of both the RealMedia application workload and wireless channel errors. Several factors contribute to these good results. First, although RealVideo is typically Variable-Bit-Rate (VBR) at the application layer, it is often streamed as Constant-Bit-Rate (CBR) at the network layer, reducing burstiness and thus the chances of packet losses due to buffer overflow in the network path. Second, while the wireless channel has bursty error characteristics, MAC-layer retransmission in 802.11b hides most errors from higher-layer protocols. Finally, the application layer's NACK-based error control is effective in recovering missing packets when needed. Our results demonstrate the viability of multimedia streaming on current and future wireless LANs.

5.4   Wireless Web Performance

Two of the most popular Internet technologies from the past ten years are the World Wide Web and wireless networks. A natural step in the wireless Internet evolution is the convergence of these technologies to form the "wireless Web": the wireless classroom, the wireless campus, the wireless office, and the wireless home. In fact, the same technology that allows Web clients to be mobile (i.e., wireless network cards) also enables the deployment of wireless Web servers.

Mobile Web servers play a useful role in *short-lived networks*. A short-lived (or *portable*) network is created spontaneously, in an *ad hoc* fashion, at a particular location in response to some event (scheduled or unscheduled). The network operates for some short time period (minutes to hours), before being disassembled, moved, and reconstituted elsewhere. Examples of deployment scenarios for short-lived networks are sporting events, disaster recovery sites, press conferences, conventions and trade shows, and classroom area networks. The potential for entertainment applications (e.g., media streaming, home networking, multiplayer gaming) is also high. In many of these contexts, an ad hoc wireless network (with a wireless Web server as an information repository) provides a suitable solution.

In recent work [BOW 2003,BW 2003,Ola 2003], we have explored the feasibility of wireless Web servers. In [BW 2003], we present simulation results that are validated with empirical measurements from wireless Web server usage in a

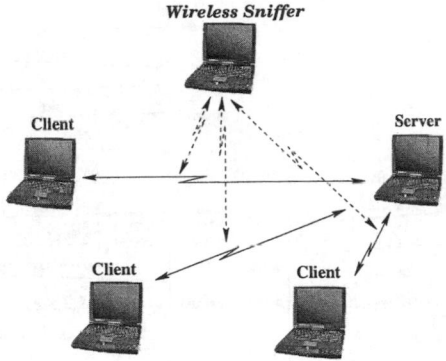

Fig. 7. Experimental Setup for Wireless Web Server Benchmarking

classroom environment. These measurements are then augmented with laboratory tests to determine experimentally the upper bounds on achievable performance [BOW 2003,Ola 2003]. In particular, we focus on the performance capabilities of an Apache Web server running on a laptop computer with an IEEE 802.11b wireless LAN interface.

The experimental setup for our measurements is illustrated in Figure 7. We study in-building Web performance for wireless Web clients. All mobile computers are configured in ad hoc mode, since no existing network infrastructure is assumed. The clients download content from the wireless Web server. A wireless network analyzer is used to collect and analyze traces from the experiments, with traffic analysis spanning from the Medium Access Control (MAC) layer to HTTP at the application layer.

Our experiments focus on the HTTP transaction rate and end-to-end throughput achievable in an ad hoc wireless network environment, and the impacts of factors such as number of clients, Web object size, and persistent HTTP connections. The results show the impacts of the wireless network bottleneck, either at the client or the server, depending on the Web workload. Persistent HTTP connections offer significant improvements both in throughput and in fairness for mobile clients accessing content from a wireless Web server.

There are three main observations from our wireless Web server experiments:

– *TCP is an extremely "chatty" protocol for wireless Web access.* An example of its behaviour is shown in Figure 8(a). Downloading a single 1 KB Web object using HTTP/1.0 requires 10 TCP packets on the network, with 6 sent by the client and 4 by the server. Only 2 of these packets carry actual user-level data: the client's GET request that specifies the desired URL, and the server's HTTP response with the Web object data. The other packets are TCP control packets to establish, maintain, update, and close TCP connection state information. This 80% protocol overhead has dire performance impacts when WLAN channel access is the bottleneck. In our experiments, the Web server can achieve only about 100 HTTP/1.0 transactions per second for 1 KB objects. The user-level throughput is below 1 Mbps.

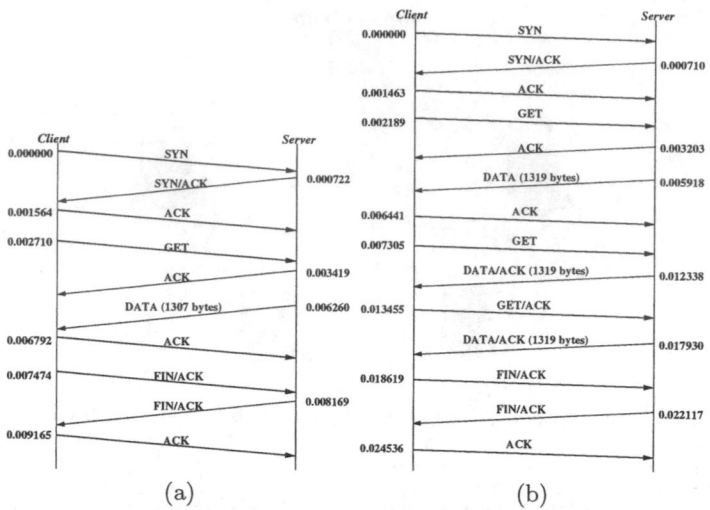

Fig. 8. Example of a 1 KB HTTP Transaction on IEEE 802.11b WLAN: (a) HTTP/1.0; (b) HTTP/1.1

Persistent-HTTP connections [PM 1995] improve performance by a factor of 3 to 5 in the WLAN environment. The performance advantages arise because the TCP connection handshaking packets are amortized over multiple HTTP transactions. For example, the first HTTP transaction within the TCP connection in Figure 8(b) requires 4 TCP packets, rather than 10. The second HTTP transaction requires only 2 TCP packets, since it takes advantage of TCP ACK piggybacking on outbound data packets in each direction. The same observation applies for the third HTTP transaction to retrieve another embedded object in the Web page. This TCP efficiency dramatically reduces the demand on the wireless channel access protocol, leading to much faster HTTP response time and better network throughput.

- *The wireless network bottleneck manifests itself differently, depending on the Web workload.* For small Web objects, the bottleneck is at the client: there is a finite limit on the HTTP request rate that can be achieved before packets are lost from the link-layer transmit queue at the client. For large Web objects, the bottleneck is at the server, since it sends more packets (and larger packets) than the client.
- *TCP behaviour is erratic under overload.* We have observed two anomalous TCP phenomena. For multiple clients requesting small Web objects from the server, the bottleneck at the server's outgoing transmit queue can lead to TCP packet loss. The loss of TCP connection handshake packets can severely affect clients, even leading to unfairness. For large Web objects, multiple clients can cause *network thrashing*, wherein the WLAN is so busy sending TCP data packets and retransmissions that clients eventually timeout and abort their HTTP transfers, drastically reducing the goodput of the network.

All three of these problems manifest themselves acutely in an IEEE 802.11b WLAN environment [BOW 2003].

6   Summary and Outlook

Wireless Internet technologies have progressed tremendously in the past five years, and will continue to reshape the networking landscape in the years ahead. The IEEE 802.11b wireless LAN standard has been a major part of the success story. This WiFi standard provides a flexible and cost-effective solution for wireless network access, while supporting mobile users in either the infrastructure mode or the ad hoc mode of operation. The user-level achievable throughput on IEEE 802.11b WLANs is approximately 6 Mbps, about an order of magnitude faster than on previous generation wireless technologies.

The future of wireless Internet is even brighter. The IEEE 802.11a standard promises up to 54 Mbps of physical-layer transmission capacity, in a much less crowded portion of the electromagnetic spectrum. This technology will offer about 32 Mbps of user-level throughput for TCP/IP networking applications. Next-generation wireless networking technologies promise greater software support, easier network configuration, better network security, and lower prices.

Wireless network access technologies will soon become an invisible part of our ubiquitous computing infrastructure. Nevertheless, protocol performance issues will continue to be a problem. Understanding these protocol performance problems, and solving them, will be important to maximize the benefits of Wireless Internet technologies.

Acknowledgements. Financial support for this research was provided by iCORE (Informatics Circle of Research Excellence) in the Province of Alberta, and by the Natural Sciences and Engineering Research Council of Canada, through NSERC Research Grant OGP0121969. The author is grateful to iCORE, NSERC, the Canada Foundation for Innovation (CFI), and the University of Calgary for help in establishing the Wireless Internet Performance Laboratory (WIPL) and the Experimental Laboratory for Internet Systems and Applications (ELISA).

The author thanks his iCORE research team and graduate students for their many contributions to Wireless Internet protocol performance research. Special thanks go to Guangwei Bai and Kenny Oladosu for their work on wireless Web servers, and for contributing several of the diagrams in this tutorial article. Tianbo Kuang and Nayden Markatchev initiated most of the work on wireless media streaming, and Fang (Shelly) Xiao has spent many hours studying wireless TCP behaviours. For all of these efforts, I am grateful.

References

[AB 2004]   K. Al-Begain. "Performance Models for 2.5/3G Mobile Systems and Networks". In M. Calzarossa and E. Gelenbe, editors, *Performance Tools and Applications to Networked Systems*, volume 2965 of *Lecture Notes in Computer Science*. Springer, 2004.

[ANSI 1999]   ANSI/IEEE Standard 802.11b, "Part 11: Wireless LAN Medium Access Control (MAC) and Physical Layer (PHY) Specificiations: Higher-Speed Physical Layer Extension in the 2.4 GHz band", 1999.

140   C. Williamson

[BOW 2003]   G. Bai, K. Oladosu, and C. Williamson, "Portable Networks: Prototype and Performance", submitted for publication, 2003.

[BW 2003]   G. Bai and C. Williamson, "Simulation Evaluation of Wireless Web Performance in an IEEE 802.11b Classroom Area Network", *Proceedings of the Third International Workshop on Wireless Local Networks* (WLN 2003), Bonn, Germany, pp. 663-672, October 2003.

[BB 1995]   A. Bakre and B. Badrinath, "I-TCP: Indirect TCP for Mobile Hosts", *Proceedings of the 15th International Conference on Distributed Computing Systems* (ICDCS), Vancouver, BC, pp. 136-143, May 1995.

[BVBP 2002]   A. Balachandran, G. Voelker, P. Bahl, and P. Rangan, "Characterizing User Behavior and Network Performance in a Public Wireless LAN", *Proceedings of ACM SIGMETRICS*, Marina Del Rey, CA, pp. 195-205, June 2002.

[BK 1998]   H. Balakrishnan and R. Katz, "Explicit Loss Notification and Wireless Web Performance", *Proceedings of IEEE GLOBECOM*, November 1998.

[BPSR 1997]   H. Balakrishnan, V. Padmanabhan, S. Seshan, and R. Katz, "A Comparison of Mechanisms for Improving TCP Performance over Wireless Links", *IEEE/ACM Transactions on Networking*, Vol. 5, No. 6, pp. 756-769, December 1997.

[BB 1997]   B. Bennington and C. Bartel, "Wireless Andrew: Experience Building a High Speed, Campus-Wide Wireless Data Network", *Proceedings of ACM MOBICOM*, Budapest, Hungary, pp. 55-65, September 1997.

[BN 2004]   A. Boukerche and S. Nikoletseas. "Algorithmic Design for Communication in Mobile Ad hoc Networks". In M. Calzarossa and E. Gelenbe, editors, *Performance Tools and Applications to Networked Systems*, volume 2965 of *Lecture Notes in Computer Science*. Springer, 2004.

[FPAY 2003]   S. Fahmy, V. Prabhakar, S. Avasarala, and O. Younis, "TCP over Wireless Links: Mechanisms and Implications", Technical Report CSD-TR-03-004, Purdue University, 2003.

[FF 1996]   K. Fall and S. Floyd, "Simulation-based Comparisons of Tahoe, Reno, and SACK TCP", *ACM Computer Communication Review*, Vol. 26, No. 3, pp. 5-21, July 1996.

[Flo 2001]   S. Floyd, "A Report on Recent Developments in TCP Congestion Control", *IEEE Communications*, Vol. 39, No. 4, pp. 84-90, April 2001.

[FZLG 2003]   Z. Fu, P. Zerfos, H. Luo, S. Lu, L. Zhang, and M. Gerla, "The Impact of Multihop Wireless Channel on TCP Throughput and Loss", *Proceedings of IEEE INFOCOM*, San Francisco, CA, April 2003.

[GLGL 2004]   E. Gelenbe, R. Lent, M. Gellman, P. Liu, and P. Su. "CPN and QoS Driven Smart Routing in Wired and Wireless Networks". In M. Calzarossa and E. Gelenbe, editors, *Performance Tools and Applications to Networked Systems*, volume 2965 of *Lecture Notes in Computer Science*. Springer, 2004.

[Jac 1988]   V. Jacobson, "Congestion Avoidance and Control", *Proceedings of ACM SIGCOMM*, Stanford, CA, pp. 314-329, August 1988.

[JPS 2003]   J. Jun, P. Peddabachagari, and M. Sichitiu, "Theoretical Maximum
                Throughput of IEEE 802.11 and its Applications", *Proceedings of
                the 2nd IEEE International Symposium on Network Computing and
                Applications* (NCA'03), Cambridge, MA, pp. 249-256, April 2003.

[KLW 2004]   A. Klemm, C. Lindemann, and O.P. Waldhorst. "Peer–to–Peer
                Computing in Mobile Ad Hoc Networks". In M. Calzarossa and
                E. Gelenbe, editors, *Performance Tools and Applications to Net-
                worked Systems*, volume 2965 of *Lecture Notes in Computer Science*.
                Springer, 2004.

[KE 2002]   D. Kotz and K. Essein, "Analysis of a Campus-Wide Wireless Net-
                work", *Proceedings of ACM MOBICOM*, Atlanta, GA, September
                2002.

[KW 2002]   T. Kuang and C. Williamson, "RealMedia Streaming Performance
                on an IEEE 802.11b Wireless LAN", *Proceedings of IASTED Wire-
                less and Optical Communications Conference* (WOC 2002), Banff,
                AB, pp. 306-311, July 2002.

[KW 2003]   T. Kuang and C. Williamson, "A Bidirectional Multichannel MAC
                Protocol for Improving TCP Throughput in Multihop Wireless Ad
                Hoc Networks", submitted for publication, 2003.

[KXW 2003]   T. Kuang, F. Xiao, and C. Williamson, "Diagnosing Wireless
                TCP Performance Problems: A Case Study", *Proceedings of SCS
                SPECTS Conference*, Montreal, PQ, July 2003.

[LLL 2002]   V. Li, Z. Liu, and S. Low, "Enhancing TCP Performance over Wi-
                reless Networks", *Proceedings of the IST Mobile and Wireless Tele-
                communications Summit*, pp. 85-89, Thessaloniki, June 2002.

[MTG 2004]   S. Mueller, R.P. Tsang, and D. Ghosal. "Multipath Routing in
                Mobile Ad Hoc Networks: Issues and Challenges". In M. Calza-
                rossa and E. Gelenbe, editors, *Performance Tools and Applications
                to Networked Systems*, volume 2965 of *Lecture Notes in Computer
                Science*. Springer, 2004.

[Ola 2003]   K. Oladosu, *Performance and Robustness Testing of Wireless Web
                Servers*, M.Sc. Thesis, University of Calgary, August 2003.

[PM 1995]   V. Padmanabhan and J. Mogul, "Improving HTTP Latency", *Com-
                puter Networks and ISDN Systems*, Vol. 28, pp. 25-35, December
                1995.

[Pent 2000]   K. Pentikousis, "TCP in Wired-Cum-Wireless Environments",
                IEEE Communications Surveys and Tutorials, Vol. 3, No. 4, Fourth
                Quarter 2000.

[RFC 1996]   RFC 1945: "Hypertext Transfer Protocol – HTTP/1.0",
                www.ietf.org/rfc/rfc1945.txt

[RFC 1997]   RFC 2616: "Hypertext Transfer Protocol – HTTP/1.1",
                www.ietf.org/rfc/rfc2616.txt

[SB 2004]   D. Schwab and R. Bunt, "Characterizing the Use of a Campus Wi-
                reless Network", to appear, *Proceedings of IEEE INFOCOM*, Hong
                Kong, April 2004.

[SS 2002]   H. Singh and P. Singh, "Energy Consumption of TCP Reno, TCP
                NewReno, and SACK in Multihop Wireless Networks", *Proceedings
                of ACM SIGMETRICS*, Marina Del Rey, CA, pp. 206-216, June
                2002.

[Ste 1994]   W. Stevens, *TCP/IP Illustrated, Volume 1: The Protocols*, Addison-
                Wesley, 1994.

[Tan 2003]   A. Tanenbaum, *Computer Networks*, Fourth Edition, Prentice-Hall, 2003.

[TB 1999]   D. Tang and M. Baker, "Analysis of a Metropolitan-Area Wireless Network", *Proceedings of ACM MOBICOM*, Seattle, WA, pp. 13-23, August 1999.

[TB 2000]   D. Tang and M. Baker, "Analysis of a Local-Area Wireless Network", *Proceedings of ACM MOBICOM*, Boston, MA, pp. 1-10, August 2000.

[WTLS 2002]   S. Wu, Y. Tseng, C. Lin, and J. Sheu, "A Multi-Channel MAC Protocol with Power Control for Multi-Hop Mobile Ad Hoc Networks", *Computer Journal*, Vol. 45, No. 1, 2002.

Performance Models for 2.5/3G Mobile Systems and Networks

Khalid Al-Begain

Mobile Computing and Networking Research Centre
School of Computing, University of Glamorgan
CF37 1DL, Pontypridd, Cardiff, Wales, UK
kbegain@glam.ac.uk

Abstract. Performance models for mobile networks have been proposed in large numbers in the literature. The majority of those models were devoted to model voice and data traffic on GSM (2G) mobile networks with General Packet Radio Services (GPRS or 2.5G) . In most cases, these models were based on Markovian assumptions or studied using simulation. Considering 3G mobile systems, the picture is far not the same. The technology behind the 3G systems became much more complex and, therefore, analytical models became rather more difficult to develop. In this area, simulation models seem to be dominant. The aim of the paper is to present some models developed to provide analytical and numerical performance evaluation for 2.5G (GSM/GPRS) and 3G (UMTS) systems. Three models are presented from which two are for GPRS systems under different channel sharing schemes; namely the Complete Partitioning (CPS) and the Partial Sharing (PSS) schemes. The third is for interference limited UMTS cell. For these models, sophisticated analytical and numerical solutions are either presented or proposed. In the first model, a generalised solution based on the principle of maximum entropy (ME) is presented for the performance modelling and evaluation of a wireless GSM/GPRS cell supporting bursty multiple class traffic of voice calls and data packets under CPS cell capacity handling scheme.
In the second part, a queueing model is proposed for similar system but with PSS capacity handling scheme. The difficulty of this model rises from the fact that the actual data bit rate depends on the state of the system including the number of active voice calls. For this case, earlier numerical solutions based on Markov chains is briefly discussed. Furthermore, analytical solutions based on Markovian and general queueing models with random environments are proposed.
In the last part of the paper, a more sophisticated 3G mobile system is considered where the system capacity depends on the interference levels in the system which are influenced by a number of factors including the number of ongoing calls and their geographical distribution over the cell area. An approximation based on discretization of the distance is proposed in order to be able to present a model based on multi-dimensional Markov chain which is then described and solved using MOSEL package.

Keywords: GSM/GPRS, UMTS, wireless mobile networks, performance models, generalised exponential (GE) distribution, maximum entropy (ME) principle.

M.C. Calzarossa and E. Gelenbe (Eds.): MASCOTS 2003, LNCS 2965, pp. 143–167, 2004.
© Springer-Verlag Berlin Heidelberg 2004

1   Introduction

Performance evaluation of mobile systems and networks has attracted a huge amount of research in the last two decades. Performance models have followed every stage of technological developments in mobile systems and networks starting from the Global System for Mobile (GSM) telecommunications (known as 2nd Generation or 2G)[1]. The introduction of the General Packet Radio Services (GPRS) (known as GSM Phase 2+ or 2.5G)[2,3,4] which allowed data services over GSM has added new challenge to the performance modelling to include multiple classes of traffic with different characteristics. Finally, the technological switch in the Third Generation (3G) mobile networks [5] has added further complexity to the system behaviour that resulted in much less flexibility in the modelling possibility of those systems. One main reason for the later is that the capacity of the 3G systems depends on the interference levels in the network which depend in turn on the actual traffic situation in the network.

In these performance studies, wide range of methodologies has been used. However, most of the published studies in the field are based on simulation modelling and/or the numerical solution of Markov models covering different traffic scenarios, usually at voice call level, with single or multiple data classes. Analytical models based on queueing network or matrix geometric solutions have also appeared for different scenarios but in limited numbers.

Simulation is an efficient tool for studying detailed system behaviour but it becomes costly, particularly as the system size increases. Markov models on the other hand provide more flexibility in producing numerical results for many interesting performance measures. Nevertheless, the numerical solution of Markov models may suffer from several drawbacks, such as

- state space explosion limiting the analysis to only small mobile systems, generally consisting of one cell
- restrictive assumptions of independent Poisson arrival processes for all types of homogeneous and uniformly distributed traffic with exponentially distributed call duration (which, if multiplexed, can be bursty and correlated).

Analytical models, if they exist, provide a powerful tool for performance evaluation. They are preferable in many cases due to there capability to provide cost effective and fast results. However, such models rarely exist for real systems and can only be achieved through a number of sometimes "strong" assumptions. There are number of analytical methodologies used in mobile network performance modelling e.g., Queueing networks based solutions [6,7] or Matrix Geometric solutions [8].

This paper devises a performance evaluation framework for the performance modelling and evaluation of 2.5 and 3G wireless mobile systems supporting bursty multiple class traffic of voice calls and data packets. This work is based on a number of modelling projects by the author in collaboration of several research groups in Germany, Hungary and the UK.

Earlier proposed models are based on resource network management parameters of GSM technology where the capacity of radio interference in the wireless cell is divided into discrete channels and operates in circuit-switched mode (e.g., [9,10,11]). Extensions of these models have been made to capture the packet-switched behaviour introduced by the GPRS which has been added to GSM to allow data packet communication with higher bit rates than those provided by a single GSM channel (e.g., [14,15,16]). More recently, Foh et al [17] proposed a single server infinite capacity queue for modelling GPRS in a Markovian environment and applied matrix geometric methods for the evaluation of performance metrics.

Let us note that several papers in this volume deal with QoS and performance of wireless networks, including [18,19,20,21] so that, taken with these other references, the present paper can provide a broad and comprehensive view for the practitioner and researcher.

This paper is organised as follows. Section 2 describes the basic operation of the GSM/GPRS system and defines the different voice and data sharing schemes. Section 4 presents a detailed analysis for the GSM/GPRS system with CPS scheme using Maximum Entropy (ME) approximation of an open queueing network that applies the Generalised Exponential distribution for the arrival and service processes. The second model for GSM/GPRS with PSS scheme is then introduced and discussed in Section 5. In Section 6, a sophisticated model for a 3G UMTS wireless cell is presented. Finally, the paper ends with some concluding remarks.

2 The GSM/GPRS System

Figure 1 shows a simplified block diagram reflecting the most important components of a GSM/GPRS system architecture. The User Equipment (UE) can generate traffic consisting of GSM voice calls and GPRS data packets. Both types of traffic will go through the same air interface or Base Station (BS) but each of them will have its own Um and Gb radio interfaces. The traditional GSM voice call connections will use the original flow path - through the Base Station Controller (BSC) and the Mobile Station Controller (MSC) - to the fixed Public Land Mobile Network (PLMN). On the other hand, GPRS data packets will use the added Serving GPRS Support Node (SGSN) which in turn will transfer the data - through the Gateway GPRS Support Node (GGSN) - to Public Data Network (PDN). The diagram of Figure 1 can be used, under different voice call/data packet handling schemes, to identify the spots where congestion is likely to build up leading to the formulation of suitable queueing network models (QNMs).

Resources for GPRS traffic can be reserved statically or dynamically whereas a combination of both is possible. Different cell capacity partitioning schemes for handling the transmission of voice calls/data packets can be defined. In such environment, partitions of the available bandwidth may be created for GSM and GPRS traffics but not for individual data services. For GPRS traffic, a complete partition is used for different data services. However, some data packets may be allocated higher priority and, therefore, they can be given higher share of the

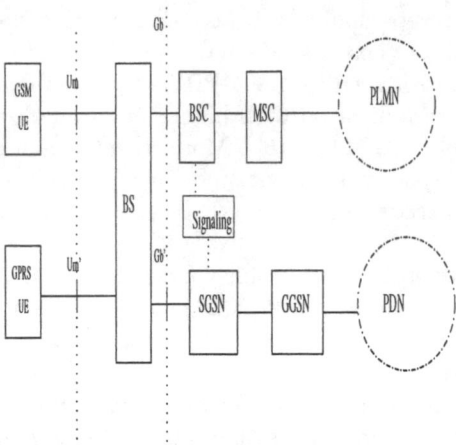

Fig. 1. A simplified block diagram of a GSM/GPRS system architecture

available bandwidth. Whenever voice calls and data packets share bandwidth, voice service is always given the highest service and space priorities, as appropriate. Three main GSM/GPRS call/data handling schemes are described as follows [22]:

- *Complete partitioning scheme (CPS)* divides the total cell capacity to serve simultaneously GSM and GPRS traffics. As a consequence, the GSM and GPRS systems can be analysed separately.
- *Partial sharing scheme (PSS)* allocates C_{data} channels for data traffic and the remaining $C_{shared} = C_{total} - C_{data}$ channels are shared by voice calls and data packets with preemptive service priority for voice calls.
- *Complete sharing scheme (CSS)* enables all available channels to serve at the same time both voice calls and data packets with voice calls having preemptive priority over data packets.

Although CPS will not clearly give the best utilisation for radio resources, it has the advantage of requiring simpler management policy and implementation. Moreover, a definite capacity for GPRS under an efficient Connection Admission Control (CAC) algorithm can guarantees some QoS. Note that CPS is the limiting case of PSS under high loads (mainly voice) whilst the CSS is in fact a special case of PSS where $C_{data} = 0$.

In the followings, two models will be presented for the CPS and PSS call handling scheme with different solution methodologies.

3   Performance Model for GSM/GPRS with CPS

A GSM/GPRS wireless cell architecture with a CPS related system of queues can be seen in Figure 2. Potential spots of congestion and delays are often attributed to the constraints imposed by BSC and SGSN dealing with GSM voice and GPRS data traffics, respectively. Moreover, data packets may experience access delays when they attempt to reach the SGSN node via BS.

A number of buffer-less channels may be allocated to the GSM partition which can be clearly seen as a loss system for multiple class voice calls. Moreover, the GPRS traffic is packet based and, therefore, each connection will attempt to use the complete available bandwidth. The GPRS partition involves two queueing systems in tandem, namely an access queue and a transfer queue where all data connections may belong to different classes and share the total capacity of the data partition.

These classes may have their own characteristics such as maximum or minimum data rates, delay sensitivity, service discrimination, arrival rates, inter-arrival-time variability and transferable file (data) length.

Note that the access queue models the Packet Control Unit (PCU)/Servicing GPRS Support Node (SGSN) buffers of a GPRS network with downlink traffic and the logical queue of data call request for transmission in the up-link stream

In the context of this paper only the up-link streams are considered.

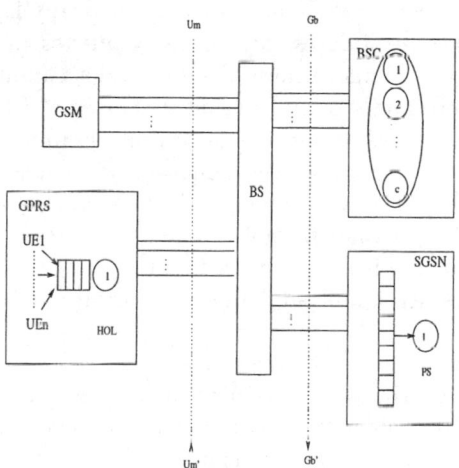

Fig. 2. A GSM/GPRS wireless cell architecture with a CPS system of queues

The model presented here is based on the works published in [23] and [24]. In this model, a Compound Poisson Process (CPP) with geometrically distributed batches is used to represent both external and internal arrival processes (or, equivalently, Generalised Exponential (GE-type)[25,26] inter-arrival times) of voice calls and data packets per class. Moreover, the GE distribution is used in all cases to model the call durations or packet channel transmission times. Note that a batch arrival process is a most suitable model of bursty multiplexed connections (belonging to various classes with different minimum capacity demands) being accepted into the mobile system (if there is enough service capacity at the moment of their arrival).

The GPRS part of the system is modelled by a tandem system consisting of two queues; the first queue (the access queue) is a logical queue representing the packets awaiting for transmission in the UE organised through the signalling channel. This queue is modelled as Head-of-Line (HOL) as the arrival of higher priority packet will not interrupt the ongoing transmission but the new packet will be served next. The second queue represents the transmission queue. This queue is modelled as Processor Sharing (PS) since all admitted data connections will share the available bandwidth according to the priority for the data class. Physically, the GPRS system is capable of allocating all available channel to one connection (subject to some battery restrictions). Also, in case of multiple connections, one time slot (channel) can be shared by eight different connections [1]. An open QNM corresponding to the GSM/GPRS wireless cell architecture with a CPS system of queues (c.f., Figure 2) can be seen in Figure 5. The model describes a trivial QNM as the GSM and GPRS channel capacity partitions compose two independent queueing systems which can be studied separately.

The GSM partition can be clearly modelled by a classical pseudo Birth-and-Death GE/GE/c/c loss system with multiple voice call classes. The GPRS partition on the other hand can be modelled as an open $GE/GE/1/N_1/HOL \rightarrow GE/GE/1/N_2/PS$ system of access and transfer queues in tandem with finite capacities N_1 and N_2 and under priority HOL and discriminatory PS scheduling disciplines, respectively, and a Complete Buffer Sharing (CBS) management rule.

More specifically, a transfer queue holds a finite number of data connections which are served according to a discriminatory PS rule, where the available service capacity is shared evenly amongst all data calls belonging to the same class. However, in the presence of multiple data classes with different priority levels, the service capacity is shared according to discrimination rates favouring higher priority classes. An admitted data call is initially held in a finite capacity HOL access queue.

A data packet at the head of the access queue will be blocked if the transfer queue is full. Upon the termination of an active data call at the transfer queue, the blocked data call at the access queue is polled into the transfer queue (within a very short time required for signalling) and immediately shares in a discriminatory PS fashion the available capacity.

Note that under PS scheduling discipline, N_2 represents the maximum number of connections sharing simultaneously the available service capacity at the transfer queue.

3.1 Maximum Entropy Analysis of an Open QNM of a GSM/GPRS Cell under CPS

The analytic methodology is illustrated by focusing on the characterisation of a product-form approximation and related queue-by-queue decomposition algorithm for the open QNM of a GSM/GPRS cell with a single class of GSM voice calls and a multiple class of GPRS data packets under blocking with Repetitive

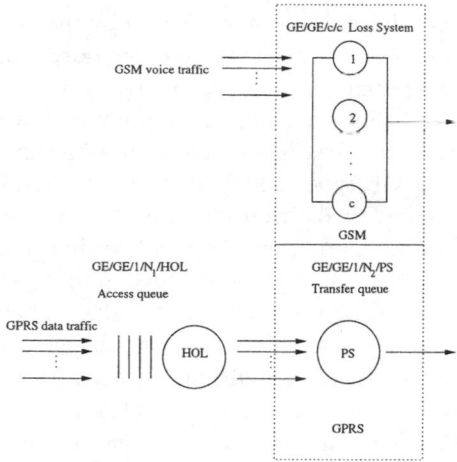

Fig. 3. A QNM of the Wireless GSM/GPRS with CPS

Service (RS) [12,13] and CBS rule. Moreover, it carries out the more detailed ME analysis of each queue of the corresponding tandem system of access and transfer queues. Throughout this section, the ME analysis and flow approximation formulae are based on the following assumptions:

- The Compound Poisson Process (CPP) (or, equivalently, the GE distribution) is used to model the arrival process (or, inter-arrival time) of both GSM voice calls and GPRS data packets
- The GE/GE/c/c loss system with a single class of voice calls is employed to model the delays of GSM partition
- The open tandem GE/GE/1/N1/HOL $\rightarrow$ GE/GE/1/N2/PS queueing system is used to model the finite capacity access and transfer queues of the GPRS partition with HOL and discriminatory PS scheduling disciplines, RS blocking and CPS buffer management rule.

Note that the justification of the aforementioned assumptions are addressed in the next section.

3.2   The GE-Type Distribution

The GE distribution is of the form (c.f.,[25,26])

$$F(t) = P(X \le t) = 1 - \tau e^{-\tau v t}, t \ge 0 \tag{1}$$

where $\tau = 2/(C^2 + 1)$, X is the inter-event time random variable and $\{1/v, C^2\}$ are the mean and squared coefficient of variation (SCV) of the inter-event time distribution, respectively. Moreover, the underlying counting process of the GE distribution is a CPP with geometrically distributed batch sizes with mean $1/\tau = (C^2 + 1)/2$.

The choice of the GE distribution is motivated by the fact that measurements of actual inter-arrival or service times may be generally limited and so only few parameters can be computed reliably. Typically, only the mean and variance may be relied upon, and thus, in a mixed time domain, a choice of a distribution which implies least bias (i.e., introduction of arbitrary and, therefore, false assumptions) is that of GE-type model. In the context of telecommunication networks (e.g., ATM), the GE distribution is particularly applicable in cases of traffic with low level of correlation or where smoothing schemes are introduced at the adaptation level (e.g., for a stored video source) with the objective of minimising or even eliminating the problem of traffic correlation.

Under renewality assumptions, the GE distribution is most appropriate to model the inter-arrival times of of multiple data packets or voice calls which arrive simultaneously. In this context, the burstiness of the arrival process is characterised by the SCV of the inter-arrival time or, equivalently, the size of the incoming bulk. In addition, the GE distribution is the extremal (limiting) case of Hyper-exponential-2 (H2) or Coxian-2 (COX-2) distributions [26] with the same first two moments as the GE. Thus, the GE model is also suitable to approximate the arrival of very small messages consisting of a single voice call or data packet. In this case the ME solutions, via the use of the extremal GE model, will provide pessimistic performance metrics in relation to those based on the use of the H2 and Cox2 distributions [26]. Moreover, IP packet length distribution is known to be non-exponential and should at least be described by the mean, $1/v$, and SCV, C^2. This is because IP packets are restricted by the underlying physical network, such as Ethernet and ATM, and, thus, they have different packet lengths, typically 1500 bytes and 53 bytes, respectively.

3.3   ME Product-Form Approximation and Flow Formulae

Consider the open GE-type QNM of Section 3.1 for a GSM/GPRS cell with a single class voice calls and multiple classes of data packets under CPS. The form of the ME joint state probability $P(\mathbf{k}), \mathbf{k} = (\mathbf{k}_1, \mathbf{k}_2, k_3)$, where $\mathbf{k}_j$ is a state vector $(k_{j1}, \ldots, k_{jR})$ and k_{ji} is the number of calls of class i at queue j for $i = 1, \ldots, R$ and $j = 1$ (access queue), $j = 2$ (transfer queue), $j = 3$ (loss system), subject to normalisation and the existence, in general, of the marginal constraints of server utilisation, busy server probability, mean queue length and full buffer state probability per class, can be clearly established by applying the method of Lagrange's undetermined multipliers and is given by

$$P(\mathbf{k}) = P_1(\mathbf{k}_1)P_2(\mathbf{k}_2)P_3(k_3), \qquad (2)$$

where $P_1(\mathbf{k}_1)$, $P_2(\mathbf{k}_2)$ and $P_2(k_3)$ are the marginal joint state (or, queue length) probabilities of the GE/GE/1/N_1/HOL access queue, GE/GE/1/N_2/PS transfer queue and GE/GE/c/c loss system, respectively. The product form approximation allows the decomposition of the QNM into the three aforementioned queues, each of which can be solved in isolation by carrying out ME analysis at the queue level in conjunction with GE-type flow formulae at the network level for the first two moments of the flow processes into the network. Both GE/GE/1/N1/HOL and GE/GE/1/N2/PS queues can be solved in isolation by carrying out ME analysis at the level of each queue in conjunction with GE-type flow formulae at the level of the tandem queueing system for the calculation of the first two moments of the related inter-departure and inter-arrival time distributions. Note that the involvement of the GE/GE/c/c loss system with a single class of voice calls is trivial as this system corresponds to the separate GSM partition. This loss system has been analysed by Kouvatsos and Xenios [27] and it will not be further considered in this paper.

The mean, $1/\lambda_{d1i}$ and SCV, C_{d1i}^2, of the GE-type inter-departure-time process can be approximately determined (c.f., [25]) by

$$1/\lambda_{d1i} = 1/\hat{\lambda}_{1i}, \; C_{d1i}^2 = 2L_{1i}p_{1i}(0) - \hat{C}_{a1i}^2 \left(1 - 2p_{1i}(0)\right), i = 1, 2, ..., R \qquad (3)$$

where $\{\hat{\lambda}_{1i}, \hat{C}_{a1i}^2, \; i = 1, 2, \dots, R\}$ are the mean rate and SCV of the effective inter-arrival time of class i stream going through the access queue 1, $\{L_{1i}, \; i = 1, 2, \dots, R\}$ are the marginal mean queue lengths and $\{p_{1i}(0), \; i = 1, 2, \dots, R\}$ are the marginal probabilities that there are no data packets of class i in the corresponding infinite capacity access queue, respectively. The effective mean rate, $\hat{\lambda}_{1i}$, and SCV, $\hat{C}_{a1i}^2$, are clearly given by, given by:

$$\hat{\lambda}_{1i} = \lambda_{a1i}(1 - \pi_{1i}), i = 1, 2, ..., R \qquad (4)$$
$$\hat{C}_{a1i}^2 = \pi_{1i} + (1 - \pi_{1i})C_{a1i}^2, i = 1, 2, ..., R \qquad (5)$$

where λ_{a1i} and C_{a1i}^2 are the actual mean arrival rate and SCV of the overall inter-arrival time at queue 1. Note that the corresponding parameters of queue 2 can be determined by

$$\lambda_{2i} = \lambda_{d1i}, \; C_{a2i}^2 = C_{d1i}^2. \qquad (6)$$

Finally, under RS blocking, a rejected data packet of class i, $i = 1, 2, \dots, R$ at the end of subsequent 'service' repetitions at the access queue GE/GE/1/N1/ HOL will continue requesting access at the transfer queue GE/GE/1/N2/PS until the later queue is not full. To eliminate this dynamically imposed feedback and, thus, create an ordinary queueing station which can utilise the aforementioned flow formulae 3- 6, the service process of class i data packets at the access queue needs to be revised to reflect the effective (or, total) 'service' time of a job. To this end, assuming that the blocking probability π_{2i}, (i=1,2, ...,R) (i.e., the blocking probability that a data packet of class i, $i = 1, 2, \dots, R$ will be blocked by the transfer queue 2) is independent of the number of times that this packet of class i is rejected by the transfer queue 2, it follows that the effective service time received by a class i packet at the access queue 1 due to AS blocking can be determined as the random sum of GE service times over a geometric counter with parameter π_{2i} with mean rate $\hat{\mu}_{1i}$ and SCV, $\hat{C}_{s1i}^2$, given by

$$\hat{\mu}_{1i} = \mu_{1i}(1 - \pi_{2i}), i = 1, 2, ..., R \qquad (7)$$
$$\hat{C}^2_{s1i} = \pi_{2i} + (1 - \pi_{2i})C^2_{s1i}, i = 1, 2, ..., R \qquad (8)$$

respectively, where μ_{1i} and C^2_{s1i} are the corresponding parameters of the actual service time at the transfer queue 1.

Note that the system of flow formulae (3)- (8) can be numerically solved to compute the overall mean rate, λ_{2i}, and SCV, C^2_{a2i}, of the transfer queues with few iterations.

3.4   The ME Analysis of a GE/GE/1/N/{HOL or PS} Queue

In this section, the $GE(\lambda_{1i}, C^2_{a1i})/ GE(\mu_{1i}, C^2_{s1i})/1/N_1/HOL$ and $GE(\lambda_{2i}, C^2_{a2i})/ GE(\mu_{2i}, C^2_{s2i})/1/N_2/PS$ queues are analysed via entropy maximisation.

Notation

Without the loss of generality and for the sake of simplifying the notation, the subscript j, $j = 1, 2$, referring to access and transfer queues, respectively, is dropped from the notation of this section.

For each class i, $(i = 1, 2, \ldots, R)$ let at any given time
  n_i be the number of data packets at the GE/GE/1/N/{HOL,PS} queue (wait-
    ing or receiving service),
  λ_i be the arrival rate, μ_i be the service rate,
  C^2_{ai} be the SCV of inter-arrival time distribution and C^2_{si} be the SCV of
    service time distribution service rate.
$\mathbf{S} = (n_1, n_2, \ldots, n_R, \omega)$ be a joint
    queue state, where $\omega\,(1 \leq \omega \leq R)$ denotes the class of the
    current job in service (n.b., for an idle queue $\mathbf{S} \equiv \mathbf{0}$ with $\omega = 0$), $\mathbf{Q}$ be the
    set of all feasible states of $\mathbf{S}$ and $P(\mathbf{S})$ be the stationary state probability
$\mathbf{n} = (n_1, n_2, \ldots, n_R)$ be an aggregate joint queue state
    (n.b., $\mathbf{0} = (0, \ldots, 0)$),
Ω be the set of all feasible states $\mathbf{n}$, and
π_i be the blocking probability that an arrival of class i finds the queue full.

For each state $\mathbf{S}$, $\mathbf{S} \in \mathbf{Q}$, and class i, $i = 1, 2, \ldots, R$, the following auxiliary functions are defined:

$$n_i(\mathbf{S}) = \text{the number of class } i \text{ customers present in state } \mathbf{S},$$
$$s_i(\mathbf{S}) = 1, \text{if the job in service is of class } i \text{ or } 0, \text{ otherwise},$$
$$h_i(\mathbf{S}) = 1, \text{ if HOL \& } n_i(\mathbf{S}) > 0 \text{ or } 0, \text{ otherwise},$$
$$f_i(\mathbf{S}) = 1, \text{ if } \sum_{i=1}^{R} n_i(\mathbf{S}) = N, \text{ and } s_i(\mathbf{S}) = 1 \text{ or } 0, \text{ otherwise}.$$

The form of the state probability distribution, $P(\mathbf{S}), \mathbf{S} \in \mathbf{Q}$, can be characterised by maximising the entropy functional $H(\mathbf{P}) = -\sum_{\mathbf{S}} P(\mathbf{S}) \log P(\mathbf{S})$, subject to prior information expressed in terms of the normalisation and for each class i, $i =$

$1, 2, \ldots, R$, the marginal constraints of server utilisation, busy server probability (with $n_i(\mathbf{S}) > 0$), mean queue length and full buffer state probability (with $s_i(\mathbf{S}) = 1$) satisfying the flow balance equations, namely

$$\lambda_i(1 - \pi_i) = \mu_i U_i, \ i = 1, \ldots, R. \tag{9}$$

By employing Lagrange's method of undetermined multipliers, the following solution is obtained:

$$P(\mathbf{S}) = \frac{1}{Z} \prod_{i=1}^{R} g_i^{s_i(\mathbf{S})} \xi_i^{h_i(\mathbf{S})} x_i^{n_i(\mathbf{S})} y_i^{f_i(\mathbf{S})}, \forall \mathbf{S} \in \mathbf{Q}, \tag{10}$$

where Z is the normalising constant and $\{g_i, \xi_i, x_i, y_i, \ i = 1, 2, \ldots, R\}$ are the Lagrangian coefficients corresponding to the server utilisation, busy busy server probability (with $n_i(\mathbf{S}) > 0$), mean queue length and full buffer state probability (with $s_i(\mathbf{S}) = 1$) constraints per class, respectively. Aggregating $P(\mathbf{S})$ over all feasible states $\mathbf{S} \in \mathbf{Q}$, and after some manipulation, the joint aggregate ME queue length distribution $P(\mathbf{n}) \ \mathbf{n} \in \Omega$, of a stable GE/GE/1/N/HOL/CBS queue is given by:

$$P(\mathbf{0}) = \frac{1}{Z}, \tag{11}$$

$$P(\mathbf{n}) = \begin{cases} \frac{1}{Z} \dfrac{\left(\sum_{j=1}^{R} k_j - 1\right)!}{\prod_{j=1}^{R} k_j!} \left(\prod_{j=1}^{R} x_j^{k_j}\right) \left(\sum_{i=1}^{R} k_i g_i y_i^{\delta(k)}\right), \forall \mathbf{n} \in \Omega - \{\mathbf{0}\}, \ (PS) \\[12pt] \frac{1}{Z} \left(\prod_{i=1}^{R} x_i^{n_i} \xi_i^{h_i(n)}\right) \left(\sum_{j=1 \wedge n_j > 0}^{R} g_j y_i^{f_i(n)}\right), \forall \mathbf{n} \in \Omega - \{\mathbf{0}\}, \qquad (HOL) \end{cases} \tag{12}$$

where $\delta(\mathbf{n}) = 1$, if $\sum_i n_i = N$, or 0, otherwise. Utilising the ME product-form solutions (11)-(12) and applying the generating function approach of Williams and Bhandiwad (c.f., [29]), the recursive closed form expressions for the aggregate state probabilities $\{P_N(n), \ n = 0, 1, \ldots, N\}$ and marginal state probabilities $\{P_i(k), \ k = 0, 1, \ldots, N_i, \ i = 1, 2, \ldots, R\}$ can be obtained with computational cost $O(RN)$ (c.f., [30]). Moreover, the Lagrangian coefficients x_i, ξ_i and g_i can be approximated analytically by making asymptotic connections to the corresponding GE-type infinite capacity queue (c.f, [25,26]).

Moreover, focusing on a tagged data packet within an arriving bulk and applying GE-type probabilistic arguments, the blocking probabilities $\{\pi_i, i = 1, 2, \ldots, R\}$ of a GE/GE/1/N queue can be approximated by

$$\pi_i = \sum_{k=0}^{N} \delta_i(k)(1 - \sigma_i)^{N-k} P_N(k), \tag{13}$$

where $\delta_i(k) = \frac{r_i}{r_i(1-\sigma_i)+\sigma_i}$ for $k = 0$ or 1, otherwise, $\sigma_i = 2/(1 + C_{ai}^2)$ and $r_i = 2/(1 + C_{si}^2)$ (c.f., [30]).

By substituting closed form expressions for the aggregate $\{P_N(n), \ n = 0 \ldots N\}$ and blocking $\{\pi_i, \ i = 1, 2, \ldots, R\}$ probabilities into the flow balance condition (9) and after some manipulation, the recursive relationships for the Lagrangian coefficients $\{y_i, \ i = 1, 2, \ldots, R\}$, can be obtained (c.f., [31]).

3.5   Sample Numerical Results for GSM/GPRS with CPS

This section presents sample results obtained by the previous model on a case of one voice and two data services. A full set of results on this case and an extended case with three data services can be found in [23].

The two data packet classes represent typical Internet applications with different mean sizes, namely, 62.5 KBytes (class 1, e.g., web browsing) and 12.5 KBytes (class 2, e.g., email), respectively. The parameterisation also involves mean arrival rates and SCV of inter-arrival and service times. It is assumed that the GPRS partition consists of one frequency providing total capacity of 171.2 Kbps. Without loss of generality, the sample results of this paper focus on marginal performance metrics of mean response time and mean queue length per class together with the aggregate blocking probability. Focusing on the GE/GE/1/N/PS queue under discriminatory PS rule favouring class 1 (with service discrimination weight 1:5), numerical tests are carried out to verify the relative accuracy of the ME algorithm against simulation at 95 % confidence intervals based on the Queueing Network Analysis Package (QNAP-2) [32], using the same assumptions and input parameterisation as the ones used for the analytic ME solution but referring to a more general open non-product form tandem queueing network. It can be observed that the ME results are very comparable to those obtained via simulation.

As a special capability of this model, results to assess the impact at varying degrees of inter-arrival time SCVs upon ME generated mean response time of both data classes are given in Fig.4. Additionally, Figures 5-6 present the effect of varying both the inter-arrival time SCV and buffer size, N, at the GE/GE/1/N/FCFS queue the on the mean queue length of both data classes, respectively. It can be seen that the analytically established mean queue lengths deteriorate rapidly with increasing external inter-arrival-time SCVs (or, equivalently, average batch sizes) beyond a specific critical value of the buffer size which corresponds to the same mean queue length for two different SCV values. It is interesting to note, however, that for smaller buffer sizes in relation to the critical buffer size and increasing mean batch sizes, the mean queue length steadily improves with increasing values of the corresponding SCVs. This 'buffer size anomaly' can be attributed to the fact that, for a given arrival rate, the mean batch size of arriving bulks increases whilst the inter-arrival time between batches increases as the inter-arrival time SCV increases, resulting in a greater proportion of arrivals being blocked (lost) and, thus, a lower mean effective arrival rate; this influence has much greater impact on smaller buffer sizes.

4   Performance Model for GSM/GPRS with PSS

A GSM/GPRS wireless cell architecture with a PSS related system of queues can be seen in Figure 7. Operationally, the main difference between the CPS and PSS is that under a PSS there is a joint bandwidth management procedure according to which the GPRS data packets are transmitted with a data rate which depends on the number of active voice connections. This arrangement

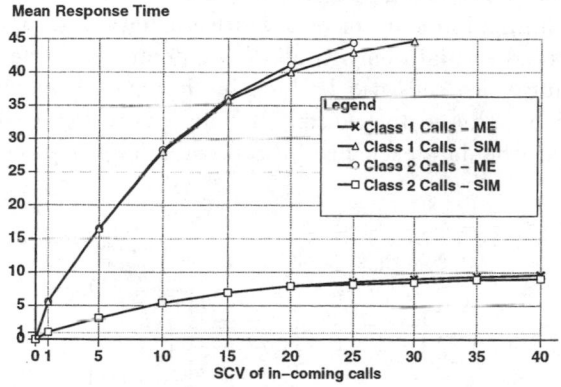

Fig. 4. Effect of varying degrees of SCV on Mean Response Time

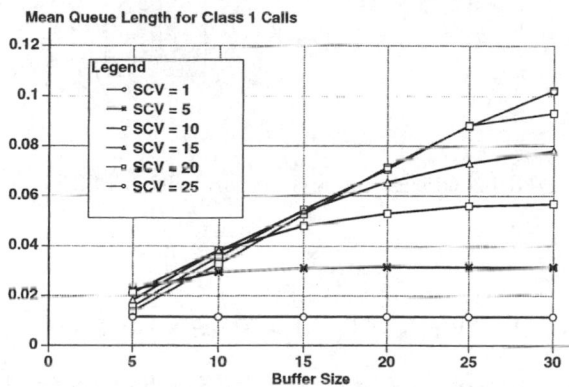

Fig. 5. Effect of varying degrees of SCV on MQLs of Class 1 at different buffer sizes

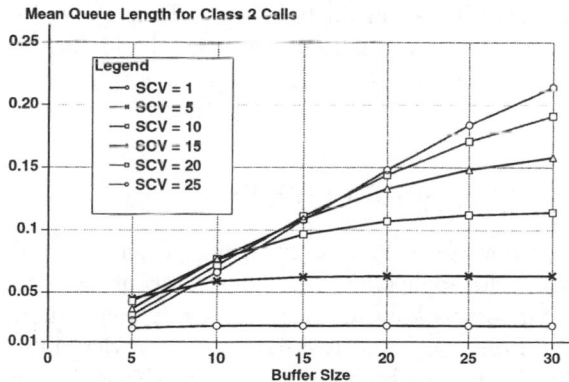

Fig. 6. Effect of varying degrees of SCV on MQLs of Class 2 at different buffer sizes

allows data to use the remaining bandwidth more efficiently. However, in order to maintain a minimum amount of bandwidth for data a given number of time slots, c, are reserved for data only. If all $N - c$ channels of the loss system are occupied incoming voice calls will be lost. In this case, a maximum of 8 data connections will be allowed to transmit but at a very low data rate which is aimed to maintain the data exchange for a possible higher protocol sessions.

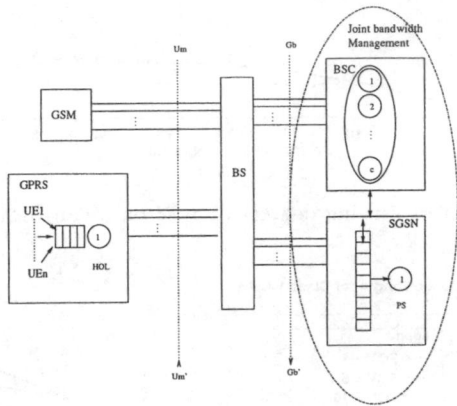

Fig. 7. A GSM/GPRS wireless cell architecture with a PSS system of queues

4.1   A QNM of the Wireless GSM/GPRS with PSS

An open QNM corresponding to the GSM/GPRS architecture with a PSS system of queues as shown in Fig. 7 can be seen in Fig. 8. It consists of the loss system for GSM and multi-class GPRS traffic and the access queue serving multiple GPRS traffic classes. The voice calls have preemptive priority over data connections and, therefore, the system can again be modelled for the voice point of view as a loss system with $N - c$ servers. On the other hand, the data connections will share the remaining bandwidth in a PS mode with total capacity varying depending on the number of active voice connections.

4.2   Analysis of GSM/GPRS with PSS

In this section, we review some of the solutions proposed for the above model and we extend to suggest the application of further solutions. Ermel et al. [33] have studied the similar model without access queue using multi-dimensional Markov chain. In this paper, one voice and two data services with different characteristics were considered; e.g., web browsing and RTTI (Road Traffic and Telematic Information). In this work, the Markov chain was described and solved using the MOSEL [34] performance evaluation language and package. Performance

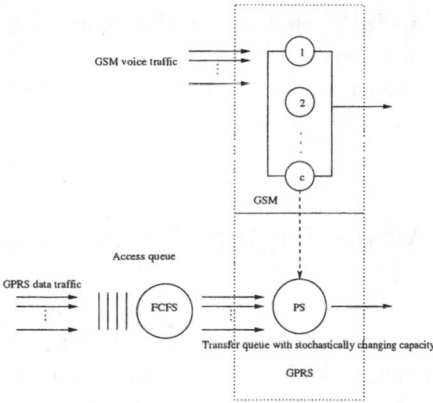

Fig. 8. A QNM of the Wireless GSM/GPRS with PSS

measures like blocking probabilities, utilisation of capacity, average data rate were derived in addition to the moments of the blocking periods for data traffic. Similar model was discussed in [15] with the access queue. In this paper, a simplified analytical (Markov) model for one data class for which the derive the mean sojourn time in the access queue. However, no mobility is considered in the paper. In [35], similar assumptions were made and service related performance measures were derived for single data service.

The above solutions were based mainly on numerical solution for the Markov chain. However, analytical solutions based on applying single queueing models with randomly changing environments seem to be appropriate for this type of model. Similar models have been proposed for the finite repair machine in random environment in [36], [37] and [38]. For the GSM/GPRS system with PSS, Foh et al [17] proposed a single server infinite capacity queue for modelling the GPRS in a Markovian environment and applied matrix geometric methods for the evaluation of performance metrics. While [39] proposed an algorithm to model the data subsystem of the GSM/GPRS system with non-homogeneous Quasi Birth-Death (QDB) process at burst level.

In all the above solutions, all arrival processes for voice and data were assumed to be Poisson and the service times for both were assumed to be exponentially distributed. Kouvatsos et al. [23], proposed an approximation of the above model in which they kept the capacity of the GPRS fixed and forwarded the blocked data connections to be served by the GSM multi-server queue with higher preemptive priority to voice call. This approximation allowed a similar set of queues based on GE-type distributions for both the access and transfer queues with multiple classes of data services. They authors proposed an analytical solution based on the Maximum Entropy similar to that presented in Section 3.

A more promising approach is seen to be based on the work of Dudin [40] and Reshetnikova [41] which was proposed for the modelling of integrated voice and data services over an ISDN channel. In this work, a general analytical and approximate solutions were given for multi-class G/G/1*/N/PS queue where the server has a Markovian random capacity. The application of the queue to the GSM/GPRS system in under investigation at the time this article has been prepared.

5   Performance Model for Interference Limited UMTS System

This section presents a performance model of a UMTS, the 3G system in Europe, cell which is significantly different from those presented for GSM/GPRS. The third generation mobile systems have a dynamic capacity that depends on the interference levels in the covered area and the number of active users in the network. Moreover, the used coding system in the UMTS implies that the received power at the base station is crucial to be equal to all active users. This leads to the fact that the distance of the mobile user (UE) from the base station Node B (the Base Station in the UMTS terminology) is also an important factor because of the signal fading [5]. The way the interference effects the capacity differs between the uplink and downlink transmission. In this paper, we consider only the uplink capacity. In [43], capacity bounds for uplink of a UMTS cell have been derived based on interference levels for both maximum number of active users and maximum distance covered by the base station. More extensive capacity analysis for UMTS systems was carried in [44] for different propagation environments. For more detailed description of the UMTS systems see [5].

5.1   Capacity Bounds

As mentioned earlier, the capacity of the UMTS system is interference limited. The interference level witnessed by the received signal of user i, I_i, at the base station is composed of multiple components as follows:

$$I_i = N_t + \chi(\mathbf{n}) + \sum_{\substack{j \neq i \\ j = 1}}^{n_1} V_j \cdot P_{Rj} \cdot \varepsilon_{ji}$$

where:
N_t is the basic thermal noise in the cell,
$\chi(\mathbf{n})$ is the total inter-cell interference caused by signals for surrounded cells,
n_1 is the number of active users in the investigated cell,
V_j is the service activity factor of user j,
P_{Rj} is the received power of user j at the base station, and
ε_{ji} is the non-orthogonality factor witnessed by user i from user j.

Assuming an ideal free space propagation of the radio signal the following equation holds [42]:

$$P_R = P_S \cdot \left(\frac{\lambda}{4\pi d}\right)^2 \cdot g_b \cdot g_m \tag{14}$$

where,
P_R is the received power of the user signal at the base station.
P_S is the transmitted power at the mobile user terminal.
λ is the wave length.
d is the distance between the UE and Node B.
g_m is the Antenna gain of the Mobile Equipment.
g_b is the Antenna gain of the Node B.

Due to the limited transmission power of the mobile equipment, P_{Smax}, the maximum distance of a mobile UE from Node B can be given as [45]:

$$d_{max} = \frac{\lambda}{4\pi} \cdot \sqrt{\frac{P_{Smax}}{P_R} \cdot g_b \cdot g_m} \tag{15}$$

For the single service case the received power P_R depends on the number of active users n and the interference levels in the cell. Assuming the antenna gains g_b and g_m to be 1, the maximum distance between UE and Node B has been derived in [45] as:

$$d_{max}(n) - \frac{\lambda}{4\pi} \cdot \sqrt{\frac{P_{Smax}}{N_t} \cdot (S - (n-1) \cdot \frac{\varepsilon \cdot V}{F})} \tag{16}$$

where
S is the service factor representing the quality of the service required, and
F defines an upper bound on the inter-cell interference level tolerated in the cell.

If we assume that the cell covers an ideal circle with the cell radius r, the maximum number of active users within a single UMTS cell can be calculated as [45]:

$$n_{max} = \left[\left[S - \left(\frac{(4\pi r)^2 \cdot N_t}{P_{Smax} \cdot L_{pce} \cdot \lambda^2}\right)\right] \cdot \frac{F}{\varepsilon \cdot V}\right] + 1 \tag{17}$$

A detailed derivation of the above expressions and their generalisation to different propagation environments can be found in [44].

5.2 Call Admission Control

Based on the above defined bounds, a call admission control (CAC)algorithm can be defined. Given n existing active connections in the cell and a new call at distance d_{new} from the Node B, assume that N represents the number of available codes (channels). The simplified operation of CAC algorithm is as follows:

1. Assume that (n+1)th new user is asking for a service.
2. IF $n + 1 > min(N, n_{max})$ THEN reject call, else continue with Step 3.
3. Calculate $d_{max}(n + 1)$ as given in Equation (16),
4. IF $d_{new} > d_{max}$ THEN reject call, else continue with Step 5.
5. Check for all existing active connections $i = 1..n$, IF for any connection i the distance $d_i > d_{max}$ THEN reject call, otherwise accept new call.

Note: Step 5 aims to protect existing connection from interruption because of the admission of a new call.

The algorithm in this form is dedicated for one cell because Step 5 aims to protect existing connection from interruption because of the admission of a new call.

5.3   The Queueing Model

In this section, we develop a simplified but still sophisticated model for a single UMTS cell with single service class applying the above CAC algorithm. In modelling UMTS cell, t Subsequently, closed form expressions for state and blocking probabilities are obtained. Typical numerical examples are included to investigate the relative accuracy of the ME solution against simulation and to assess the effect of external GE-type bursty traffic upon the GPRS performance. Extensions of the work towards the study of a GE/GE/c/N/PR queue with preemptive resume (PR) priorities and its implications towards the performance modelling and evaluation of GSM/GPRS wireless cells with PSS and ASS protocols as discussed.

Therefore, the following approximation is adopted: assume that the area of the cell is virtually divided into, say, Z concentric zones (rings) with equal areas. We assume that the inner border of Zone i, d_i represents the distance of all connections in Zone i. Let n_i denote the number of connections in Zone i. It is obvious that $\sum_{i=1}^{Z} n_i = n$, where n is the total number of active connections in the cell. It is important to note that all connections in the cell will share the same set of available codes. However, the CAC algorithm will treat a new connection differently based on its location.

Given the above, we can now define the system state as the vector,

$$\mathbf{n} = (n_1, n_2, \ldots, n_Z)$$

with the condition that $\sum_{i=1}^{Z} n_i = n \leq min\{N, n_{max}\}$, where N is the total number of available codes and n_{max} is the maximum number of connection based on the interference levels and calculated using Equation 17. It is obvious that there must be a trade-off between the accuracy of the model and minimising the size of the state space, in choosing the number of zones, as every new zone add a new dimension to the state space.

Since the geographical distribution of the traffic in the cell is relevant, we introduce the vector $(\alpha_1, \alpha_2, \ldots, \alpha_Z)$, where $\sum_{i=1}^{Z} \alpha_i = 1$, to represent the geographical distribution of the traffic over the area of the cell. In other words, if the overall arrival rate of traffic to the whole cell is λ then the arrival rate on Zone i will be equal to $\alpha_i\lambda$. This distribution vector allows for the consideration of different scenarios. Accordingly, the model that represents the UMTS can be shown as in Figure 9.

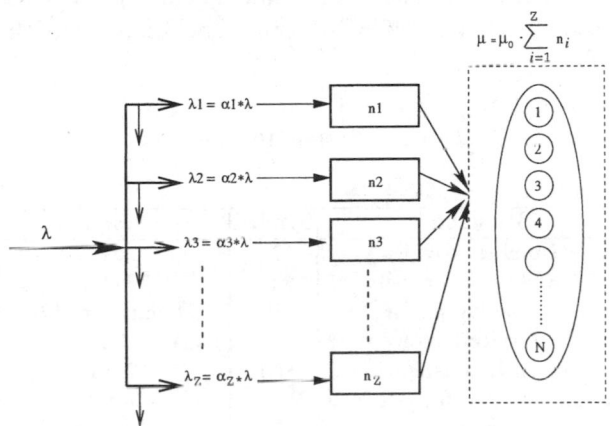

Fig. 9. Queueing model of UMTS cell with Z virtual zones (where μ_0 is the service rate for one call)

Applying the CAC algorithm described in the previous section, it is important to note that a new connection to a certain (outer) zone may be blocked while another call may be accepted. This kind of behaviour makes an analytical solution of the model rather complex.

Regarding the traffic arrival process, it is assumed that new connection requests are following the Poisson distribution with arrival rate λ. The call duration is also assumed to be exponentially distributed random variable. This is usually not an optimal assumption especially for data connections. However, this assumption has been made in order to be able to have a numerical solution for the model. A possible justification is that we consider the burst level rather than packet level. A burst in this case consists of a number of buffered packets that are to transmitted continuously. As the air interface of the UMTS systems seems to be the slowest part of the network, this justification can be acceptable. This assumption will be released when the model is validated using simulation where the more realistic Pareto distribution is used with the same mean service time.

Given the above assumptions, the model can be described using a Continuous-Time Markov Chain (CTMC) with a state $\mathbf{n} = (n_1, n_2, \ldots, n_Z)$. The set of all permissible states will then give the state space $\mathbb{N} = \{\mathbf{n_i}\}$.

The example in this paper does not consider mobility which is part of another more detailed research.

The Markov model is written and analysed using the MOSEL language [34]. MOSEL (Modelling, Specification and Evaluation Language) is a high level model description language. The model described with MOSEL can be solved with any performance evaluation technique. In this study the solution was based on the numerical solution associated with the SPNP package [46].

For validation purposes,we have developed a much more detailed simulation model of the system written in C. Using the steady state probabilities of the Markov chain, performance measures like burst blocking probabilities and system utilisation are expressed. The MOSEL results are compared with the obtained simulation results.

Table 1. System Parameters

Parameter	Symbol	Values
Number of codes	N	64
Radius of the cell	r	5000 m
Spreading factor	SF	64 chips/symbol
Service factor	S	32
Signal to noise ratio	SNR	2 db
Maximum transmission rate	P_{Smax}	125 mW
Basic noise	N_0	-80 dBm
Interference factor	ε	0.5
Service rate	$1/\mu$	100s
Input Traffic load	λ/μ	25, 30, ... 70 Erlang

5.4   Sample Numerical Results for the UMTS Cell

We consider a UMTS cell with parameters as given in Table 1. Furthermore, we assume $Z=4$.

Figures 10 and 11 show the blocking probability and the utilisation, respectively, versus input traffic in the single cell scenario for different interference factors and uniform traffic distribution over the cell area ($\alpha_i = 1/Z$ for $i = 1..Z$). Accordingly, a call near to Node B is rejected if its acceptance will interrupt far connections. Figures 12 and 13 show the blocking probability and the utilisation using non-uniform traffic distribution in a single scenario cell, in this case the traffic is assumed to be concentrated in the centre of the cell and decrease linearly towards the border of the cell. It can been noticed that non-uniform traffic will dramatically decrease the blocking probability since most of the traffic will be concentrated on the area around the base station. While in the uniform traffic, all the zones have the same percentage of traffic. Similarly, the utilisation is better in the non-uniform traffic espaecially at higher intereference conditions.

6   Conclusions

A general framework on the performance modelling and evaluation of GSM/ GPRS and 3G UMTS wireless mobile networks is presented in this paper. First,

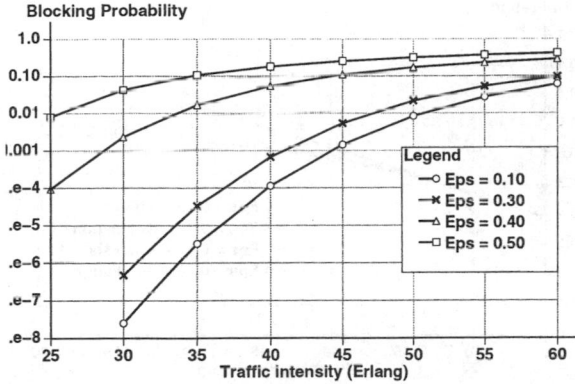

Fig. 10. Blocking Probability versus input traffic - Uniform Traffic

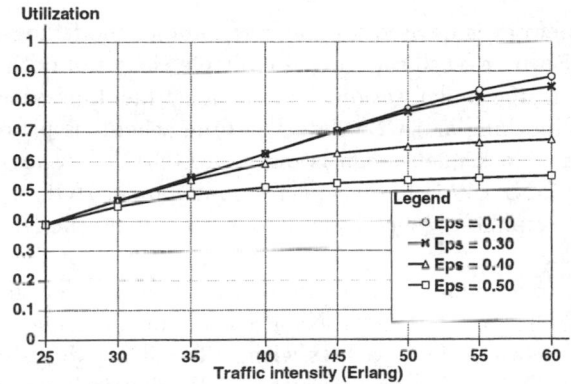

Fig. 11. Utilisation versus versus input traffic - Uniform Traffic

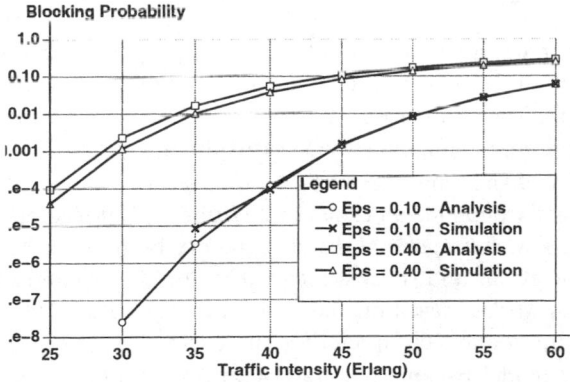

Fig. 12. Blocking Probability versus input traffic - Non-uniform Traffic

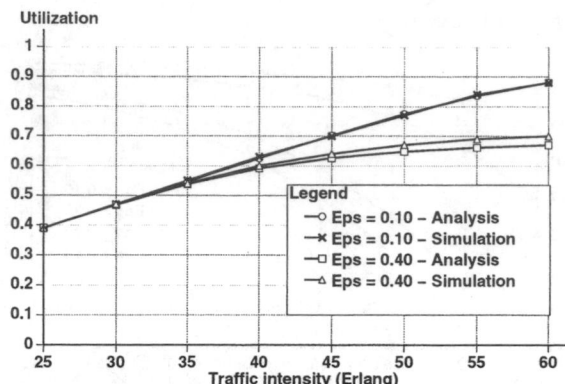

Fig. 13. Utilisation versus versus input traffic - Non-uniform Traffic

an analytic framework is devised for the performance modelling and evaluation of a GSM/GPRS wireless cell supporting multiple classes of both voice calls and data packets of traffic under various cell capacity handling schemes. An open QNMs with RS blocking and CPS data handling scheme is presented based on GE-type building block queues consisting of a GE/GE/c/c loss system and a GE/GE/1/N_1/{HOL or PS} finite capacity queue under HOL or discriminatory PS scheduling disciplines, respectively, under a complete buffer sharing (CBS) management rule.

The analytic methodology is illustrated by focusing on the study of the GE/GE/1/N_1/HOL $\to$ GE/GE/1/ N_2/PS tandem queueing system under a CPS for traffic classes of data packets with different arrival rates, inter-arrival and service times SCVs, data packet sizes and PS discriminatory service levels. The solution, based on the principle of ME, leads into the decomposition of the tandem queueing system into individual building block queues each of which is analysed in isolation by applying queueing theoretic concepts and entropy maximisation.

In the second part, a similar model has been proposed for the GSM/GPRS system with PSS capacity handling scheme. A review on some published solutions is given for the evaluation of this model. Solutions based on queueing systems with randomly changing service rates are suggested.

The third part of the paper presents a sophisticated model of an UMTS cell in which the capacity depends on the actual traffic and interference levels within the cell and the network in general. First, capacity bounds for both the number of admitable connections and the distance of the admitable connections are given based on previous works. Based on these bounds a simple CAC is suggested for the sake of building a queueing model for the system.

The queueing model became possible by applying an approximation for the dicretisation of the mobile user distance from the mobile station. The resulting queueing model was then solved using a multi-dimensional Markov chain.

The models included in this paper are the result of several research projects and, therefore, more details on each of the models can be found in the appropriate references. It is important to note that it was not the intention of the author to give a comprehensive survey on all published models in the very wide topics but only to report on models developed in projects in which the author is directly involved.

References

1. T.S. Rappaport: *Wireless Communications*, Prentice Hall, Englewood Cliffs, NJ, 1996.
2. ETSI GSM Specification 02.60, Digital cellular telecommunications system (Phase2+); General Packet Radio Service; Service description, Stage 1, 1999.
3. ETSI GSM Specification 03.60, Digital cellular telecommunications system (Phase2+); General Packet Radio Service; Service description, Stage 2, 1999.
4. ETSI GSM Specification 03.64: Digital cellular telecommunications system (Phase2+); General Packet Radio Service; Overall description of the GPRS radio interface, Stage 2, 1999.
5. ETSI/TR 101 112 V3.1.0: Universal Mobile Telecommunications System (UMTS); Selection Procedures for the choice of Radio Transmission Technology, Nov. 1997.
6. L. Kleinrock: *Queueing Systems, Volume I: Theory*, John Wiley & Sons, 1975.
7. L. Kleinrock: *Queueing Systems, Volume II: Computer Applications*, John Wiley & Sons, 1976.
8. M.F. Neuts: *Matrix Geometric Solutions on the Stochastic Models: An Algorithmic Approach*, Baltimore, MD: The Johns Hopkins University Press, 1981.
9. K. Begain, G. Bolch, M. Telek: Scalable Schemes for Call Admission and Handover Handling in Cellular Networks with Multiple Services. *Journal on Wireless Personal Communications*, 15, Kluwer Academic Publishers, 2000, pp. 125-144.
10. K. Begain: Scalable Multimedia Services on GSM-like Networks: An Analytical Approach, in Proc.*ITC Specialist Seminar on Mobile Systems and Mobility*, Lillehammer, Norway, March 2000. pp.73-84.
11. M. Ajmone Marsan, S. Marano, C. Mastroianni, M. Meo: Performance Analysis of Cellular Mobile Communications Networks supporting Multimedia Services, *Mobile Networks and Applications* 5, 2000, pp. 167-177.
12. H.G. Perros: *Queueing Networks with Blocking*, Oxford University Press, 1994
13. S. Balsamo, V. De Nitto Persone, R. Onvural: *Analysis of Queueing Networks with Blocking*, Kluwer academic publishers, 2001.
14. K. Begain, M. Ermel, T. Mueller, J. Schueller, M. Schweigel: Analytical Call Level Model of GSM/GPRS Network, in *SPECTS'00, SCS Symposium on Performance Evaluation of Computer and Telecommunication Systems*, Vancouver, BC, Canada, July 16-26, 2000.
15. R. Litjens, R. Boucherie: Radio Resource Sharing in GSM/GPRS Network. *ITC Specialist Seminar on Mobile Systems and Mobility*, Lillehammer, Norway, March 2000. pp. 261-274.
16. W. Ajib, P. Godlewski: Effect of Circuit-Switched Transmissions over GPRS Performance, in Proc. *ACM Intl Workshop on Modelling and Simulation of Wireless and Mobile Systems as part of MobiCom'00*, 2000, pp.20-26.
17. C.H. Foh, B. Meini, B. Wydrowski, M.Zuerman: Modeling and Performance Evaluation of GPRS, in *Proc. of IEEE VTC 2001*, Rhodes, May 2001, Greece, pp. 2108-2112.

18. A. Boukerche and S. Nikoletseas: Algorithmic Design for Communication in Mobile Ad hoc Networks. In M. Calzarossa and E. Gelenbe, editors, *Performance Tools and Applications to Networked Systems*, volume 2965 of *Lecture Notes in Computer Science*. Springer, 2004.

19. E. Gelenbe, R. Lent, M. Gellman, P. Liu, and P. Su: CPN and QoS Driven Smart Routing in Wired and Wireless Networks. In M. Calzarossa and E. Gelenbe, editors, *Performance Tools and Applications to Networked Systems*, volume 2965 of *Lecture Notes in Computer Science*. Springer, 2004.

20. P. Lorenz: IP–Oriented QoS in the Next Generation Networks: Application to Wireless Networks. In M. Calzarossa and E. Gelenbe, editors, *Performance Tools and Applications to Networked Systems*, volume 2965 of *Lecture Notes in Computer Science*. Springer, 2004.

21. C. Williamson: Wireless Internet: Protocols and Performance. In M. Calzarossa and E. Gelenbe, editors, *Performance Tools and Applications to Networked Systems*, volume 2965 of *Lecture Notes in Computer Science*. Springer, 2004.

22. M. Ermel, K. Begain, T. Mueller, J. Schueller, M. Schweigel: Analytical Comparison of Different GPRS Introduction Strategies, in *Proc. ACM Intl Workshop on Modelling and Simulation of Wireless and Mobile Systems as part of MobiCom'00*, 2000, pp.3-10.

23. D.D. Kouvatsos, I. Awan, K. Al-Begain: Performance Modelling of GPRS with Multiclass Traffic, *IEE Proceedings on Computers and Digital Technology*, Vol. 150, No. 2. March 2003. pp. 75-85.

24. K. Al-Begain, I. Awan, D. Kouvatsos: Analysis of GSM/GPRS Cell with Multiple Data Service Classes, *Wireless Personal Communications* **25**, Kluwer, 2003. pp. 41-57.

25. D.D. Kouvatsos, P.H. Georgatsos, N.M. Tabet-Aouel: A Universal Maximum Entropy Algorithm for General Multiple Class Open Networks with Mixed Service Disciplines, *Modelling Techniques and Tools for Computer Performance Evaluation*, eds. R. Puigjaner and D. Potier, Plenum, pp 397-419, 1989.

26. D.D. Kouvatsos: Entropy Maximisation and Queueing Network Models, *Annals of Operation Research*, Vol. 48, pp. 63-126, 1994.

27. D.D. Kouvatsos, N.P. Xenios: MEM for Arbitrary Queueing Networks with Multiple General Servers and Repetitive-Service Blocking, *Performance Evaluation*, Vol. 10, pp 169-195, 1989.

28. I.U. Awan, D.D. Kouvatos: Maximum Entropy Analysis of Arbitrary Queueing Network Models with multiple servers, PR Service Priority and finite capacity queues *Research Report* RS-05-02, Performance Modelling and Engineering Research Group, Department of Computing, University of Bradford, May (2002).

29. A.C. Williams, R.A. Bhandiwad: A Generating Function Approach to Queueing Network Analysis of Multiprogrammed Computers, *Networks* **6**, 1976, pp. 1-22.

30. R.M. Bryant, E. Krzesinskim, M. Seetha, K.m. Chandy: The MVA Priority Approximation, *ACM Transactions on Computer Systems*, Vol. 2, pp.335-359, 1984.

31. D.D. Kouvatsos, I.U. Awan: Open Queueing Networks with RS-Blocking and Multiple Job Classes, Research Report RR-08-01, Performance Modelling and Engineering Research Group, Department of Computing, Bradford University, August, 2001.

32. M. Veran, D. Potier: QNAP-2, A Portable Environment for Queueing Network Modelling Techniques and Tools for Performance Analysis, D. Potier (ed.), North Holland, pp. 25-63, 1985.

33. M. Ermel, T. Mueller, J. Schueller, M. Schweigel, K. Begain: Performance of GSM networks with General Packet Radio Services, *Performance Evaluation*, Vol. 48, 2002.pp. 285-310.
34. K. Begain, G. Bolch, H. Herold: *Practical Performance Modeling, Application of the MOSEL Language*, Kluwer Academic Publishers, Boston, 2001.
35. Y.R. Haung, Y.B. Lin, J.M. Ho: Performance Analysis for Voice/Data Integration on a Finite Buffer Mobile System, *IEEE Transactions on Vehicular Technology*, Vol. 49, No. 2. 2000.
36. D.P. Gaver, P.A. Jacobs, G. Latouche: Finite Birth and Death Models in Randomly Changing Environment, *Journal of Advances in Applied Probability*, Vol. 16, 1984, pp.715-731.
37. J. Sztrik: Modelling of a Multiprocessor System in Randomly Changing Environment, *Performance Evaluation*, Vol. 17, 1993, pp. 1-11.
38. J. Sztrik: Machine Interference Problem with a Random Environment, *European Journal on Operational Research*, Vol. 65, 1993, pp.259-269.
39. M. Mehdavi: *Queueing Analysis and Evaluation of Mixed Services for Mobile TDMA Cellular Networks*, PhD Thesis, University of Sheffield, 2003.
40. A. Dudin: Analysis of the Characteristics of the Data Transmission Process in an ISDN Channel, *Automatic Control and Computer Sciences*, Vol. 21, No. 6, Allerton Press, 1987, pp.42-49.
41. N.D. Reshetnikova: Approximate Analysis of Data Transmission Through an ISDN Channel, *Automatic Control and Computer Sciences*, Vol. 25, No. 1, Allerton Press, 1991, pp.64-70.
42. W.C. Jakes: *Microwave Mobile Communications*, IEEE Press, 1993.
43. J. Schueller, K. Begain, M.Ermel, T. Mueller, M. Schweigel: Performance Analysis of a Single UMTS Cell, in *Proc. of the European Wireless Conference*, Dresden, Germany, 2000, pp. 281-286.
44. A. Zreikat: *Performance Modelling of Mobile Networks*, PhD Thesis, University of Bradford, 2003.
45. A. Zreikat, K. Al-Begain: Interference based CAC for Up-link Traffic in UMTS Networks, in *Proc. of the World Wireless Congress 2002*, San Francisco, 2002, pp. 298-303.
46. G. Ciardo, R. Fricks, J.K. Muppala, K. Trivedi: *Manual of SPNP Package Version 4.0*, Duke University, Durham, NC, USA, 2000.

IP-Oriented QoS in the Next Generation Networks:
Application to Wireless Networks

P. Lorenz

University of Haute Alsace
34, rue du Grillenbreit
68008 Colmar - France
lorenz@ieee.org

Abstract. The challenge of new communication architectures is to offer QoS in Internet networks. This paper investigates the QoS aspects in the NGN (Next Generation Networks), and especially in the wireless networks. Voice and video multimedia applications will be developed and used when QoS mechanisms will exist. Therefore new functions must be developed to guarantee performance, offer security, avoid jitter and to allow the respect of real-time constraints.

1 Introduction

To offer Quality of Service (QoS) in a network, there are a lot of parameters that should be taken into account, such as bandwidth, latency, jitter, packet loss, packet delay. A given application does not take into account all parameters with the same priority. For video applications, the most important QoS parameter is based on the bandwidth. For Voice over IP (VoIP) applications, the most important QoS parameter is based on latency with end to end delay no larger than 200 ms.

The QoS can be linked to the:

- network level: in this case the QoS depend of the network policy and of the used mechanisms such as filters, rerouting in the core of the network, control access at the corners of the network.
- application level: in this case it is the application which improve the QoS and there is no link with the network infrastructure.

The Classes of Service (CoS) classify the services in different classes and manage each type of traffic with a particular way. ETSI (European Telecommunications Standards Institute) has introduced four CoS: Class 1 for Best Effort until Class 4 for QoS guaranteed. Many SLA offers three CoS: Premium (for 15% of network resources), Olympic (for 80% of network resources) and Best Effort.

M.C. Calzarossa and E. Gelenbe (Eds.): MASCOTS 2003, LNCS 2965, pp. 168–186, 2004.
© Springer-Verlag Berlin Heidelberg 2004

Internet is increasing exponentially and the bandwidth double every 6 months. In 2001, there was 180 million users and in 2005, there will be 500 millions users. In 2005, the data traffic will be 20 times the voice traffic and there will be more wireless voice traffic than wired traffic. 90% of the Internet traffic is based on TCP and 10% on UDP protocol (75% for WWW applications, 3% for the Emails, 4% for FTP and 7% for the News).

2 QoS Mechanisms

There are many QoS mechanisms that can be used to offer QoS in the network. In the last years, we have observed a growth of the networks capacity with the development of Wavelength Division Multiplexing (WDM) technologies. In 2005, WDM networks will offer 1000 Wavelengths and 1000 Tbit/s in the core of the networks.

The different types of QoS mechanisms can:

- Provide a fair service with Best Effort algorithms.
- Maximize the bandwidth allocation to the source receiving the smallest allocation (max-min allocation of bandwidth). This algorithm allows to decrease the bandwidth allocated to other source.
- Drop the packets if congestion occurs in routers when the buffer is full (tail drop algorithm) or when the buffer occupancy increases too much (RED: Random Early Detection algorithm).
- Use congestion control mechanisms in end systems to inform the source about network congestion with ICMP or tagged packets with ECN (Explicit Congestion Notification) protocols. In this case, all routers should implement the congestion control mechanisms.
- Divide the output buffers in N queues and introduce a scheduler (processor sharing or round robin algorithms).
- Classify the IP flows at different layers: the edge routers perform classification/marking and the backbone routers rely on marking.
- Support n drop priorities to offer a minimum bandwidth service with n RED algorithms running in parallel (weighted RED algorithm).
- Introduce a weight to each queue (Generalized Processor Sharing/Weighted Round Robin algorithms).

The new communication networks must offer QoS and mobility. The two major possibilities are:

- QoS mechanisms based on signalization and routers. It is the solution used by the telecommunication world or
- Overprovisioning the network for new applications such as TV on demand, telephony IP. Overprovisioning is not a global solution but is an asset for traffic engineering and to QoS in Internet network.

In the core of the network, the architectures can be based on these two following models:

- With signalization (such as the SS7, X25/ATM, Internet/Telecom networks). These networks offer good QoS but theses solutions are expensive: an UMTS access point costs about 15000 $.
- Without signalization (such as the Arpanet, 1st and 2nd Internet generation networks). These networks offer no QoS but these solutions are cheap: a Wi-Fi access point costs about 100 $.

The mechanisms to offer QoS can be based on the connection-oriented ATM protocol or on the connectionless-oriented IP protocol. Each solution has his own advantage and offer different QoS guaranties.

The different network architectures can be represented as follows:

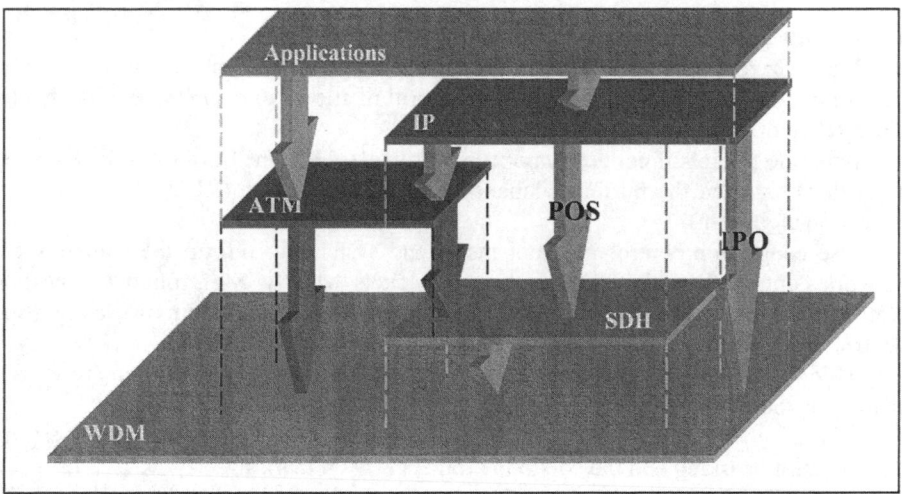

Fig. 1. Representation of the different networks architectures

We will now present the ATM (Asynchronous Transfer Mode) protocol that is the first protocol which offer QoS.

3   ATM Protocol

ATM is a connection-oriented protocol, which offer real QoS guaranties. The QoS is negotiated during the establishment of the connection and depends of the available resources.

ATM is based on VC (Virtual Channels) and on VP (Virtual Paths). VP and the VC can be represented as follow:

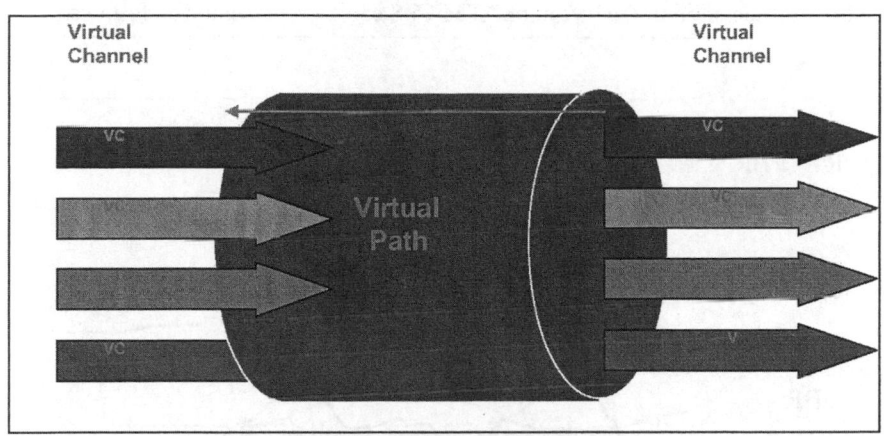

Fig. 2. Representation of VP and VC

The representation of VP and VC in an ATM switched architecture can be represented as follows:

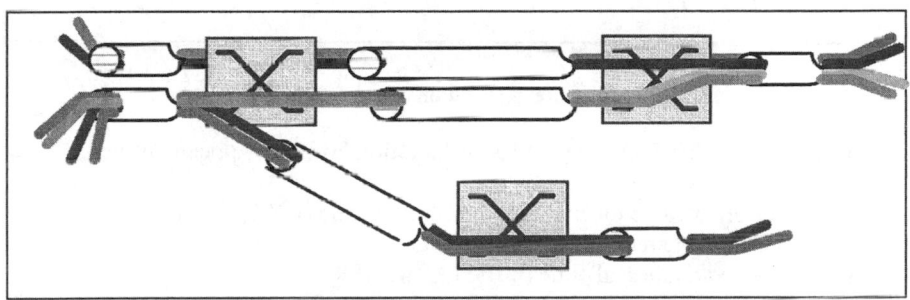

Fig. 3. Representation of VP and VC in an ATM switched architecture

In ATM networks, there are six CoS:

- CBR (Constant Bit Rate), which guarantees a constant bit rate for applications such as vidcoconferencing, telephony.
- RT-VBR (Real-Time Variable Bit Rate) for transmissions with a variable rate for applications requiring real-time constraints, such as MPEG transmissions.
- NRT-VBR (Non-Real-Time Variable Bit Rate) for transmissions with a variable rate for applications requiring no real-time constraints, such as multimedia transfers.
- ABR (Available Bit Rate) for transmissions of traffic using the remaining bandwidth or based on bursty traffic. ABR guaranties always a minimum rate.
- GFR (Guaranteed Frame Rate) for applications, which accept to loose sometime some services.
- UBR (Unspecified Bit rate), which offer no rate guaranty and no congestion indication. UBR is a Best Effort CoS.

The representation of the different ATM CoS can be represented as follows:

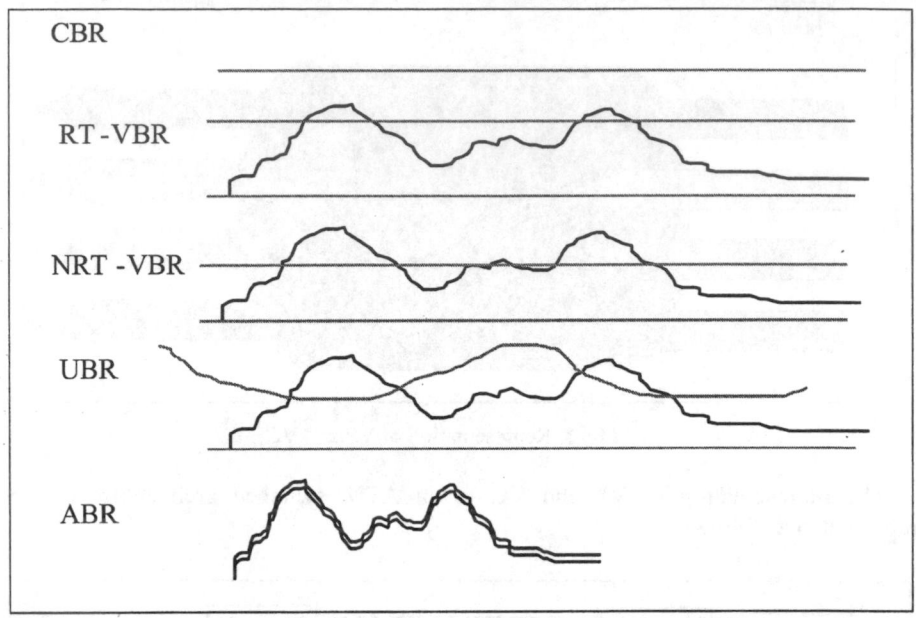

Fig. 4. Representation of the ATM CoS

For theses CoS, different AAL (ATM Adaptation layer) have been defined:

- AAL1: for oriented connection and real-time traffic (CBR)
- AAL2: for variable real time traffic (VBR)
- AAL3/4: for variable real-time traffic (ABR, GFR)
- AAL5: for reliable or non-reliable services and unicast or multicast traffic (UBR).

In ATM networks, QoS is provided by the signalization and stream controls mechanisms. The major QoS parameters used by ATM networks are:

- CTD: Cell Transfer Delay
- CMR Cell Misinsertion Ratio
- CLR: Cell Loss Ratio
- CER: Cell Error Ratio
- PCR: Peak Cell Rate
- MCR: Minimum Cell Rate
- CVDT: Cell variation Delay Tolerance
- SCR: Sustainable Cell Rate
- BT: Burst Tolerance
- CDV: Cell Delay Variation

The ATM stream control mechanisms are based on:

- CAC (Connection Admission Control) that determines if a connection can be accepted or not.
- Usage Parameter Control/Network Parameter Control (UPC/NPC) that controls the traffic and the conformity of a connection.
- The Resource Management mechanisms that optimize the traffic.

An example of an ATM control mechanism can be represented as follows:

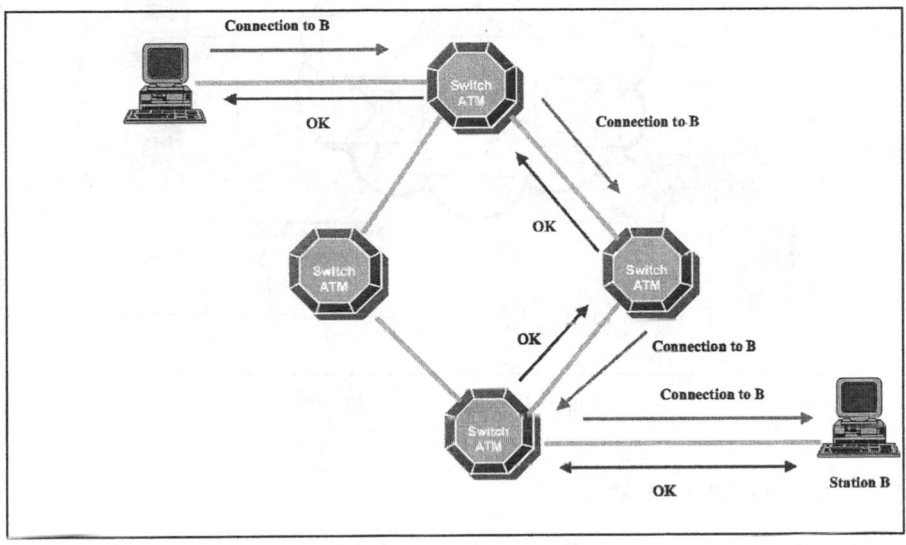

Fig. 5. ATM control mechanism

There are three major protocols used to manage IP over ATM: LAN Emulation (LANE), Classical IP and Multi Protocol Over ATM (MPOA).

3.1   Classical IP

Classical IP has been defined by IETF. It introduces the notion of LIS (Logical IP Subnetwork), which is an emulation of an IP subnetwork. Each IP station registers a connection with an ARP (Address Resolution Protocol) server. The ARP server does the translation between IP and ATM addresses.

Classical IP allows multicast traffic through the MARS (Multicast Address Resolution Server) protocol. The problem is that only IP protocol is supported and that the connections between two ARP servers must be done through additional protocols such as NHRP (Next Hop Resolution Protocol).

The classical IP architecture can be represented as follows:

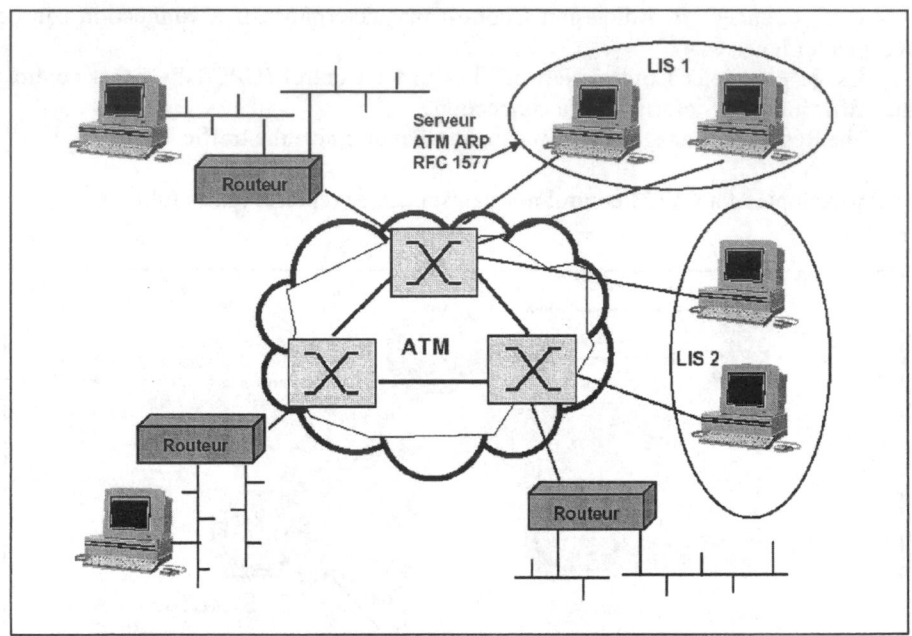

Fig. 6. Classical IP architecture

3.2 LAN Emulation (LANE)

The LANE protocol supports all types of protocol (IP, IPX, …). The problem is that it cannot be used in wide area networks and routers are always necessary. Three servers and one client compose LANE:

- LEC (LAN Emulation Client) is a station with an ATM address and a MAC address.
- LES (LAN Emulation Server) is used for the translation between the 6 octets of the IP address and the 20 octets of the ATM address.
- LECS (LAN Emulation Configuration Server) used for the automatic configuration. It allows to find the LES ATM address.
- BUS (Broadband Unknown Server) used for the management of multicast traffics, broadcast and unknown frames.

The interconnection of an ATM network with an IP network can be represented as follows:

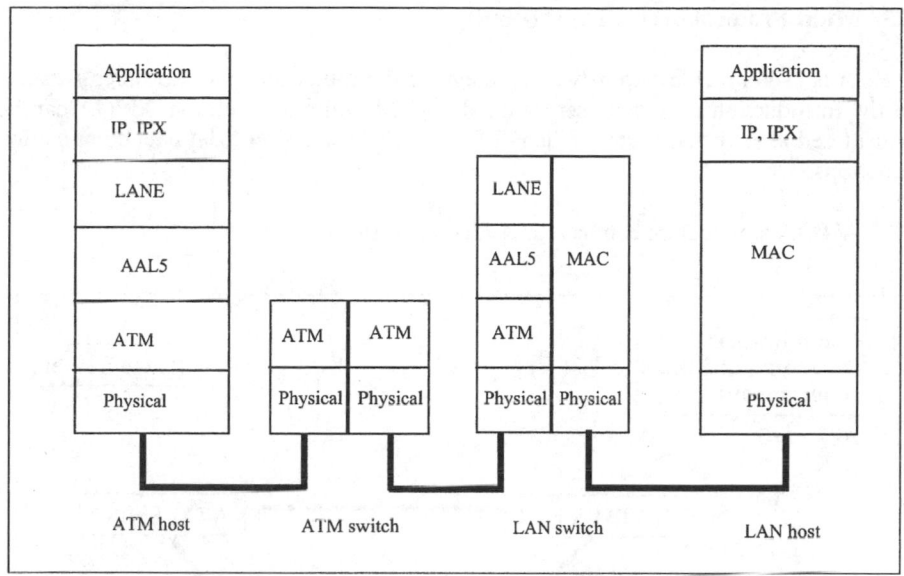

Fig. 7. Interconnection of an ATM networks with an IP network

The LANE architecture can be represented as follows:

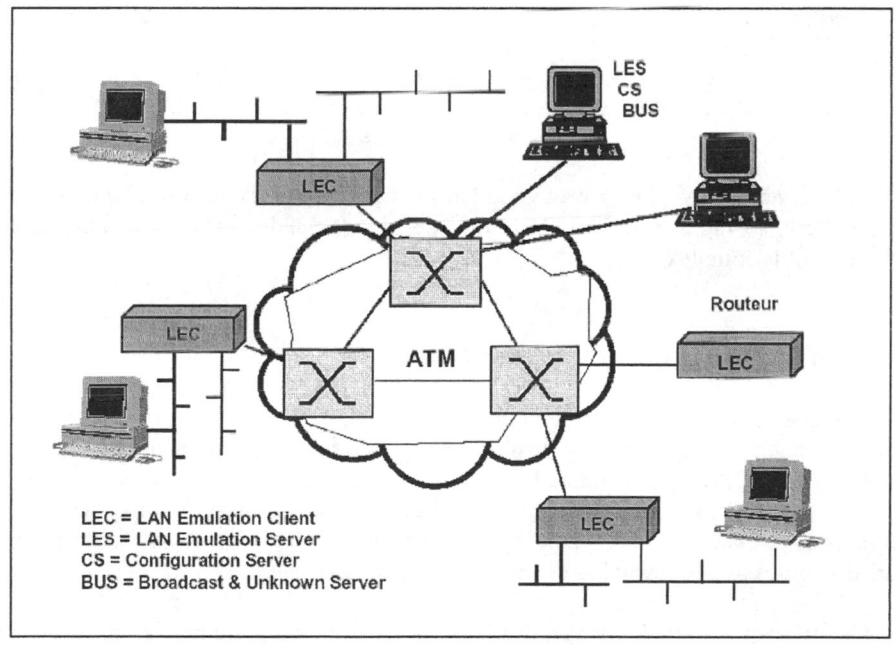

LEC = LAN Emulation Client
LES = LAN Emulation Server
CS = Configuration Server
BUS = Broadcast & Unknown Server

Fig. 8. LANE architecture

3.3 Multi Protocol over ATM (MPOA)

MPOA is used in wide area network and avoids the router bottleneck problems thanks to the introduction of a route server for the ATM address resolution. MPOA can be considered as a virtual router, which divide data transmission from data computation functions.

A MPOA architecture can be represented as follows:

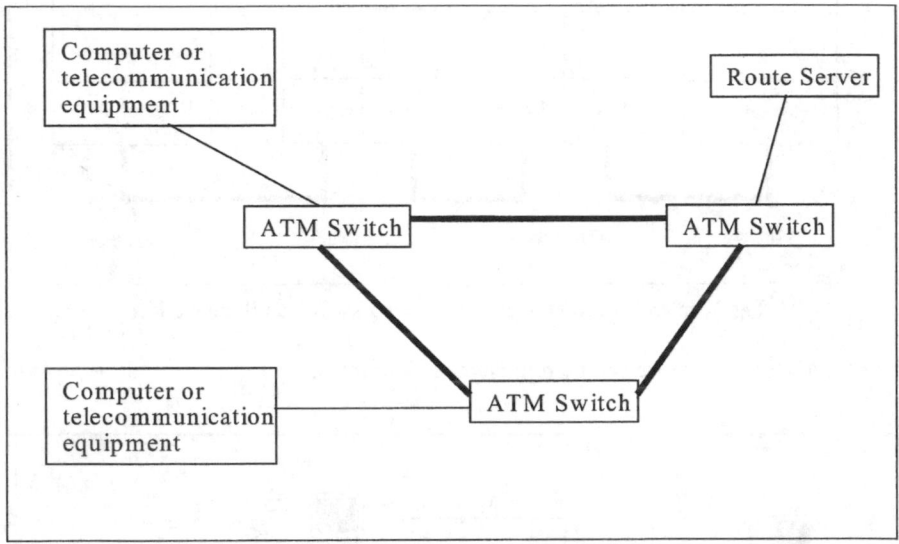

Fig. 9. MPOA architecture

ATM is necessary when a very good QoS is needed, but the disadvantage of ATM is that there is a big overhead. ATM is an expensive technology and the support of the IP protocol is complex.

4 New Communication Architectures

The first mechanisms allowing QoS have been developed in 1996 with proprietary solutions such as Tag Switching (Ipsilon), IP Switching, Net Flow Switching (Cisco), ARIS (IBM), IP Navigator (Cascade), ...

The signalization steps are separated from the data transmissions steps. The signalization management is done by the routers to offer control and management functionalities and the switches do only the data transmission.

The IP switching solution, represented in the following figure, is based on the separation of routing and switching functionalities:

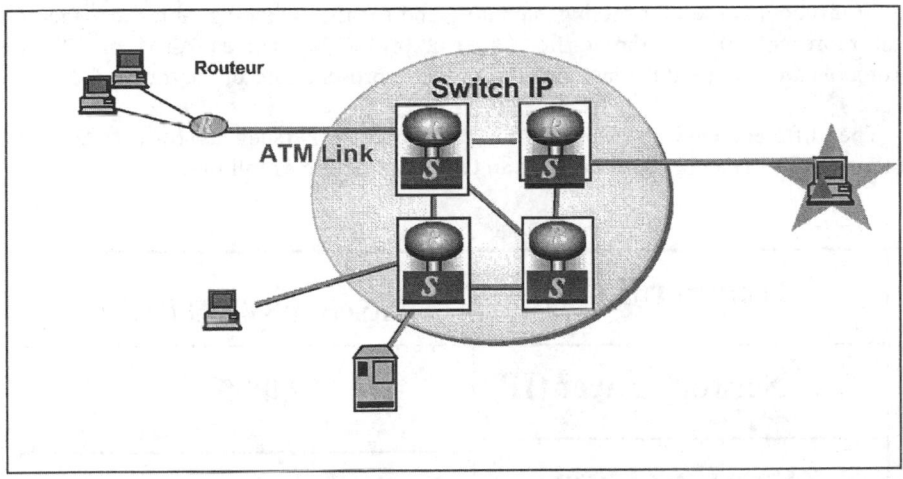

Fig. 10. IP Switching architecture

In a given communication, only the first packet is routed (if the first packet is unknown from the switch) and the other packets of the application are switched and no more routed as represented in following Fig. 11:

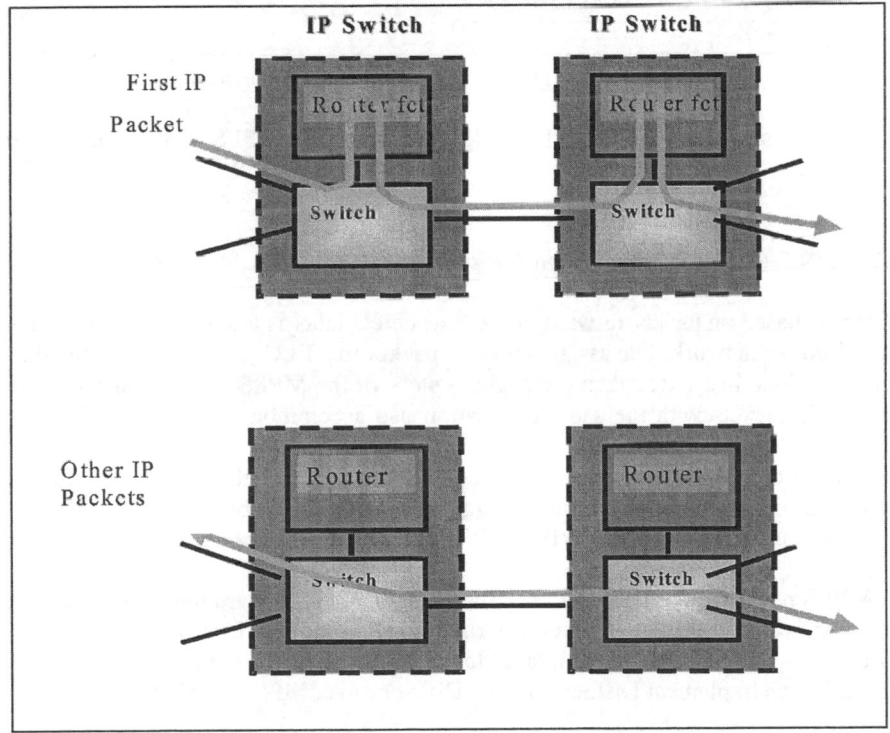

Fig. 11. Representation of IP Switch mechanisms

IP protocol is used for routing, signaling and for the switching tables management that represent 20% of the traffic. Layer 2 protocols such as ATM or Ethernet protocols are only used the fast forwarding that represent 80% of the traffic.

The different QoS mechanisms for an IP network (such as MPLS, DiffServ, IntServ, RSVP) can be represented in an OSI architecture as follows:

Transport Layer	IntServ, RSVP, DiffServ
Network Layer (IP)	MPLS
Data Link Layer (Ethernet, FR, ATM, PPP)	
Physical Layer (Sonet/SDH, optical fiber, 802.17: Resilient Packet Ring)	

Fig. 12. Description of the different IP QoS protocols

We will now describe the different IP QoS protocols MPLS, RSVP, IntServ and DiffServ.

4.1 MPLS (Multi Protocol Label Switching)

MPLS is based on packet forwarding. A four octets label is assigned when the packet enter into the network. The assignment of a packet to a FEC (Forwarding Equivalence Class) is done just once when the packet enters in the MPLS network at the ingress node. All packets with the same destination use a common route and at the egress node the labels are removed.

The label is inserted between the layer 2 header and the IP header. Existing protocols are extended to enable to piggyback on MPLS labels. The IP protocol is switched instead of routed and RIP, OSPF or BGP protocols can still be used.

MPLS nodes (called LSR or Label Switching Router) forward the packets based on the label value. MPLS combines L3 routing (IP) and L2 forwarding (ATM with VPI/VCI or Ethernet with a shim label located between MAC and VLAN addresses). The LSR can implement DiffServ, with a DiffServ over MPLS architecture.

A LSP (Label Switched Paths) is a sequence of routers. The signalization protocol LDP (Label Distribution Protocol) manages the information exchange between the

LSR to establish a LSP and associates a FEC for each LSP. A LSR sent periodically a LDP Hello Message.

With MPLS it is possible to introduce a path protection/restoration with the introduction of an alternate route and the use of RSVP as Label Distribution Protocol. With CR-LDP (Constraint-based Routing LDP), the LSR establishes LSPs satisfying to a set of constraints. MPLS supports IP QoS models and can be used to build VPNs. It supports all types of traffic by defining a trunk for each pair of ingress/egress router.

GMPLS (Generalized MPLS) integrates ATM, Ethernet, FR, TDM, optical networks. It is possible to do traffic engineering, to control network resources and the input load to offer performance optimization. In this case, a new model based on CR-LDP protocol is necessary.

4.2 Ressource ReSerVation Protocol (RSVP)

RSVP is a signalization protocol used to establish unidirectional flows in IP networks. RSVP is used by routers to deliver QoS and to reserve resources in each node along a path. RSVP sends periodic refresh messages to maintain a state along a reserved path. A bandwidth is reserved for a given flow and requires resources reservation and releasing at regular intervals.

The establishment/maintain of unidirectional flows in IP networks is done through the PATH and RESV messages. The RSVP messages are encapsulated inside IP packets. RSVP supports MPLS, multicast and unicast traffics.

Signalization management can be done by COPS (Common Open Policy Service) protocol. There is an exchange between a policy server (Policy Decision Point) and an edge router (Policy Enforcement Point). In PEP, it is possible to have a Local PDP (LPDP).

There is two policy management based on:

- outsourcing policy model : the PDP decides if a request can be accepted or not (for example a RSVP request).
- provisioning policy model: the PDP decides what politic should be installed in routers.

PDP includes bandwidth, security and mobility brokers, authentication and billing servers. COPS can use IPsec for authentication and VPN to offer secure communications.

PDP can be connected to a:
- LDAP server to accept or not a new user,
- PIB (Policy Information Base), which is a database with all network politics,
- bandwidth broker that manage the available resources,
- mobility broker,
- security broker.

4.3   Integrated Services (IntServ)

IntServ is based on traffic control mechanisms and on the signalization protocol RSVP. The reservation is done at the router level. The problem is that:

- There is a poor scalability because the amount of state increase proportionally with the number of flows.
- All routers must implement RSVP.
- There is no policy for reservation control.
- Stations must support signalization.

Therefore, RSVP is only for small networks. The three CoS for IntServ are:

- Guaranteed Service (Premium service), for applications requiring fixed delay bound (CBR, RT-VBR).
- Controlled-Load Service (Olympic service) for applications requiring reliable and enhanced best-effort service (NRT-VBR, GFR, ABR).
- Null service, when there is no need of time constraints, but only a better best-effort service (UBR).

4.4   Differentiated Services (DiffServ)

DiffServ is a relative-priority scheme in which the IP packets are classified and marked at the network ingress to create several packet classes. The selected type of service is indicated inside each IP packet. DiffServ scalability comes from the traffic aggregation with the use of aggregate classification state in the core of the network. To share the bandwidth, DiffServ offers a hierarchy of different flows and is similar to MPLS, but more adapted for MAN.

Complex mechanisms depend of the number of services implemented in boundary nodes. SLA can be used between the client and the provider to specify for each service the amount of traffic that can be sent. The three CoS for DiffServ are:

- Expedited Forwarding (Premium service) for fixed bit rate between the source and the destination (CBR, RT-VBR).
- Assured Forwarding (Olympic service) for bursty services. There is no QoS guaranteed but only a low loss probability (ABR, GFR, nrt-VBR).
- Bulk Handling for applications requiring no QoS such as file transfer or mail (UBR).

4.5   Conclusion

The integration of QoS mechanisms is easier in small networks, because large networks ingrate a lot of heterogeneous domains.

DiffServ is less complex and more easy to be implemented than IntServ, but it gives less accurately and less QoS flow differentiation. DiffServ is located in the core of the network between the routers and IntServ at the periphery of the networks.

IntServ works on micro-flows, it is a complex technology based on a "hard" approach of QoS. DiffServ is an evolution of IP service, the load control is done at aggregate level by the network and not at flow level by TCP. MPLS is another evolution of IP service: it is a generic connection orientation that increases of routing functionalities.

To summarize, in the LAN it is IntServ that offers the best approach, in the MAN it is DiffServ (or IntServ) and in the WAN it is MPLS.

5 QoS in Wireless Networks

There are many different types of wireless networks: cellular networks, mobile networks, data transmission networks and satellites networks.

For example, the wave radio-electrical networks penetrate the buildings and can be used for large distances. The wave infrared networks are used for small distance and do not penetrate the buildings. The micro-wave frequency networks do not penetrate the buildings and are used for networks no larger than 80 km. Light wave networks are based on lasers that are quickly absorbed by the rain or the snow.

There are a multiple access techniques: FDMA (Frequency Division Multiple Access) developed for analogical networks, TDMA (Time Division Multiple Access) developed for numerical networks, CDMA (Code Division Multiple Access) developed for third generation networks or techniques based on demand assignment through reservation mechanisms. In the last case, there is no waste of bandwidth, because the protocol minimizes the wasted bandwidth. The protocol assigns bandwidth on demand and avoids collisions wasted bandwidth.

5.1 Satellites

There are three types of satellites:

- LEO (Low Earth Orbit),
- MEO (Medium Earth Orbit),
- GEO (Geostationary Earth Orbit).

The frequencies used by the satellites use:

- Ku band (10 GHz to 18 GHz),
- C band (4 GHz to 6 GHz) for the connections between terrestrial stations and satellites,
- Ka band (20 GHz to 30 GHz).
- V band (40 GHz to 50 GHz) for future applications.

The LEO satellites are located between 500 and 2000 km. The communication delays are 0.01 second and the maximum rate is 155 Mbit/s. To cover the world 50 satellites are necessary and one satellite covers the skylink during 15 minutes.

LEO based on 800 MHz, offer a 300 kbit/s rate and can be used for localization such as the GPS (Global Positioning System) system.

LEO based on 2 GHz, offer a 10 kbit/s rate used essentially for telephony applications.

LEO based on 20 GHz to 30 GHz, offer 155 Mbit/s rate for multimedia applications.

The different major LEO projects are:

- Iridium (Motorola) composed by 66 satellites and by 6 emergency satellites located at an altitude of 780 km. Iridium is out of service since 1998.
- Globalstar (France Telecom, Daimler-Benz Aérospace) composed by 48 satellites and by 8 emergency satellites located at an altitude of 1414 km.
- Teledesic (Microsoft and de Craig McCaw) planned for 2006. It will be composed by 288 satellites located at an altitude of 1375 km with a download rate of 64 Mbit/s and an upload rate of 2 Mbit/s.
- Skybridge (Alcatel Space) composed by 80 satellites located at an altitude of 1469 km.

MEO satellites are located at an altitude between 5000 km and 20000 km for communication delays of 0.1 second. A communication can remain one hour and 12 satellites are necessary to cover the earth.

GEO satellites are located at an altitude of 36600 km with communication delays of 0.27 second. The duration of GEO satellites are between 15 and 20 years and three satellites can cover the world.

Today, we can observe the development of pico-satellite (1 kilo) located at an altitude of 340 km and of HAPS (High Altitude Stratospheric Platform) with the following projects:

- Proteus airplane (Awacs) that will offer a bandwidth of 164 kbit/s for a 100 km diameter.
- Airship located at an altitude of 23 km (Sky Station project) with a rate of 10 Mbit/s in the 48 GHz band.

5.2   2G and 3G Networks

In this section, we will present the second generation (2G) and the third generation (3G) wireless networks.

The pico-cells are used for distances between 5 and 50 meters, the micro-cells for distances between 50 and 500 meters and the macro-cells for distances between 0.5 and 10 km.

The 2G USA technologies are composed by:

- D-AMPS (Advanced Mobile Phone System) based on 800 MHz or 1900 MHz frequencies and on TDMA,
- PCS 1900 (Personal Communication Services) based on 1900 MHz frequencies and on TDMA,
- IS-95 based on CDMA.
- IS-136 based on TDMA

The 2G Japan networks are based on PDC (Personal Digital Cellular) numerical technology and in Europe the 2G networks are based on GSM, DCS and PCS networks. These networks offer a 10 kbit/s transmission rate.

In wireless networks, the Public Land Mobile Network (PLMN) is composed by the:

- Base Station Subsystem (BSS) that manage radio resources with the:
    * Mobile Station,
    * Base Transceiver Station (BTS),
    * Base Station Controller (BSC).

- Network and Switching Subsystem (NSS) that manage network resources with the:
    * Visitor Location Register (VLR) for mobiles localization,
    * Home Location Register (HLR) that contain subscription informations,
    * Mobile Switching Center (MSC).

- Operation Sub-System (OSS) for the administration and management of the network.

The BSC establishes the communications with the Mobile services Switching Center (MSC). When the best BTS is selected, the mobile asks for a logical signaling channel to the BSC, which manages the communications synchronization.

GPRS (General Packet Radio Service) is a 2.5G networks that offers a maximum rate of 170 kbit/s. It is based on packet switching, the cost of the communication depends only of the amount of data and no more on duration. The evolution of a 2G network to a 2.5G network can be done without modification of the BSS, because 2.5 networks use the same frequency and it can reuse the BTSs and the BSCs.

In GPRS networks, there is a need of two additional routers:

- SGSN (Serving GPRS Support Node) for the resources, sessions, taxation and mobility management,
- GGSN (Gateway GPRS Service Node) for IP networks interconnections.

EDGE (Enhanced Data rate for GSM Evolution) is also a 2.5G network that will offer a 384 kbit/s rate. E-GPRS (Enhanced GPRS) will apply EDGE to GPRS to offer services similar to UMTS (Universal Mobile Telecommunication System) services.

IMT2000 (International Mobile Telecommunication 2000) is essentially composed by UMTS and CDMA2000 systems.

UMTS (Universal Mobile Telecommunication System) is a 3GPP (Third Generation Partnership Project) developed by the:
- SMG (Special Mobile Group) of ETSI (European Telecommunication Standard Institute),
- ARIB (Association of Radio Industries and Business) and TTC (Telecommunication Technology Association) Japanese organizations,
- TTA (Telecommunication Technology Association) Korean organization,
- T1P1 American committee.

CDMA 2000 is an American evolution of the IS-95 standard.

The 3G networks will integrate in a same network, the cellular network, the wireless network, the data transmission network, the intelligent terminals and multimedia services such as bandwidth on demand. Two billions users are expected in 2010 for 3G networks. VHE (Virtual Home Environment) will enable to obtain their usual proprietary services and his usual environment thanks to the introduction of smart cards. The 2 Mbit/s rate will be offered in the cities, 144 kbit/s in the countryside and 30 kbit/s for a global mobility with the use of satellites. The problem is that new frequencies and new infrastructures are necessary for the networks.

The fourth generation networks will be developed for 2010 and will use the 30 GHz frequencies.

5.3 Wireless LAN (WLAN)

WLAN networks are composed by different versions of IEEE 802.11 standards. The majors IEEE 802.11 standards are:

- IEEE 802.11b (WiFi – Wireless Fidelity): frequency of 2.4 GHz, rate of 11 Mbit/s for a maximum distance of 100 meters. This protocol is based on CDMA/CA.
- IEEE 802.11g: frequency of 2.4 GHz, rate of 54 Mbit/s.
- IEEE 802.11a (WiFi 5): frequency of 5 Ghz, rate of 54 Mbit/s.
- IEEE 802.11i: developed to manage security aspects.
- IEEE 802.11e: developed to manage QoS aspects.
- IEEE 802.11f: developed to manage handover aspects.
- IEEE 802.11n: based on power control will offer a rate of 400 Mbit/s.

IEEE 802.20 standard Mobile Broadband Wireless Access (MBWA) will replace in the future the WiFi protocols. MBWA is based on 3.5 GHz frequency and will offer a 1 Mbit/s rate in a 1 km diameter around the access point.

5.4 Wireless Personal Area Networks (WPAN) IEEE 802.15

WPAN networks are composed by a lot of different versions of IEEE 802.15. The majors IEEE 802.15 standards are:
  - IEEE 802.15.1: Bluetooth WPAN has a rate of 1 Mbit/s and use the 2400 MHz frequency in a 10 meters diameter around the access point.
  - IEEE 802.15.3: Ultra WideBand (UWB) is a wireless technology for transmitting digital data over a wide spectrum of frequency bands with very low power and a 400 Mbit/s rate.
  - IEEE 802.15.4: standard developed for the communications between toys and sensors with a 200 kbit/s rate.

We can observe that Bluetooth is developed for phone and computer while the IEEE 802.11standard (Wifi) is developed only for computers communications.

5.5 Wireless Local Loop (WLL) and WMAN (Wireless Metropolitan Area Network) IEEE 802.16

The WMAN (Wireless Metropolitan Area Network) use the 10 GHz and the 11 GHz frequencies. The majors WLL IEEE 802.16 standards are:

  - LMDS (Local Multi-point Distribution Service) based on:
    + Bi-directional transmissions point to multi-points.
    + Rate of 1 Gbit/s in the 28 GHz to 31 GHz band.
    I 1 km to 2 km distances.
    + Directive antenna without shadow area.
  MMDS (Multi-channel Multi-point Distribution Service) based on:
    + Unidirectional video.
    + Distances of 10 km in the 2.5 GHz to 2.7 GHz band for example for rural areas without CATV.

6 Conclusion

There are a lot of wireless networks, each wireless network offer different types of QoS based on a specific protocol.

GSM standard was initially based on ISDN, GPRS on Frame Relay and UMTS on ATM/AAL2 protocol. Now, the second generation of UMTS and CDMA2000 standards are based on the IP protocol, which is the major protocol to offer QoS in wired and wireless networks.

References

1. Aljadhai, A., and Znati, T.F., Predictive mobility support for QoS provisioning in mobile wireless environments, IEEE Journal on Selected Areas in Communications, Volume 19, Issue 10, Oct. 2001, pp. 1915-1930.

2. Becchetti, L. et al, Enhancing IP service provision over heterogeneous wireless networks: a path toward 4G, IEEE Communications Magazine, Volume 39, Issue 8, Aug. 2001, pp. 74-81.
3. Carmel, B. et al, PiNet: wireless connectivity for organizational information access using lightweight handheld devices, IEEE Personal Communications, Volume 8, Issue 4, Aug. 2001, pp. 18-23.
4. Chakrabarti, S. and Mishra, A., QoS issues in ad hoc wireless networks, IEEE Communications Magazine, Volume 39, Issue 2 , Feb. 2001, pp. 142-148.
5. Chi-Jui H. et al., Call admission control in the microcell/macrocell overlaying system, IEEE Transactions on Vehicular Technology, Volume 50, Issue 4 , July 2001, pp. 992-1003.
6. Dixit, S. et al., Resource management and quality of service in third generation wireless networks, IEEE Communications Magazine, Volume 39, Issue 2, Feb. 2001, pp. 125-133.
7. Frodigh, M., et al., Future-generation wireless networks, IEEE Personal Communications, Volume 8, Issue 5, Oct. 2001, pp. 10-17.
8. Gutierrez, J.A. et al. IEEE 802.15.4: a developing standard for low-power low-cost wireless personal area networks, IEEE Network, Volume 15, Issue 5, Sept.-Oct. 2001, pp. 12-19.
9. Hao F., et al., "QoS Routing for Anycast Communications: Motivation and an Architecture for DiffServ Networks", IEEE Communications Magazine, June 2002.
10. Jamalipour, A.; Tung, T., The role of satellites in global IT: trends and implications, IEEE Personal Communications, Volume 8, Issue 3, June 2001, pp. 5-11.
11. Leung, K.K., et al, Link adaptation and power control for streaming services in EGPRS wireless networks, , IEEE Journal on Selected Areas in Communications, Volume 19, Issue 10, Oct. 2001, pp. 2029-2039.
12. Misra, A. et al, Autoconfiguration, registration, and mobility management for pervasive computing, IEEE Personal Communications, Volume 8, Issue 4, Aug. 2001, pp. 24-31.
13. Ramjee, R, et al., IP-based access network infrastructure for next-generation wireless data IEEE Personal Communications, Volume 7, Issue 4, Aug. 2000, pp. 34-41.
14. Sari, H., A multimode CDMA with reduced intercell interference for broadband wireless networks, IEEE Journal on Selected Areas in Communications, Volume 19, Issue 7, July 2001, pp. 1316-1323.
15. Striegel A. and Maniaran G., "Managing Group Dynamics and Failures in QoS Multicasting", IEEE Communications, June 2002.
16. Varshney, U. and Jain, R., Issues in emerging 4G wireless networks, Computer, Volume 34, Issue 6, June 2001, pp. 94-96.
17. Wang, W. and Akyildiz, I.F., A new signaling protocol for intersystem roaming in next-generation wireless systems, IEEE Journal on Selected Areas in Communications, Volume 19, Issue 10, Oct. 2001, pp. 2040-2052.
18. Webb, W., Broadband fixed wireless access as a key component of the future integrated communications environment, IEEE Communications Magazine, Volume 39, Issue 9, Sept. 2001, pp. 115-121.
19. Yu-Chee Tseng and Cheng-Chung Tan, Termination detection protocols for mobile distributed systems, IEEE Transactions on Parallel and Distributed Systems, Volume 12, Issue 6, June 2001, pp. 558-566.
20. Zhang, T., et al., IP-based base stations and soft handoff in all-IP wireless networks, IEEE Personal Communications, Volume 8, Issue 5, Oct. 2001, pp. 24-30.
21. Zhang Z.-L., et al., "On Scalable Design of Bandwidth Brokers," IEICE Transaction on Communications, E84-B (8), Aug. 2001.

Peer-to-Peer Computing in Mobile Ad Hoc Networks

Alexander Klemm, Christoph Lindemann, and Oliver P. Waldhorst

University of Dortmund
Department of Computer Science
August-Schmidt-Str. 12
44227 Dortmund, Germany
http://www4.cs.uni-dortmund.de/~Lindemann/

Abstract. Establishing applications based on the popular peer-to-peer (P2P) paradigm in mobile ad hoc networks (MANET) requires the employment of efficient mechanisms for communication and information exchange among peers. In this paper, we show that P2P systems designed for the wireline Internet cannot be deployed in MANET in a straightforward way. As a major shortcoming, these systems construct an overlay network consisting of application-layer connections that do not reflect the network-layer topology of the MANET. Furthermore, overlay construction and maintenance does not consider node mobility. We present two solutions to cope with these shortcomings: P2P communication based on cross-layer cooperation and P2P information exchange by epidemic dissemination. As an example for a system that implements the first concept, we describe the Optimized Routing Independent Overlay Network (ORION). As an example for a system that implements the second concept, we describe Passive Distributed Indexing (PDI). Detailed simulation studies show that both ORION and PDI provide efficient building blocks for the deployment of P2P computing applications in MANET.

1 Introduction

Beside the World Wide Web, peer-to-peer (P2P) file sharing constitutes one of the most popular applications of today's Internet. Driven by users' desire for mobile and wireless communication, deploying P2P systems for mobile networks is of high interest. Mobile nodes equipped with network interfaces for communication without further network infrastructure form a mobile ad hoc network (MANET, [20]). MANET and P2P systems both exhibit a lack of fixed infrastructure and possess no a-priori knowledge of arriving and departing peers. Due to this common nature, P2P systems seem natural and attractive to be deployed for MANET. Interesting application scenarios include sharing traffic and weather data by car-to-car communication in a wide-range MANET, a groupware system for mobile e-learning applications in a local-range MANET using IEEE 802.11 [13] wireless LAN, and sharing music, jingles, video clips etc. from mobile device to mobile device via Bluetooth. However, although similar solutions can be applied to enable self-organization and communication in both P2P systems and MANET, straightforward combination of both approaches multiplies the performance problems of each individual approach.

M.C. Calzarossa and E. Gelenbe (Eds.): MASCOTS 2003, LNCS 2965, pp. 187–208, 2004.
© Springer-Verlag Berlin Heidelberg 2004

The operation of most P2P systems in the wireline Internet depends on application-layer connections among peers, forming an application-layer overlay network. In general, these connections are static, i.e., a connection between two peers remains established as long as both peers dwell in the system. We identify the maintenance of static overlay connections as the major performance bottleneck for deploying a P2P system in a MANET. We find that the overlay network topology does not reflect the underlying MANET topology, neither in terms of connection layout nor in connection lifetime. Whereas the overlay network topology is static in the timescale of a node's dwell time in the system, the MANET topology changes much more frequent due to node mobility. This induces significant control overhead for connection maintenance, resulting in increasing network traffic and decreasing communication efficiency. Motivated by this observation, we present two different approaches for implementing efficient P2P communication in MANET that do not suffer from the shortcomings of static application-layer overlay connections. Such approaches can be classified as (1) P2P communication based on cross-layer cooperation, or (2) P2P information exchange by epidemic dissemination.

As example for an approach following paradigm (1), we describe the Optimized Routing Independent Overlay Network (ORION, [17]), which provides all communication mechanisms required to implement a MANET-based P2P file sharing system. ORION comprises of an algorithm for construction and maintenance of an application-layer overlay network that enables routing of all types of messages required to operate a P2P file sharing system, i.e., queries, responses, and file transmissions. Overlay connections are set up on demand and maintained only as long as necessary, closely matching the current topology of the underlying network. ORION follows a cross-layer approach, i.e., it combines application-layer query processing with the network-layer process of route discovery, substantially reducing control overhead and increasing search accuracy compared to an off-the-shelf approach utilizing a P2P system for the wireline Internet based on TCP and a state-of-the-art MANET routing protocol. Additionally, the ORION overlay network enables low overhead file transfer, as well as an increasing probability for successful file transfers compared to the off-the-shelf approach.

As an example for an approach following paradigm (2), we describe Passive Distributed Indexing (PDI, [18], [19]), which implements epidemic information dissemination among nodes in a MANET. PDI constitutes a general-purpose distributed lookup service for mobile P2P applications, which supports resolution of application-specific keys to application-specific values. Key-to-value resolution is similar to the functionality provided by wireline P2P systems, e.g., Gnutella [7], CAN [23], and Chord [26]. As building block, PDI stores index entries in form of (key, value) pairs in index caches located at each mobile device. Index entries are propagated by epidemic dissemination, i.e., they are transmitted between devices that get in direct contact, similar to the spread of an infections disease. By exploiting node mobility, such contacts occur randomly between arbitrary devices. PDI can resolve most queries locally without sending messages outside the radio coverage of the inquiring node. Thus, PDI effectively reduces network traffic for the resolution of keys to values for applications possessing a sufficiently high degree of temporal locality in their query streams as known for web content, Internet search engines and P2P file sharing systems [28], [27]. To foster information dissemination for devices with limited radio transmission range, PDI features a bandwidth-efficient message relaying mechanism. For keeping index caches coherent, PDI features two novel consistency mechanisms implementing explicit and implicit invalidation.

The remainder of this paper is organized as follows. Section 2 discusses the shortcomings of P2P systems using static application-layer overlay connections for the deployment of a MANET-based P2P system. Section 3 presents ORION as an example for a P2P system using a cross-layer approach, while Section 4 presents PDI as an example for a system based on epidemic information dissemination. Finally, concluding remarks are given.

2 Approaches for P2P Systems in MANET

2.1 Wireline Peer-to-Peer Systems

File sharing constitutes the most popular application of P2P technology in the wireline Internet. In general, a P2P file sharing system consists of two building blocks: a search algorithm for transmitting queries and search results and a file transfer protocol for downloading files matching a query. Whereas most file sharing systems transfer files directly between peers using TCP connections, efficient searching is an active area of research for wireline P2P systems.

The P2P system Gnutella [7] released by AOL affiliate Nullsoft in 2000 constitutes the first system implementing a fully distributed file search. Queries are broadcasted to all peers using an application-layer overlay network. Despite its poor performance, Gnutella gained rapidly increasing popularity. Several recent approaches substantially improved the scalability of the search process by reducing the number of messages generated to resolve a user query. Aberer et al. propose P-Grid, a virtual binary search tree, to route query messages to a number of nodes, which are responsible to answer these queries [1]. Each peer in a P-Grid has to maintain application-layer connections to two other peers. Other state-of-the-art P2P systems like CAN [23] and Chord [26] implement distributed hash tables. These systems allow queries for keys, which are resolved by routing the query to a peer storing the value matching this key. For systems with n peers, a query can be resolved involving $O(\log n)$ (or $O(n^{\alpha})$ for $\alpha < 1$) intermediate routing hops. Each node in CAN has to maintain connections to $2d$ neighbors, where d is a configuration parameter. Chord maintains connections to $O(\log n)$ neighbors. Thus, like Gnutella, all state-of-the-art P2P systems build overlay networks with static connections between participating peers.

2.2 Shortcomings of an Off-the-Shelf Approach for MANET

Deploying an off-the-shelf P2P file sharing system on top of a MANET routing protocol, e.g. Dynamic Source Routing (DSR, [15]) or Ad hoc On-demand Distance Vector routing (AODV, [21]), seems a straightforward approach towards deployment of P2P applications in MANET. However, though attractive because of its off-the-shelf

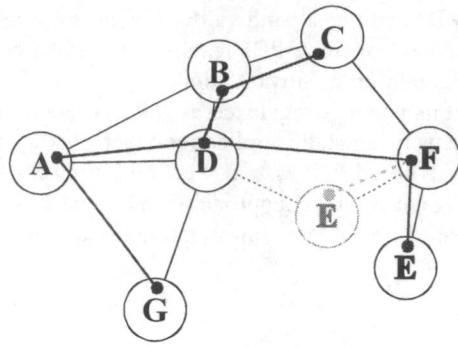

Fig. 1. Link-layer versus application-layer connections.

nature, this approach reveals several shortcomings due to the dynamics of the mobile environment. Recall that each host participating in the static overlay system has to maintain an application-layer connection to at least one peer for all state-of-the-art approaches. The maintenance of a static application-layer connection between two arbitrary peers in a mobile environment reveals several shortcomings, as illustrated in Figure 1. Mobile nodes are drawn as circles named by capital letters A to G. If two nodes can communicate with each other using a wireless connection, a thin line connects the corresponding circles. E.g., there is a direct connection between nodes D and G, while nodes A and G cannot communicate directly over a wireless link. We refer to a direct wireless connection as link-layer connection, while a sequence of link-layer connections as provided by a routing protocol is denoted as network-layer connection. Independent of direct connectivity, nodes can establish application-layer connections, which form the application-layer overlay network. These connections are shown as thick lines in Figure 1. Note that an application-layer connection can be established across several hops, i.e., spanning several link-layer connections. E.g., nodes A and G establish an application-layer connection, which uses the sequence of link-layer connections A-D-G for IP communication. By analyzing scenarios as shown in Figure 1 we identify the following shortcomings of static overlay systems for MANET:

Inefficient usage of link-layer connections: Most overlay construction mechanisms are unaware of link-layer connections and changing node positions. Thus, a message sent on an application-layer connection between two nodes may traverse a particular link-layer connection several times. E.g., consider a message sent from node C to node F in Figure 1. Until it finally reaches node F, the query will be forwarded using the application-layer connections C–B, B–D and D–F. Considering the underlying network-layer connections, we find that the message will traverse the link-layer connections D-B and B-C twice. This phenomenon, already observed for connections between fixed hosts in wireline networks [24], has a dramatic impact in wireless environments due to scarce radio bandwidth and energy resources. Furthermore, many responses for identical documents matching a query may traverse the same link-layer connection, wasting resources without extending the search results. E.g., consider a scenario where all nodes apart from E store an identical document matching a query sent out by E. In this case, the same information is transmitted over the link-layer connection F-E six times. As we will show in Section 2.3, inefficient usage of link-layer connections will result in significantly increasing transmitted message volume.

High Routing Overhead: Maintaining an application-layer connection between two nodes in a mobile environment requires some efforts on the network-layer. Node movement forces the routing protocol to permanently update and repair broken routes, generating a high amount of route request, reply and repair messages. E.g., consider the application-layer connection between nodes D and F in Figure 1, which has been established over the network-layer connection D-E-F when node E was located at the position indicated in gray. When node E moves to the final position, a route discovery must be initiated to maintain a network route for the application-layer connection between D and F. Note that although network routes may change in wireline networks, this event is even more common in MANET, due to node mobility. As we will show in Section 2.3, routing overhead is a serious problem for systems maintaining static application-layer connections.

Frequent loss of application-layer connections: In general, the ad hoc routing protocol may fail to maintain a connection between two nodes due to node movement. In the worst case, this will lead to a disconnection of a single node from all application-layer neighbors. E.g., consider that the link-layer connections A-B and A-D in Figure 1 fail. In this case, G will lose the application-layer connection to A, and with it the entire connectivity to the network, although it remains connected to D on the link-layer. Recall that if a peer loses all application layer connections to other peers, it has to employ a mechanism to discover initial peers. Although peers may well lose connection to the application-layer network in wireline P2P systems, this scenario is much more common in MANET, again, due to node mobility. As we will show in Section 2.3, traffic generated by initial peer discovery will result in significant control overhead.

2.3 Performance Evaluation of the Off-the-Shelf Approach

To illustrate the impact of each of the shortcomings described above, we conduct a simulation study of an off-the-shelf approach using a P2P system from the wireline Internet which utilizes TCP and a MANET routing protocol. Since almost all known P2P file sharing systems use static connections to build an overlay network we consider Gnutella as example for such P2P systems. To complete the off-the-shelf approach, we employ TCP-SACK as transport-layer protocol on top of the MANET routing protocol DSR. Performance results are obtained using the Network Simulator ns-2 [6]. For a detailed description of the simulation environment we refer to [17]. We simulate 40 mobile nodes moving in an area of 1000 m × 1000 m according to the random waypoint mobility model [2] with a maximum speed of 2 m/s and a pause time of 50 s between two successive epochs of movement. This mobility model is well suited to mimic the movement of pedestrians. Each mobile node shares 100 files and generates queries for files shared by remote peers in a mean interval of 120 s. In our simulations, we focus on the search algorithm and do not consider file transfers, as Gnutella transfers files without employment of the overlay network using direct TCP connections.

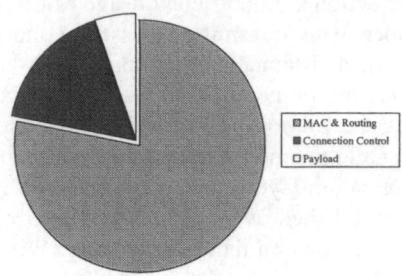

Fig. 2. Breakdown of message volume for an off-the-shelf P2P system in MANET

During the simulations, we record the volume of all messages transmitted by the mobile nodes. We brake down the total message volume into three categories depending on the message types. Gnutella query and response messages are classified as payload, as they constitute the only message types that are used for searching. Gnutella connection request, ping and pong messages are used to establish and maintain overlay connections and are therefore classified as connection control. All messages used by DSR for establishment and maintenance of network routes as well as the request-to-send (RTS), clear-to-send (CTS) and acknowledgement generated by the IEEE 802.11 medium access control mechanisms are classified as mac & routing.

The breakdown of message volume for a system consisting of 40 nodes is presented in Figure 2. We find that the payload constitutes only 5.2% of the transmitted message volume. Connection control for maintenance of application-layer connectivity constitutes 16.6% of the volume. The major share of 78.2% is consumed by MAC & routing. These results confirm the argumentation presented in Section 2.2. We conclude from Figure 2 that for successful deployment of P2P systems in MANET more sophisticated approaches than employment of off-the-shelf components are required.

3 P2P Communication Using Cross-Layer Information Transfer

3.1 Basic Concepts

As discussed in Section 2.2, unawareness of the overlay network construction algorithm to the actual network- and link-layer conditions is a major source of overhead when employing an off-the-shelf P2P system in MANET. As a consequence, efficient deployment of P2P systems in MANET requires close cooperation between the different layers in the network stack. For instance, it is easy to see that the application-layer can benefit from knowledge of network-layer routes by adjusting overlay routes accordingly.

Several current research projects try to exploit cross-layer information transfer. Conti, Giordano, Maselli, and Turi proposed a reference-architecture for cross-layer design of a MANET protocol stack [3]. The architecture uses a shared memory area called "network status", which serves as a repository of network information collected by all network protocols. In the context of P2P communication in MANET, Chen, Shah and Nahrstedt proposed a system for implementing data advertising, lookup and replication based on cross-layer information [4]. In their approach, the network-layer provides a node profile to a middleware-layer, which stores information about a nodes position, velocity, and energy resources. Similar, the middleware-layer provides a data profile to the network-layer that contains information on the quality-of-service requirements of each data flow. Node and data profiles are communicated via shared memory, requiring the employment of a proactive link-state routing protocol that is capable to communicate with the middleware-layer. Schollmeier, Gruber, and Niethammer presented the Mobile Peer-to-Peer Protocol (MPP) suite that spans form the network to the application-layer [25]. The suite tries to reuse existing protocols as far as possible. It uses an enhanced version of DSR called EDSR. P2P applications can register with EDSR and communicate information using a message passing mechanism called Mobile Peer Control Protocol (MPCP). Information passed via MPCP are used to establish application-aware routes that can be employed for data transfers using standard mechanisms, e.g., the hypertext transfer protocol HTTP.

In the remainder of this section, we describe a special-purpose approach for P2P file sharing tailored to MANET denoted Optimized Routing Independent Overlay Network (ORION, [17]). ORION employs cross-layer information and comprises of an algorithm for construction and maintenance of an application-layer overlay network that enables routing of all types of messages required to operate a P2P file sharing system, i.e., queries, responses, and file transmissions. ORION operation does not depend on the deployment or support of any MANET routing protocol. Note that some data structures and mechanisms used by ORION are also provided by reactive MANET routing protocols, e.g., AODV or DSR. Opposed to [3], [4], [25], ORION does not define or employ a reference-architecture for extensive cross-layer information exchange, but implements all required functionality on the application-layer. The only cross-layer information employed by ORION is a link-failure notification from the MAC layer, which is known from implementations of reactive MANET routing protocols. Thus, ORION minimizes the usage of cross-layer information as far as possible, providing maximum independence from other protocols in the network stack.

3.2 The Optimized Routing Independent Overlay Network

To implement ORION, each mobile device maintains a local repository, consisting of a set of files stored in the local file system. ORION provides searching capabilities for all files in the repository. We assume that each file is associated with a unique file identifier. Additionally, ORION maintains two routing tables, a response routing table and a file routing table. Similar to the routing tables used by AODV, the response routing table is used to store the node from which a query message has been received as next hop on the reverse path. Thus, a node is able to return responses to the inquiring node without explicit route discovery. The file routing table is a data structure that stores alternative next hops for file transfers based on the file identifier. ORION updates both the response routing table and the file routing table during query processing. Limiting the maximum size and applying the Least Recently Used (LRU) replacement policy control the memory consumption for both routing tables.

For the query phase, ORION defines two types of messages. A QUERY message contains a query string, which consists of one or more keywords. A message of type RESPONSE contains unique identifiers of one or more files matching a query. To enable controlled message forwarding, each ORION message contains a field SRC, representing the unique identifier of the mobile device that generated the original message. Furthermore, a field SEQ contains a sequence number unique and strictly increasing for each mobile device. The rollover of SEQ numbers is handled as in [21]. When a mobile device relays a message as described below, SRC and SEQ are preserved. Thus, storing the largest SEQ entry received from each source device can prevent relaying a message more than once.

To illustrate the operation of the ORION search algorithm, we consider the four-node scenario shown in Figure 3. There, the boxes near the mobile devices show parts of the local repository (first box), the response routing table (second box), and the file routing table (third box), respectively. Assume node A issues a query matching files 1, 2, 3 and 4 (see Figure 3a). The QUERY message is distributed by link-layer flooding, i.e., application data is piggybacked on a link-layer broadcast message. This technique is borrowed from the simple multicast and broadcast protocol for MANET [14]. On its way through the network the QUERY message sets up reverse paths to node A in the response routing tables of all intermediate nodes. This technique is borrowed from the route discovery process in AODV, in which the flooded route request message sets up the reverse route for the route reply message.

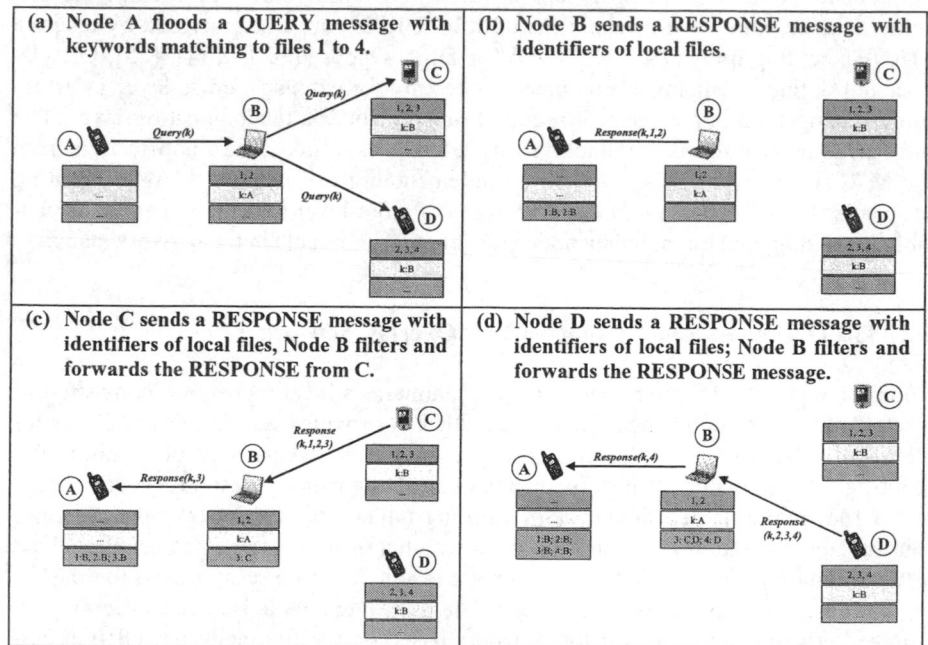

Fig. 3. Illustration of data structures and messages used by the ORION search algorithm

After searching the local repository, node B sends a RESPONSE message containing the identifiers of files 1 and 2 to the next hop in node A's direction, i.e., directly to node A (see Figure 3b). Additionally, node C sends a RESPONSE message with the identifiers of files 1, 2, and 3 in node A's direction, i.e., to node B. Before forwarding the message to node A, node B examines the contained result. File 3 is so far unknown to node B. Therefore, node C is stored as feasible next hop for this file in the file routing table. Node C is not stored as next hop for files 1 and 2, because node B stores these files itself. Therefore, node B will never request any of these two files from node C. Instead of relaying the complete RESPONSE message to node A, node B sends a reduced RESPONSE message to node A containing only the new file identifiers (see Figure 3c). Similar to node C, node D answers to the query with the identifiers of matching local files with a RESPONSE message to the next hop in the direction to node A, i.e., to node B. As before, node B stores node D's files in the file routing table and sends an own additional RESPONSE message to node A. Because node B has already forwarded responses for files 2 and 3, the new RESPONSE message only contains the identifier of file 4 (see Figure 3d). Note that node B not only stores node C as next hop for file 3, but also node D, adding redundancy to the file routing table without additional transmission cost. After the query phase, node A may chose one of the four matching documents for download.

For efficient file transfer, ORION incorporates a transfer protocol tailored to the characteristics of both the MANET and the file sharing application. As first building block, the ORION transfer protocol utilizes the routes given by the file and response routing tables for transmission of control- and data packets. Recall that the file routing table may store several redundant paths to copies of the same file. Due to changing network conditions, the sender of a file might change during a file transfer. Therefore, it is essential to keep the complete control over the transfer at the receiver. For transfer, a file is split into several blocks of equal size. Since the maximum transfer unit of the mobile network is assumed to be equal between all neighboring nodes, the block size can be selected such that the data blocks fit into a single packet. The receiver sends a DATA_REQUEST message for one of the blocks along the path given by the file routing tables. Once the DATA_REQUEST reaches a node storing the file in it's local repository, the node responds with a DATA_REPLY message, containing the requested block of the file. This message is routed back to the requesting node via the same path as the RESPONSE message in the query phase. After receiving the DATA_REPLY message, the receiver continues with a request for the next block until the complete file has been transferred. A scheduling mechanism described below prevents ORION from waiting indefinitely for lost packets.

To illustrate the operation of the file transfer mechanism, suppose node A in Figure 3d decides to download the file with the file identifier 3. The DATA_REQUEST message for the first part of the file is sent to the next hop in direction to the node storing file 3, i.e., node B. Node B cannot deliver the file itself, thus, it relays the request to the best-suited next hop towards a node storing file 3. From node B's point of view, node C is best suited, because in the query phase node C has responded first to the query message. Node C stores a copy of file 3 and, thus, sends a DATA_REPLY message containing the requested part of file 3 to the next hop in direction to node A as identified by the response routing table. Node B relays the DATA_REPLY message to node A. Subsequently, node A sends a DATA_REQUEST for the next block of file 3. This process continues until the complete file has been transferred to node A. In this (ideal) scenario, the file transfer implicitly uses the optimal route without any overhead for retransmissions due to route failures.

ORION transfers control and data packets on the best-suited route chosen from a set of redundant routes. Selecting an alternative route provides an efficient mechanism to locally resolve link failures. Consider again the example in Figure 3d. Suppose the link between node B and node C fails during the file transfer, e.g., because node C moves out of node B's transmission range. As soon as B recognizes the link failure, it deletes node B in its routing tables and forwards subsequent DATA_REQUEST messages to the now best-suited next hop to a node possessing file 3, i.e. node D. Thus, the link failure can locally be resolved by node B without involving other nodes. Note that if a sender becomes unavailable, ORION transparently changes to another sender. In the off-the-shelf approach, a change of the sender must be explicitly triggered by the P2P application. We have shown in extensive simulation studies that changing the sender transparently outperforms explicit changes of the sender in terms of transmission overhead and reliability for a P2P file sharing application [17].

The timely recognition of link failures is necessary to avoid delays and unnecessary data transmissions. Building upon a cross-layer approach, ORION uses feedback from the link-layer as the second building block for an efficient file transfer. Feedback can be provided by link-layer notification if available, e.g., as in IEEE 802.11. Otherwise, packet receipt must be acknowledged by sending explicit application-layer packets or by passive acknowledgements [15]. In the case of a send failure on the link-layer, the ORION application is notified and can immediately chose an alternative next hop as described above.

As a further feature, the route maintenance algorithm utilizes ROUTE_ERROR messages. These ROUTE_ERROR messages indicate that a node could not forward a DATA_REQUEST message, because there are no next hop entries for the requested file left in its file routing table. To illustrate the usage of ROUTE_ERROR messages, we refer again to Figure 3d. Suppose that node B looses the link-layer connections to both node C and node D and receives further DATA_REQUEST messages from node A. In that case, node B sends a ROUTE_ERROR message for file 3 to node A. Subsequently, node A deletes node B as next hop for file 3 in its files routing table. If node A's file routing table contains another next hop for file 3, subsequent DATA_REQUESTs are sent to this node. Otherwise node A can either cancel the file transfer or start a special search phase called re-query. Opposed to an ordinary query, a re-query does not search for keywords, but for a file identifier. The messages transmitted during a re-query are equal to the messages during an ordinary query, thus updating the routing tables for the specific file.

Recall that the ORION transfer protocol as described so far will not continue a file transfer, if a DATA_REQUEST or DATA_REPLY message is lost. To cope with packet losses, the ORION transfer protocol incorporates a packet scheduling and loss-recovery mechanism as third building block. So far, the ORION specification does not elaborate on performance and fairness optimization using sophisticated flow control and congestion avoidance as for example provided by the transmission control protocol, TCP. Recall that TCP tries to transmit data packets with a small delay jitter as a continuous stream. In a file sharing application, the order in which the blocks of a file are received is irrelevant, as long as each block is received at least once. Therefore, ORION maintains a list of pending blocks, which is initialized with all blocks of a particular file. The blocks are requested in a round-robin fashion, until they are successfully transmitted. To determine the minimal request rate, the receiver keeps track of the average round trip time, i.e., the time elapsed between sending a DATA_REQUEST and receiving the corresponding DATA_REPLY. The time between two successive requests is equal to the average round trip time, unless a DATA_REPLY is received before. That is, if a packet is lost or delayed, the next packet will be requested at most one round trip time later. Due to the round-robin selection of blocks, the ORION transfer protocol re-requests a lost or delayed block after all other pending blocks have been requested. Therefore, a delayed block will reach the receiver with high probability before it is requested a second time.

3.3 Performance Evaluation of ORION

To illustrate the effectiveness of ORION's search algorithm and transfer protocol, we conduct a simulation study comparing the performance of ORION with an off-the-shelf approach utilizing a P2P system for the wireline Internet, TCP, and a state-of-the-art MANET routing protocol. Again, the off-the-shelf approach comprises of Gnutella, TCP-SACK and DSR. We label corresponding curves as ORION and Off-the-Shelf, respectively. As in Section 2.3, performance results were obtained using the Network Simulator ns-2. For a detailed description of the simulation environment we refer to [17]. Similar to the experiments described above, mobile nodes move on an area of 1 km^2 according to the random waypoint mobility model at pedestrian speed. For the evaluation of search performance, each mobile node shares 100 files and generates queries for files shared by remote peers in a mean interval of 120 s. To evaluate download performance, a selected peer performs 200 successive downloads of different files, each of size 3 MB.

An important performance measure from the user's point of view is the quality of the results received in response to a query. Therefore, we consider search accuracy, i.e., the fraction of received unique files in relation to the number of all files matching a query. In Figure 4, we plot search accuracy as a function of number of nodes participating in the mobile file sharing application. We find that search accuracy for both applications increase with the number of nodes. This is clearly due to the increasing connectivity. However, the off-the-shelf approach has searching performance worse than ORION and also shows decreasing search accuracy for more than 40 mobile nodes. We conclude from Figure 4 that the off-the-shelf approach fails to provide sufficient search accuracy due to the shortcomings of static overlay connections discussed in Section 2.2.

To illustrate the benefit of changing the sender of a file during a transmission, we investigate the average overhead for a successful file transmission in percent of the total file size of 3 MB. Figure 5 plots the overhead as a function of the number of nodes. For few nodes, in most cases files are transferred between direct neighboring nodes. In this case, most overhead is control overhead. We find that the off-the-shelf approach transmits more than twice the file size over the wireless medium. As shown

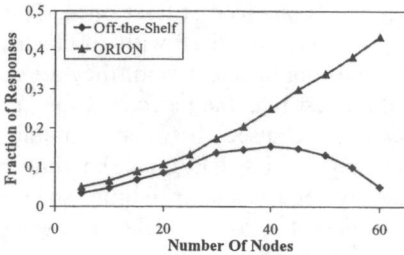

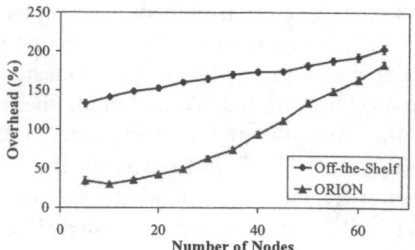

Fig. 4. Search accuracy for different systems sizes

Fig. 5. Overhead for the transfers of a 3 MB file.

by an analysis of the ns-2 trace files, the main reason for this are retransmissions of packets delayed by the route maintenance of DSR, as well as longer transfer routes. Route length increases if a direct connection between the sending and receiving node fails, while a longer route via an intermediate node exists. Recall that in this case, DSR will recover from route failure globally, using the longer route. The scheduling algorithm of ORION reduces the amount of retransmissions of delayed packets. Furthermore, in case of route failure, ORION will resume the transmission from another nearby node using the file routing table. These features lead to a total overhead below 50% for few nodes. For a growing number of nodes, overhead increases for both the off-the-shelf approach and ORION, as the average length of a network route increases. However, ORION stays superior to the off-the-shelf for the same reasons. From Figure 5 we conclude that ORION provides a file transfer mechanism that is highly efficient in bandwidth usage.

4   P2P Information Exchange by Epidemic Data Dissemination

4.1   Basic Concepts

As discussed in Section 2.2, route changes, weak connectivity or even disconnected operation hampers the implementation of searching functionality using static overlay connections in mobile and wireless environments. The smart collaboration of mobile devices in an ad hoc fashion constitutes an attractive alternative for implementing an effective distributed lookup service for such scenarios. As shown by Grossglauser and Tse node mobility does not only hinder communication in MANET, but also supports cost-effective information exchange by epidemic dissemination [9]. To implement epidemic information dissemination, information is transmitted when nodes get in direct contact, similar to the transmission of an infectious disease between individuals. Mathematical models for the spread of epidemic diseases have been widely studied. There exist applications of epidemic algorithms in various fields of computer science, e.g., for the maintenance of replicated databases [5].

As a first approach to epidemic information dissemination in mobile environments, Papadopouli and Schulzrinne introduced seven degrees of separation (7DS), a system for mobile Internet access based on Web document dissemination between mobile users [22]. To locate a Web document, a 7DS node broadcasts a query message to all mobile nodes currently located inside its radio coverage. Recipients of the query send response messages containing file descriptors of matching Web documents stored in their local file caches. Subsequently, such documents can be downloaded with HTTP by the inquiring mobile node. Downloaded Web documents may be distributed to other nodes that move into radio coverage, implementing an epidemic dissemination of information.

Using a related approach, Goel, Singh, Xu and Li proposed broadcasting segments of shared files using redundant tornado encoding [8]. Their approach enables nodes to restore a file, if a sufficient number of different segments have been received from one or more sources. In [16], Khelil, Becker, Tian, and Rothermel presented an analytical model for a simple epidemic information diffusion algorithm inspired by the SPIN-1 protocol [11]. Based on these results, Häner, Becker, and Rothermel present the Negotiation-based Ad Hoc Data Dissemination Protocol (NADD) [10], which uses active data advertisements to negotiate data transmissions between neighboring peers. Both systems [8] and [10] implement a push model for information dissemination. That is, shared data is advertised or even actively broadcasted without a node requesting it. Recently, Hanna, Levine, and Mamatha proposed a fault-tolerant distributed information retrieval system for P2P document sharing in MANET [12]. Their approach distributes the index of a new document to a random set of nodes when the document is added to the system. The complete index of a document, i.e., all keywords matching it, constitutes the smallest unit of disseminated information.

In the remainder of this section, we describe a general-purpose distributed lookup service for mobile applications, denoted Passive Distributed Indexing (PDI, [18], [19]). PDI supports resolution of application-specific keys to application-specific values. As building block, PDI stores index entries in form of (key, value) pairs in index caches located at each mobile device. Index entries are propagated by epidemic dissemination. As for PDI, a sufficiently high degree of locality in user queries is essential for all related approaches [22], [8], [10], and [12]. Opposed to 7DS [22], PDI implements document search based on a distributed index rather than immediate document sharing. In fact, PDI disseminates individual index entries whereas 7DS disseminates entire documents. Opposed to [8] and [10], PDI implements a pull model and does not advertise data yielding a substantial reduction of traffic over wireless links. In contrast to [12], PDI disseminates pairs of a single key and one matching value rather than the entire index of a document. Furthermore, opposed to all related work, PDI cannot only be employed for sharing documents, but also for numerous other mobile applications.

4.2 Passive Distributed Indexing

In this section, we describe the functionality of PDI. We assume a system consisting of several mobile nodes, e.g. mobile users equipped with notebooks or PDAs and wireless network interfaces as illustrated in Figure 6. All mobile nodes collaborate via a shared application that uses a distributed lookup service. Radio coverage is small

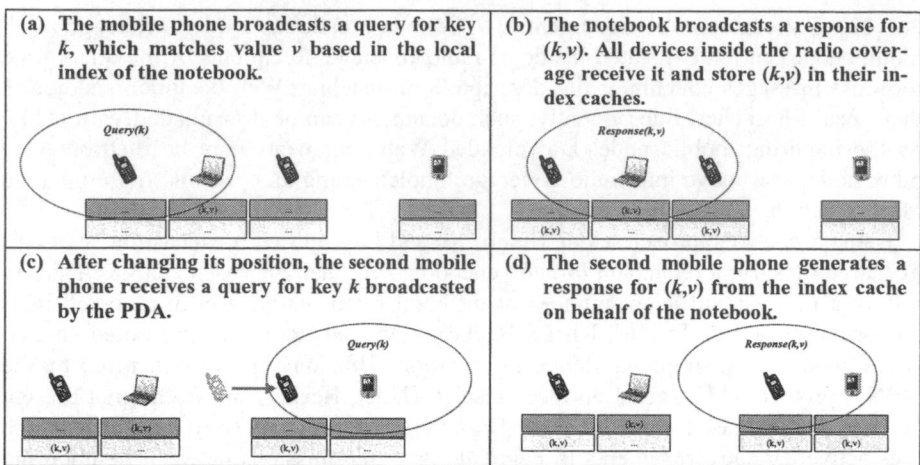

Fig. 6. Illustration of epidemic information dissemination with PDI.

compared to the area covered by all nodes, so that most nodes cannot contact each other directly. Thus, communication may be performed using several intermediate hops as in MANET. Subsequently, we assume IEEE 802.11 as underlying radio technology. However, we would like to point out that PDI could be employed on any radio technology that enables broadcast transmissions inside a node's radio coverage.

PDI implements a general-purpose lookup service for mobile applications. In general, PDI stores index entries in the form of pairs (k,v). Keys k and values v are both defined by the mobile application. For example, in case of file sharing, keys are given by keywords derived from the file name or associated meta data. Values are given by references to files in form of URIs. Opposed to distributed hash-table systems [23], [26], PDI does neither limit the number of keys matching a value nor the number of values matched by a key. However, some mechanisms implemented in PDI require that a value is unique in the system, i.e., it is only added to the system by a single node. This can be easily achieved by extending the application specific value v by a unique node identifier i_n for node n. For example, the node identifier i_n may be derived from the node's IP address or the MAC address of the radio interface. For ease of exposition, we will abbreviate the unique value given by (v, i_n) pairs just by v.

A node n may contribute index entries of the form (k,v) to the system by inserting them in a local index. In Figure 6, the local index is drawn as the first box below each mobile device. We refer to such an index entry as supplied. The node n is called the origin node of an index entry. For example, the notebook shown in Figure 6 is the origin node of the index entry (k,v). A key k matches a value v, if (k,v) is currently supplied to the PDI system. Each node in the system may issue queries in order to resolve a key k to all matching values v_i (see Figure 6a). A node issuing a query is denoted as *inquiring node*.

Queries are transmitted by messages containing the key to resolve. Query messages are sent to the IP limited broadcast address 255.255.255.255 and a well-defined port. Thus, all nodes located inside the radio coverage of the inquiring node receive a query message. Each of these nodes may generate a response message. A response message contains the key from the query and all matching values from either the local index or a second data structure called index cache. To enable epidemic data dissemination, PDI response messages are sent to the IP limited broadcast address 255.255.255.255 and a well-defined port, too. Thus, all mobile nodes within the radio coverage of the responding node will overhear the message (Figure 6b). Not only the inquiring node but also all other mobile nodes that receive a response message extract all index entries and store them in the index cache (see Figure 6b). In Figure 6, index caches are drawn as the second box below mobile devices. Index entries from the index cache are used to locally resolve queries, if the origin nodes of matching values reside outside the radio coverage of the inquiring node (see Figures 6c and 6d). Obviously, the index cache size is limited to a maximum number of entries adjusted to the capabilities of the mobile device. The replacement policy least-recently-used (LRU) is employed if a mobile device runs out of index cache space. Note that information is disseminated to all other nodes that are in direct contact, similar to the spread of an infectious disease. Due to the movement of nodes and overhearing response messages of neighboring nodes, index entries are disseminated within the network without costly global communication. In fact, PDI builds and maintains an index distributed among mobile node of the MANET in a passive way.

Note that PDI might return no results to a query for key k, even if an index entry (k,v) is currently supplied. This may occur, if the corresponding query neither reaches the origin node of the index entry (k,v), nor another node storing (k,v) in its index cache. We refer to such unresolved query as false miss. For the application using PDI, there are two ways to resolve false misses: First, the application can tolerate and simply ignore them. Second, the application has to resort to a centralized lookup service via the cellular infrastructure of a mobile network as fallback. Since typical query streams for lookup services possess a high degree of temporal locality, e.g. as observed for P2P file sharing systems [27] and Internet search engines [28], PDI will produce only few false misses for such applications.

Recall that all PDI messages are send to the limited broadcast address and received by all nodes located inside the radio coverage of the sender. Depending on the transmission range of the wireless network interfaces, this may considerably limit the number of nodes that receive a message. To overcome the limitations of small transmission ranges, packet forwarding is a common technique in mobile ad hoc networks. Using packet forwarding, packets are exchanged between source and destination using one ore more intermediate hops as relays. Several approaches for multi-hop communication in MANET have been proposed. Sophisticated protocols for unicast or multicast communication use control messages to set up sequences of nodes (or routes) that are used by the actual data packets. Using such stateful approach is out of scope for a distributed lookup service as described above for two reasons. First, communication occurs infrequently, so that most routes will only be used for a single query. Second, size of PDI messages is roughly equal to the size of control messages, resulting in a high relative overhead for route setup. A stateless approach to one-to-many communication constitutes flooding. For flooding a packet through the network, each node that receives the packet will forward it exactly once. Since every neighbor of the node will receive the packet and forward it, uncontrolled flooding can generate an infinite number of duplicate transmissions of the same packet. To avoid duplicate

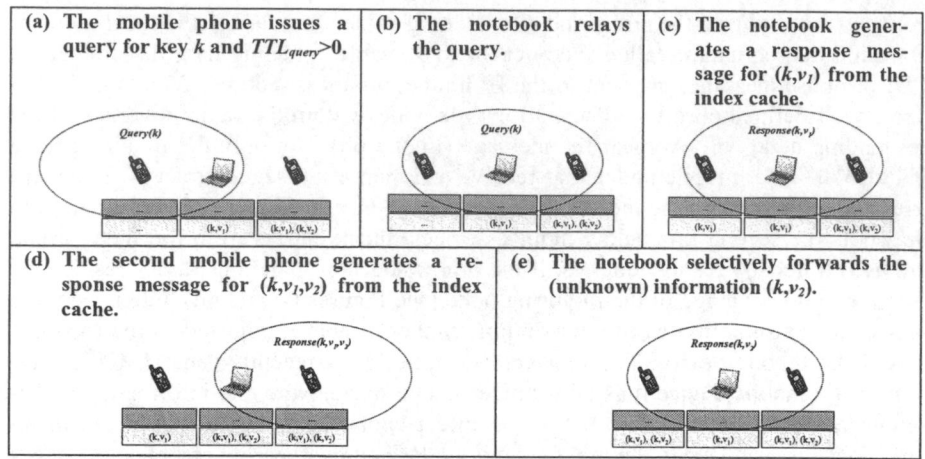

Fig. 7. Message forwarding in PDI

transmissions, a unique tag is included in every packet, e.g., given by a unique source identifier (ID) and a sequence number. Thus, a node can check on packet reception whether it has received this packet already and discard duplicates. Furthermore, the scope of flooding may be limited by including a time-to-live (TTL) field into each packet, which indicates the maximum number of hops the packet may be relayed. A node will decrement the TTL field on packet reception, and will not forward the packet if TTL is equal to 0.

PDI includes a customized flooding mechanism to extend the dissemination of information beyond the radio coverage of the inquiring node. The mechanism is illustrated in Figure 7. Query messages are flooded with a TTL TTL_{query}, which is specified by the inquiring node. For detecting duplicate messages, each massage is tagged with a unique source ID and a sequence number as described above. We will show in Section 4 that $TTL_{quer} \leq 2$ yields sufficient performance in most scenarios, thus, PDI communication remains localized despite of the flooding mechanism.

Similarly to query messages, response messages are forwarded with time-to-live TTL_{query}. Recall that the payload of query messages consists of a few keys. Thus, query messages are small and may be flooded without significantly increasing network load (see Figure 7a and 7b). In contrast, response messages can contain numerous values that may each have a considerable size, depending on the application using PDI. Therefore, flooding of complete response messages will significantly increase network load, even if the scope of flooding is limited to two hops. For the cost-efficient flooding of response messages, PDI incorporates a concept called selective forwarding. That is each node that receives a response message will search the index cache for each index entry contained in the message (see Figure 7d). If an entry is found, in most cases the node itself has already responded with this index entry (e.g. as shown in Figure 7c). Therefore, forwarding this index entry constitutes redundant information. Using selective forwarding, each relay node removes all index entries found in its local index cache from the response message, before the message is forwarded (see Figure 7e).

The basic concepts of PDI as described so far do not take into account intermittent connectivity and spontaneous departures of nodes; circumstances under which all information previously supplied become stale. Examples of these cases include node failure or nodes leaving the area covered by the system. In such cases, an implicit invalidation mechanism can achieve cache coherency. To cope with this problem, value timeouts limit the time for which any index entry (k,v) with a given value v will be stored in an index cache. By receiving a response from the origin node of (k,v), the corresponding value timeout will be reset. Let $a_{(k,v)}$ be the time elapsed since (k,v) has been extracted from a response message generated by its origin node. We define the age a_v of value v as $a_v = \min_k(a_{(k,v)})$, i.e., the time elapsed since the most recent response message of this kind was received. If at a node holds $a_v > T$ for the given timeout value T, all pairs (k,v) are removed from its index cache. PDI implements only one timeout per value v rather than an individual timeout for each index entry (k,v). This is because in most applications the fact that one index entry (k,v) for a given v expires indicates a substantial change of the value. Subsequently, all other index entries (k',v) are likely to be influenced. For example, in a file sharing system, a pair (keyword_i, URI) is removed when the file specified by URI is withdrawn from the system. Thus, all other pairs (keyword_j, URI) also become stale. Note that depending on the application the concept of value timeouts can be easily extended to individual timeout durations T_v for each value v. Such duration may be included in a response message generated by the origin node. For ease of exposition, we assume a global timeout value T for all values in the system in the remainder of this paper.

To determine the current age of a value, an age field is included in the response message for each value. This age field is set to zero in each response from the origin node. When receiving a response message, a node n extracts the age of each value and calculates the supply time s_v. That is the time at which a response for this value was generated by the origin node. Assume that the response message contains age a_v, then s_v is determined by $s_v = c_n - a_v$, where c_n denotes the local time of node n. s_v is stored

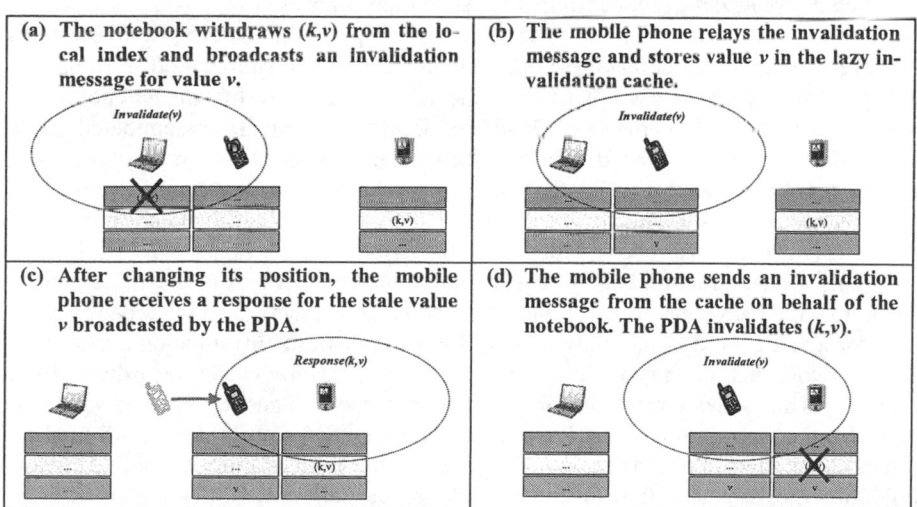

(a) The notebook withdraws (k,v) from the local index and broadcasts an invalidation message for value v.	(b) The mobile phone relays the invalidation message and stores value v in the lazy invalidation cache.
(c) After changing its position, the mobile phone receives a response for the stale value v broadcasted by the PDA.	(d) The mobile phone sends an invalidation message from the cache on behalf of the notebook. The PDA invalidates (k,v).

Fig. 8. Epidemic dissemination of invalidation messages using lazy invalidation caches.

in the index cache together with v. Note that v might already be present in the index cache with supply time s'_v. The copy in the index cache might result from a more recent response by the origin node, i.e., $s_v < s'_v$. Thus, in order to relate the age of a value to the most current response from the origin node, the supply time is updated only if $s_v > s'_v$. When a node generates a response for a cached index entry (k,v), it sets the age field for each value v to $a_v = c_n - s_v$. Note that only time differences are transmitted in PDI messages, eliminating the need for synchronizing clocks of all participating nodes.

Additional to the scenarios described above, a node produces stale index entries by modifying information. That is the case when an index entry is removed from the local index. In a worst-case scenario, a node suddenly leaves the system and all index entries supplied by the node expire at the same time. One way to handle such modification of information is to wait until the timeouts of the values in the stale index entries elapse. Depending on the application and the timeout duration T, this straightforward solution may cause severe inconsistency, especially if T is large. A more effective way to handle information modification in distributed applications constitutes the explicit invalidation by control messages. As basic idea of PDI's explicit invalidation mechanism, a node removes all index entries (k,v) from the index cache when it receives an invalidation message for value v. Flooding is a straightforward way to propagate invalidation messages. To flood an invalidation message for value v, the node removing an index entry sends the message to the limited broadcast address. All mobile nodes that receive the message will relay it exactly once, so that the message is propagated to each node that is connected to the initial node via one or more hops. Unfortunately, in mobile systems even a multi-hop connection between two nodes frequently does not exist. Subsequently, stale index entries are still contained in the index caches of nodes not reached by the invalidation message. Note that these index entries will be redistributed in the system due to the epidemic dissemination. We have shown that even repeated flooding of invalidation messages does not significantly reduce the number of hits for stale index entries [19].

This observation is consistent with [5], which reports that deleted database items "resurrect" in a replicated database environment due to epidemic data dissemination. In [5], a solution is proposed that uses a special message to testify the deletion of an item, denoted as death certificate. Death certificates are actively disseminated along with ordinary data and deleted after a certain time. In contrast, we propose a mostly passive (or "lazy") approach for the epidemic propagation of invalidation messages, which is illustrated in Figure 8. For the initial propagation of an invalidation message by the origin node, we rely on flooding as described above (Figure 8a). Each node maintains a data structure called lazy invalidation cache, which is drawn as a third box below the mobile devices in Figure 8. When a node receives an invalidation message for a value v it does not only relay it, but stores v in the invalidation cache (Figure 8b). Note that an entry for v is stored in the invalidation cache, regardless if the node stores any index entry (k,v) for v in the index cache. Thus, every node will contribute to the propagation of invalidation messages, so that distribution of information and invalidation messages is separated. To enable the epidemic propagation of the invalidation message, a node scans the invalidation cache for all values contained in an overheard response message (Figure 8c). If a value v' is found, the node will generate an invalidation message for v' itself, because the hit in the invalidation cache indicates that the index cache of a nearby node contains a stale entry (Figure 8d). The invalidation message is not flooded through the entire network, but only with a certain scope TTL_{inv} similar to forwarding query and response messages as described in Section 3.2. A node that receives a cached invalidation message will store the included value v in the invalidation cache, and remove all index entries (k,v) from the index cache. Additionally, the node checks whether it has recently received hits for v in response to an own query, which must also be invalidated and may not be passed to the application using PDI.

As the index cache size, the invalidation cache size is limited to a fixed number of values and LRU replacement is employed. We have shown that setting the invalidation cache size to a fraction below 20% of the index cache size achieves sufficient reduction of false hits assuming a reasonable rate of data modification [19]. Note that LRU replacement does neither guarantee that an invalidation cache entry is kept until all stale index entries are invalidated, nor that it is removed after a certain time, inhibiting a node indefinitely from restoring a value it has once invalidated. Increasing the invalidation cache size solves the first problem, though, doing so amplifies the second problem. To avoid this tradeoff, storing the supply time of invalidation messages similar to the supply time of values yields an efficient mechanism to decide whether a result for a value is more recent than an invalidation message.

4.3 Performance Evaluation of PDI

To illustrate the effectiveness of PDI, we conduct a simulation study using the Network Simulator ns-2. For a detailed description of simulation environment we refer to [19]. Similar to the experiments described above, mobile nodes move on an area of 1 km^2 according to the random waypoint mobility model at pedestrian speed. Each node contributes 16 values to the system and sends queries for remote keys in a mean interval of 120 s. To incorporate consistency issues due to node departures into our simulation model, a fraction of approximately 30% of the nodes leave the system in Poisson-distributed intervals. Similarly, new nodes enter the system in equally distributed intervals with empty index caches and fresh values. To take data modification into account, each value expires once in a simulation and is replaced by another value. We investigate the performance of an integrated approach combining both value timeouts and lazy invalidation caches. The duration of the value timeout is fixed to 1000s and the invalidation cache size to 128 entries.

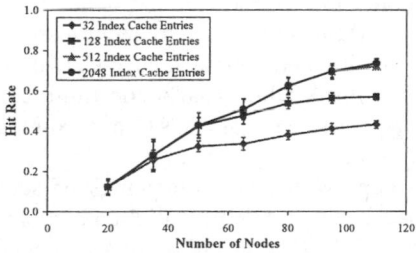
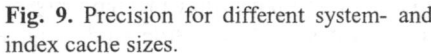
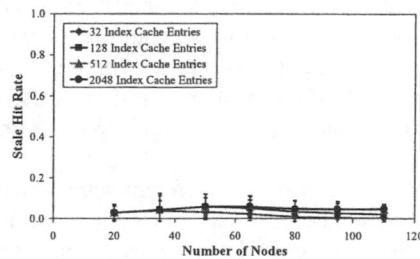

Fig. 9. Precision for different system- and index cache sizes.

Fig. 10. Coherence for different system- and index cache sizes.

We choose performance measures to evaluate the quality and the coherence of the results delivered by PDI and the impact of the introduced invalidation mechanisms. Quality is measured by the hit rate HR, i.e., $HR = H_F / K_F$ for H_F denoting the number of up-to-date hits and H_F the total number of all up-to-date matching values currently in the system. Note that hit rate can be compared to the information retrieval measure recall. Coherence is measured by the stale hit rate SHR, i.e., $SHR = H_S / (H_S + H_F)$, where H_S denotes the number of stale hits returned on a query. Note that stale hit rate is related to the information retrieval measure precision by $precision = 1 - SHR$. To evaluate the impact of the consistency mechanisms on hit rate, we investigate hit rate versus system size for such scenarios in an experiment not shown here. Figure 9 plots hit rates versus system sizes. We find that hit rate is reduced for small systems due to invalidations for up-to-date index entries by value timeouts. This leads to a decrease of at most 20%. The performance of index cache sizes of both 512 and 2048 is equal because a large cache cannot benefit from long-term correlations between requests due to the short timeout. For growing system sizes, the hit rate converges towards results without an invalidation mechanism. We conclude form Figure 9 that PDI provides sufficient hit rate in systems with reasonable node density even when consistency mechanisms are enabled.

As compensation of the reduction of hit rate, the stale hit rate is significantly reduced compared to a system without invalidation. As shown in Figures 10, the stale hit rate is highest for medium system size and sufficient large index cache sizes. The reason is that a fixed cache size of 128 entries is somewhat to large for small systems, while for large systems the natural limit of stale index entries reduces stale hit rate. We conclude from Figures 10 that the integrated approach comprising of the introduced implicit and explicit invalidation mechanisms can effectively handle both spontaneous node departures and modification of information. In fact, for large index caches, the stale hit rate can be reduced by more than 85%. That is, more than 95% of the results delivered by PDI are up-to-date.

5 Conclusion

In this paper, we showed that P2P systems for mobile ad hoc networks (MANET) cannot be deployed by using off-the-shelf P2P systems developed for the wireline Internet. As a major shortcoming, these systems build up overlay networks consisting of static application-layer connections. Such connections suffer from inefficient usage of link-layer connections, require high routing overhead and lead to frequent losses of application-layer connectivity. Simulation of a off-the-shelf approach have shown that only 5.2% of the generated traffic is used for payload, while 16.6% constitute connection control messages and 78.2% constitute messages from MAC and routing-layer.

We presented two different approaches to cope with the shortcomings of static overlay connections. As an example for approaches that implement P2P communication based on cross-layer cooperation, we presented the Optimized Routing Independent Overlay Network ORION. ORION provides all communication mechanisms required to implement a P2P file sharing system in MANET. Simulation studies have shown that ORION outperforms an off-the-shelf approach to P2P file sharing in terms of search accuracy and file transfer overhead. As an example for approaches that implement P2P information exchange by epidemic dissemination, we presented Passive Distributed Indexing PDI. PDI constitutes a general-purpose distributed lookup service for mobile applications. Simulation studies have shown that PDI resolves up to 80% of all keys correctly while providing more than 95% up-to-date results. We concluded from the simulation studies that both ORION and PDI provide efficient building blocks for the deployment of P2P applications in MANET.

References

1. K. Aberer, M. Punceva, M. Hauswirth, and R. Schmidt, Improving Data Access in P2P Systems, *IEEE Internet Computing* **6**(1), 58-67, 2002.
2. J. Broch, D. Maltz, D. Johnson, Y.-C. Hu, and J. Jetcheva, A Performance Comparison of Multi-Hop Wireless Ad Hoc Network Routing Protocols, *Proc. 6th ACM/IEEE MobiCom 98*, Dallas, TX, 85-97, 1998.
3. M. Conti, S. Giordano, G. Maselli, and G. Turi, MobileMAN: Mobile Metropolitan Ad hoc Networks, *Proc. 8th Int. Conf. On Personal Wireless Communications (PWC 2003)*, Venice, Italy, 169-174, 2003.
4. K. Chen, S. Shah, and K. Nahrstedt, Cross-layer Design for Data Accessibility in Mobile Ad Hoc Networks, *Wireless Personal Communications* **21**(1), 49-76, 2002.
5. A. Demers, D. Greene, C. Hauser, W. Irish, J. Larson, S. Shenker, H. Sturgis, D. Swinehart, and D. Terry, Epidemic Algorithms for Replicated Database Maintenance, *Proc. 6th Symp. on Principles of Distributed Computing (PODC 1987)*, Vancouver, Canada, 1-12, 1987.
6. K. Fall and K. Varadhan (editors), *The ns-2 manual*, Technical Report, The VINT Project, UC Berkeley, LBL, and Xerox PARC, 2003.
7. Gnutella Developer Forum, *Gnutella – A Protocol for a Revolution*, 2003. http://rfc-gnutella.sourceforge.net
8. S. Goel, M. Singh, D. Xu, and B. Li, Efficient Peer-to-Peer Data Dissemination in Mobile Ad-Hoc Networks, *Proc. Int. Workshop on Ad Hoc Networking (IWAHN 2002)*, Vancouver, BC, 2002.
9. M. Grossglauser and D. Tse, Mobility Increases the Capacity of Ad-hoc Wireless Networks, *IEEE/ACM Trans. on Networking* **10**, 477-486, 2002.
10. J. Hähner, C. Becker, and K. Rothermel, A Protocol for Data Dissemination in Frequently Partitioned Ad-Hoc Networks, *Proc. IEEE International Symposium on Computers and Communications (ISCC 2003)*, Antalya, Turkey, 633-640, 2003.
11. W. Heinzelman, J. Kulik, and H. Balakrishnan, Adaptive Protocols for Information Dissemination in Wireless Sensor Networks, *Proc 5th ACM/IEEE MobiCom 99*, Seattle, WA, 174-185, 1999.
12. K. Hanna, B. Levine, and R. Manmatha, Mobile Distributed Information Retrieval For Highly-Partitioned Networks, *Proc. 11th IEEE Int. Conf. on Network Protocols (ICPN 2003)*, Atlanta, GA, 2003.
13. IEEE Computer Society LAN MAN Standards Committee, *Wireless LAN Medium Access Control (MAC) and Physical Layer (PHY) Specifications*, IEEE Standard 802.11-1997, New York, NY, 1997.
14. J. Jetcheva, Y. Hu, D. Maltz, and D. Johnson, *A Simple Protocol for Multicast and Broadcast in Mobile Ad Hoc Networks*, IETF Internet Draft (work in progress), July 2001. http://www.ietf.org/proceedings/01dec/I-D/draft-ietf-manet-simple-mbcast-01.txt

15. D. Johnson, D. Maltz, and Y. Hu, *The Dynamic Source Routing Protocol for Mobile Ad Hoc Networks (DSR)*, IETF Internet Draft (work in progress), April 2003. *http://www.ietf.org/internet-drafts/draft-ietf-manet-dsr-09.txt*
16. A. Khelil, C. Becker, J. Tian, and K. Rothermel, An Epidemic Model for Information Diffusion in MANETs, *Proc. 5th ACM Int. Workshop on Modeling, Analysis and Simulation of Wireless and Mobile Systems (MSWiM 2002)*, Atlanta, Georgia, 2002.
17. A. Klemm, C. Lindemann, and O. Waldhorst, A Special-Purpose Peer-to-Peer File Sharing System for Mobile Ad Hoc Networks, *Proc. IEEE Semiannual Vehicular Technology Conference (VTC2003-Fall)*, Orlando, FL, 2003.
18. C. Lindemann and O. Waldhorst, A Distributed Search Service for Peer-to-Peer File Sharing in Mobile Applications, *Proc. 2nd IEEE Conf. on Peer-to-Peer Computing (P2P 2002)*, Linköping, Sweden, 71-83, 2003.
19. C. Lindemann and O. Waldhorst, Consistency Mechanisms for a Distributed Lookup Service supporting Mobile Applications, *3rd Int. ACM Workshop on Data Engineering for Wireless and Mobile Access (MobiDE 2003)*, San Diego, CA, 61-68, 2003.
20. IETF Working Group MANET, *Mobile Ad-hoc Networks (MANET) Charter*. *http://www.ietf.org/html.charters/manet-charter.html*
21. C. Perkins, E. Royer, and S. Das, *Ad hoc On-Demand Distance Vector (AODV) Routing*, IETF RFC 3561, 2003. *ftp://ftp.rfc-editor.org/in-notes/rfc3561.txt*
22. M. Papadopouli and H. Schulzrinne, Effects of Power Conservation, Wireless Coverage and Cooperation on Data Dissemination among Mobile Devices, *Proc. 2nd ACM MobiHoc 2001*, Long Beach, NY, 117-127, 2001.
23. S. Ratnasamy, P. Francis, M. Handley, R. Karp, and S. Shenker, A Scalable Content-Addressable Network, *Proc. ACM SIGCOMM 2001*, San Diego, CA., 149-160, 2001.
24. M. Ripeanu, Adriana Imnitchi, and I. Foster, Mapping the Gnutella Network. *IEEE Internet Computing* **6**(1), 50-57, 2002.
25. R. Schollmeier, I. Gruber, and F. Niethammer, Protocol for Peer-to-Peer Networking in Mobile Environments, *Proc. IEEE 12th Int. Conf. on Computer Communications and Networks (ICCCN'03)*, Dallas, TX, 2003.
26. I. Stoica, R. Morris, D. Karger, F. Kaashoek, and H. Balakrishnan, Chord: A Scalable Peer-to-Peer Lookup Service for Internet Applications, *Proc. ACM SIGCOMM 2001*, San Diego, CA, 149-160, 2001.
27. K. Sripanidkulchai, The Popularity of Gnutella Queries and its Implications on Scalability. *Proc. O'Reilly Peer-to-Peer and Web Services Conf.*, 2001
28. Y. Xie and D. O'Hallaron, Locality in Search Engine Queries and Its Implications for Caching, *Proc. IEEE INFOCOM 2002*, New York, NJ, 2002.

Multipath Routing in Mobile Ad Hoc Networks: Issues and Challenges*

Stephen Mueller[1], Rose P. Tsang[2], and Dipak Ghosal[1]

[1] Department of Computer Science, University of California, Davis, CA 95616
[2] Sandia National Laboratories, Livermore, CA 94551

Abstract. Mobile ad hoc networks (MANETs) consist of a collection
of wireless mobile nodes which dynamically exchange data among themselves without the reliance on a fixed base station or a wired backbone
network. MANET nodes are typically distinguished by their limited power, processing, and memory resources as well as high degree of mobility.
In such networks, the wireless mobile nodes may dynamically enter the
network as well as leave the network. Due to the limited transmission
range of wireless network nodes, multiple hops are usually needed for
a node to exchange information with any other node in the network.
Thus routing is a crucial issue to the design of a MANET. In this paper, we specifically examine the issues of multipath routing in MANETs.
Multipath routing allows the establishment of multiple paths between a
single source and single destination node. It is typically proposed in order to increase the reliability of data transmission (i e , fault tolerance)
or to provide load balancing. Load balancing is of especial importance
in MANETs because of the limited bandwidth between the nodes. We
also discuss the application of multipath routing to support application
constraints such as reliability, load-balancing, energy-conservation, and
Quality-of-Service (QoS).

1 Introduction

Mobile ad hoc networks (MANETs) consist of a collection of wireless mobile
nodes which dynamically exchange data among themselves without the reliance
on a fixed base station or a wired backbone network. MANETs have potential
use in a wide variety of disparate situations. Such situations include moving
battlefield communications to disposable sensors which are dropped from high
altitudes and dispersed on the ground for hazardous materials detection. Civi-
lian applications include simple scenarios such as people at a conference in a
hotel where their laptops comprise a temporary MANET to more complicated
scenarios such as highly mobile vehicles on the highway which form an ad hoc
network in order to provide vehicular traffic management.

MANET nodes are typically distinguished by their limited power, processing,
and memory resources as well as high degree of mobility. In such networks,

---

* This research was funded in part by a grant from Sandia National Laboratories, CA,
USA.

M.C. Calzarossa and E. Gelenbe (Eds.): MASCOTS 2003, LNCS 2965, pp. 209–234, 2004.
© Springer-Verlag Berlin Heidelberg 2004

the wireless mobile nodes may dynamically enter the network as well as leave the network. Due to the limited transmission range of wireless network nodes, multiple hops are usually needed for a node to exchange information with any other node in the network. Thus routing is a crucial issue to the design of a MANET.

Routing protocols in conventional wired networks are usually based upon either distance vector or link state routing algorithms. Both of these algorithms require periodic routing advertisements to be broadcast by each router. In distance vector routing [1], each router broadcasts to all of its neighboring routers its view of the distance to all other nodes; the neighboring routers then compute the shortest path to each node. In link-state routing [2], each router broadcasts to its neighboring nodes its view of the status of each of its adjacent links; the neighboring routers then compute the shortest distance to each node based upon the complete topology of the network.

These conventional routing algorithms are clearly not efficient for the type of dynamic changes which may occur in an ad-hoc network. In conventional networks, routers do not generally move around and only rarely leave or join the network. In an environment with mobile nodes, the changing topology will not only trigger frequent re-computation of routes but the overall convergence to stable routes may be infeasible due to the high-level of mobility.

Clearly, routing in MANETs must take into consideration their important characteristics such as node mobility. Work on single path (or unipath) routing in MANETs has been proposed in [3] [4]. In this paper, we specifically examine the issues of multipath routing in MANETs. Multipath routing allows the establishment of multiple paths between a single source and single destination node. Multipath routing is typically proposed in order to increase the reliability of data transmission (i.e., fault tolerance) or to provide load balancing. Load balancing is of especial importance in MANETs because of the limited bandwidth between the nodes.

The rest of the paper is organized as follows. In Section 2, we provide background into the area of multipath routing for wired networks. In Section 3, we present an overview of the characteristics of MANETs. We discuss techniques for supporting multipath routing in MANETs in Section 4. In Section 5, we discuss the application of multipath routing to support application constraints such as reliability, energy-conservation, and Quality-of-Service (QoS). Finally in Section 6, we provide the conclusion.

2   Background on Multipath Routing

Multipath routing has been explored in several different contexts. Traditional circuit switched telephone networks used a type of multipath routing called alternate path routing. In alternate path routing, each source node and destination node have a set of paths (or multipaths) which consist of a primary path and one or more alternate paths. Alternate path routing was proposed in order to decrease the call blocking probability and increase overall network utilization.

In alternate path routing, the shortest path between exchanges is typically one hop across the backbone network; the network core consists of a fully connected set of switches. When the shortest path for a particular source destination pair becomes unavailable (due to either link failure or full capacity), rather than blocking a connection, an alternate path, which is typically two hops, is used. Well known alternate path routing schemes such as Dynamic Nonhierarchical Routing and Dynamic Alternative Routing are proposed and evaluated in [5] [6].

Multipath routing has also been addressed in data networks which are intended to support connection-oriented service with QoS. For instance, Asynchronous Transfer Mode (ATM) [7] networks use a signaling protocol, PNNI, to set up multiple paths between a source node and a destination node. The primary (or optimal) path is used until it either fails or becomes over-utilized, then alternate paths are tried. Using a crankback process, the alternate routes are attempted until a connection is completed.

Alternate or multipath routing has typically lent itself to be of more obvious use to connection-oriented networks; call blocking probability is only relevant to connection oriented networks. However, in packet-oriented networks, like the Internet, multipath routing could be used to alleviate congestion by routing packets from highly utilized links to links which are less highly utilized. The drawback of this approach is that the cost of storing extra routes at each router usually precludes the use of multipath routing. However, multipath routing techniques have been proposed for OSPF [2], a widely used Internet routing protocol.

3 Overview of Mobile Ad Hoc Networks

In MANETs[1] communication between nodes is done through the wireless medium. Because nodes are mobile and may join or leave the network, MANETs have a dynamic topology. Nodes that are in transmission range of each other are called neighbors. Neighbors can send directly to each other. However, when a node needs to send data to another non-neighboring node, the data is routed through a sequence of multiple hops, with intermediate nodes acting as routers. An example ad hoc network is depicted in Figure 1.

There are numerous issues to consider when deploying MANETs. The following are some of the main issues.

1. **Unpredictability of environment**: Ad hoc networks may be deployed in unknown terrains, hazardous conditions, and even hostile environments where tampering or the actual destruction of a node may be imminent. Depending on the environment, node failures may occur frequently.
2. **Unreliability of wireless medium**: Communication through the wireless medium is unreliable and subject to errors. Also, due to varying environmental conditions such as high levels of electro-magnetic interference (EMI) or inclement weather, the quality of the wireless link may be unpredictable.

---

[1] In this paper, we will use the terms MANETs and ad hoc networks interchangeably.

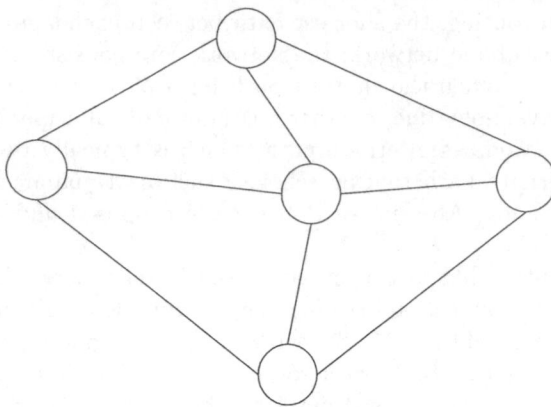

Fig. 1. An example ad hoc network, with circles representing nodes. Two nodes that are in transmission range of each other are connected by a line.

Furthermore, in some applications, nodes may be resource-constrained and thus would not be able to support transport protocols necessary to ensure reliable communication on a lossy link. Thus, link quality may fluctuate in a MANET.

3. **Resource-constrained nodes**: Nodes in a MANET are typically battery-powered as well as limited in storage and processing capabilities. Moreover, they may be situated in areas where it is not possible to re-charge and thus have limited lifetimes. Because of these limitations, they must have algorithms which are energy-efficient as well as operating with limited processing and memory resources. The available bandwidth of the wireless medium may also be limited because nodes may not be able to sacrifice the energy consumed by operating at full link speed.

4. **Dynamic topology**: The topology in an ad hoc network may change constantly due to the mobility of nodes. As nodes move in and out of range of each other, some links break while new links between nodes are created.

As a result of these issues, MANETs are prone to numerous types of faults including,

1. **Transmission errors**: The unreliability of the wireless medium and the unpredictability of the environment may lead to transmitted packets being garbled and thus received in error.

2. **Node failures**: Nodes may fail at any time due to different types of hazardous conditions in the environment. They may also drop out of the network either voluntarily or when their energy supply is depleted.

3. **Link failures**: Node failures as well as changing environmental conditions (e.g., increased levels of EMI) may cause links between nodes to break.

4. **Route breakages**: When the network topology changes due to node/link failures and/or node/link additions to the network, routes become out-of-date and thus incorrect. Depending upon the network transport protocol, packets forwarded through stale routes may either eventually be dropped or be delayed; packets may take a circuitous route before eventually arriving at the destination node.

5. **Congested nodes or links**: Due to the topology of the network and the nature of the routing protocol, certain nodes or links may become over-utilized, i.e., congested. This will lead to either larger delays or packet loss.

Routing protocols for MANETs must deal with these issues to be effective. In the remainder of this section, we present an overview of some of the key unipath routing protocols for MANETs.

3.1   Unipath Routing in MANETs

Routing protocols are used to find and maintain routes between source and destination nodes. Two main classes of ad hoc routing protocols are table-based and on-demand protocols [3] [4] [8] [9]. In table-based protocols [8] [9], each node maintains a routing table containing routes to all nodes in the network. Nodes must periodically exchange messages with routing information to keep routing tables up-to-date. Therefore, routes between nodes are computed and stored, even when they are not needed. Table-based protocols may be impractical, especially for large, highly mobile networks. Because of the dynamic nature of ad hoc networks, a considerable number of routing messages may have to be exchanged in order to keep routing information accurate or up-to-date.

In on-demand protocols [3] [4], nodes only compute routes when they are needed. Therefore, on-demand protocols are more scalable to dynamic, large networks. When a node needs a route to another node, it initiates a route discovery process to find a route. On-demand protocols consist of the following two main phases.

1. **Route discovery** is the process of finding a route between two nodes (see Figure 2)

2. **Route maintenance** is the process of repairing a broken route or finding a new route in the presence of a route failure (see Figure 3)

Most currently proposed routing protocols for ad hoc networks are unipath routing protocols. In unipath routing, only a single route is used between a source and destination node. Two of the most widely used protocols are the Dynamic Source Routing (DSR) [4] and the Ad hoc On-demand Distance Vector (AODV) [3] protocols. AODV and DSR are both on-demand protocols. Since most of the multipath routing protocols discussed in this paper are an extension of one of these two protocols, the following subsection gives a brief overview of DSR and AODV.

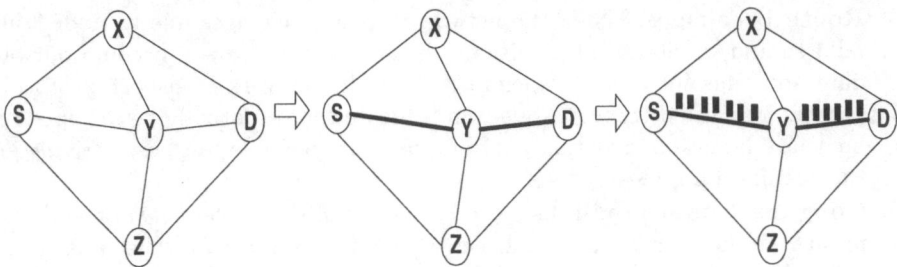

Fig. 2. An example of route discovery in an ad hoc network. In order for node S to send data to node D, it must first discover a route to node D. Node S discovers a route to node D going through node Y, and sets up the route. Once the route is established, node S can begin sending data to node D along the route.

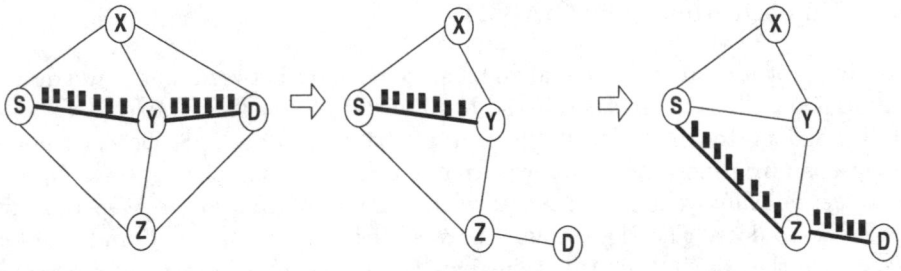

Fig. 3. An example of route maintenance in an ad hoc network. Node S sends data along an established route to node D through node Y. When node D moves out of range of node Y, this route breaks. Node S finds a new route to node D through node Z, and thus can begin sending data to node D again.

Dynamic Source Routing. DSR is an on-demand routing protocol for ad hoc networks. Like any source routing protocol, in DSR the source includes the full route in the packets' header. The intermediate nodes use this to forward packets towards the destination and maintain a route cache containing routes to other nodes.

Route discovery. If the source does not have a route to the destination in its route cache, it broadcasts a route request (RREQ) message specifying the destination node for which the route is requested. The RREQ message includes a route record which specifies the sequence of nodes traversed by the message. When an intermediate node receives a RREQ, it checks to see if it is already in the route record. If it is, it drops the message. This is done to prevent routing loops. If the intermediate node had received the RREQ before, then it also drops the message. The intermediate node forwards the RREQ to the next hop according to the route specified in the header. When the destination receives the RREQ, it sends back a route reply message. If the destination has a route to the source in its route cache, then it can send a route response (RREP) message along this route. Otherwise, the RREP message can be sent along the reverse route back to the source. Intermediate nodes may also use their route cache to reply to RREQs. If an intermediate node has a route to the destination in its cache, then it can append the route to the route record in the RREQ, and send an RREP back to the source containing this route. This can help limit flooding of the RREQ. However, if the cached route is out-of-date, it can result in the source receiving stale routes.

Route maintenance. When a node detects a broken link while trying to forward a packet to the next hop, it sends a route error (RERR) message back to the source containing the link in error. When an RERR message is received, all routes containing the link in error are deleted at that node.

Ad Hoc On-demand Distance Vector. AODV is an on-demand routing protocol for ad hoc networks. However, as opposed to DSR, which uses source routing, AODV uses hop-by-hop routing by maintaining routing table entries at intermediate nodes.

Route Discovery. The route discovery process is initiated when a source needs a route to a destination and it does not have a route in its routing table. To initiate route discovery, the source floods the network with a RREQ packet specifying the destination for which the route is requested. When a node receives an RREQ packet, it checks to see whether it is the destination or whether it has a route to the destination. If either case is true, the node generates an RREP packet, which is sent back to the source along the reverse path. Each node along the reverse path sets up a forward pointer to the node it received the RREP from. This sets up a forward path from the source to the destination. If the node is not the destination and does not have a route to the destination, it rebroadcasts the RREQ packet. At intermediate nodes duplicate RREQ packets are discarded. When the source node receives the first RREP, it can begin sending data to the destination.

To determine the relative degree out-of-dateness of routes, each entry in the node routing table and all RREQ and RREP packets are tagged with a destination sequence number. A larger destination sequence number indicates a more current (or more recent) route. Upon receiving an RREQ or RREP packet, a node updates its routing information to set up the reverse or forward path, respectively, only if the route contained in the RREQ or RREP packet is more current than its own route.

Route Maintenance. When a node detects a broken link while attempting to forward a packet to the next hop, it generates a RERR packet that is sent to all sources using the broken link. The RERR packet erases all routes using the link along the way. If a source receives a RERR packet and a route to the destination is still required, it initiates a new route discovery process. Routes are also deleted from the routing table if they are unused for a certain amount of time.

Many multipath routing protocols have been proposed in literature. Most of these protocols are an extension of either DSR or AODV. In the following section, we first outline the key advantages of multipath routing and then discuss some of the proposed multipath extensions to both DSR and AODV.

4   Multipath Routing in MANETs

Standard routing protocols in ad hoc wireless networks, such as AODV and DSR, are mainly intended to discover a single route between a source and destination node. Multipath routing consists of finding multiple routes between a source and destination node. These multiple paths between source and destination node pairs can be used to compensate for the dynamic and unpredictable nature of ad hoc networks.

4.1   Benefits of Multipath Routing

As mentioned before, multiple paths can provide load balancing, fault-tolerance, and higher aggregate bandwidth. Load balancing can be achieved by spreading the traffic along multiple routes. This can alleviate congestion and bottlenecks.

From a fault tolerance perspective, multipath routing can provide route resilience. To demonstrate this, consider Figure 4, where node S has established three paths to node D. If node S sends the same packet along all three paths, as long as at least one of the paths does not fail, node D will receive the packet. While routing redundant packets is not the only way to utilize multiple paths, it demonstrates how multipath routing can provide fault tolerance in the presence of route failures.

Since bandwidth may be limited in a wireless network, routing along a single path may not provide enough bandwidth for a connection. However, if multiple paths are used simultaneously to route data, the aggregate bandwidth of the paths may satisfy the bandwidth requirement of the application. Also, since there is more bandwidth available, a smaller end-to-end delay may be achieved. Due to issues at the link layer, using multiple paths in ad hoc networks to achieve higher bandwidth may not be as straightforward as in wired networks. Because nodes in the network communicate through the wireless medium, radio interference must be taken into account. Transmissions from a node along one path may interfere with transmissions from a node along another path, thereby limiting the achievable throughput. However, results show that using multipath routing in ad hoc networks of high density results in better throughput than using unipath routing [10]. We defer further discussion of the link layer issues in multipath routing to Section 4.3.

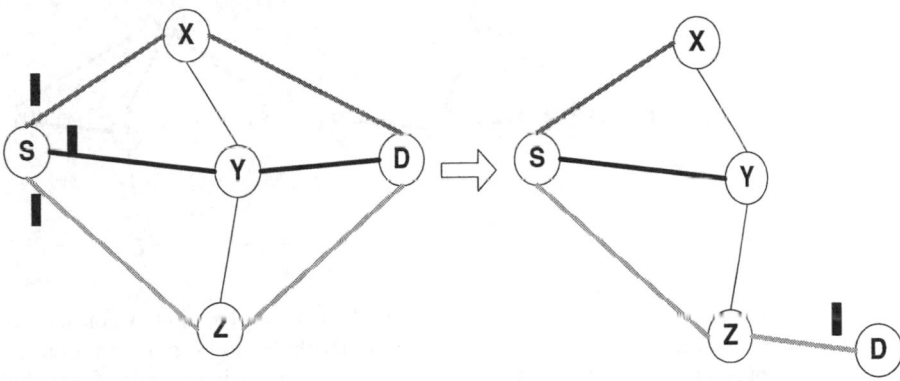

Fig. 4. Source node S routes the same packet to destination node D along the routes SXD, SYD, and SZD. When node D moves, routes SXD and SYD break, but route SZD is still able to deliver the packet to node D.

4.2   Multipath Routing Components

Multipath routing consists of three components: route discovery, route maintenance, and traffic allocation. We discuss these components in the following subsections.

Route Discovery and Maintenance. Route discovery and route maintenance consists of finding multiple routes between a source and destination node. Multipath routing protocols can attempt to find node disjoint, link disjoint, or non-disjoint routes. Node disjoint routes, also known as totally disjoint routes, have no nodes or links in common. Link disjoint routes have no links in common, but may have nodes in common. Non-disjoint routes can have nodes and links in common. Refer to Figure 5 for examples of the different kinds of multipath routes.

Disjoint routes offer certain advantages over non-disjoint routes. For instance, non-disjoint routes may have lower aggregate resources than disjoint routes, because non-disjoint routes share links or nodes. In principle, node disjoint routes offer the most aggregate resources, because neither links nor nodes are shared between the paths. Disjoint routes also provide higher fault-tolerance. When using non-disjoint routes, a single link or node failure can cause multiple routes to fail. In node or link disjoint routes, a link failure will only cause a single route to fail. However, with link disjoint routes, a node failure can cause multiple routes that share that node to fail. Thus, node disjoint routes offer the highest degree of fault-tolerance.

The main advantage of non-disjoint routes is that they can be more easily discovered. Because there are no restrictions that require the routes to be node or link disjoint, more non-disjoint routes exist in a given network than node or link

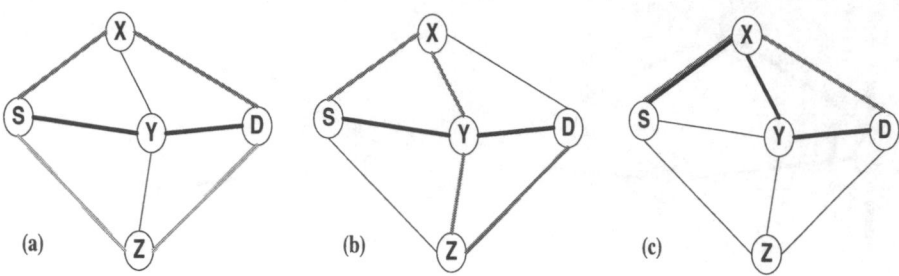

Fig. 5. Routes SXD, SYD, and SZD in (a) have no links or nodes in common and are therefore node disjoint. Routes SXYZD and SYD in (b) have node Y in common and are therefore only link disjoint. Routes SXD and SXYD in (c) have node X and link SX in common and are therefore non-disjoint.

disjoint routes. Because node-disjointedness is a stricter requirement than link-disjointedness, node-disjoint routes are the least abundant and are the hardest to find. It has been shown that in moderately dense networks, there may only exist a small number of node disjoint routes between any two arbitrary nodes, especially as the distance between the nodes increases [11]. This is because there may be sparse areas between the two nodes that act as bottlenecks. Given the trade-offs between using node disjoint versus non-disjoint routes, link disjoint routes offer a good compromise between the two. In the following subsection, we review some of the proposed multipath protocols for finding node disjoint, link disjoint, and non-disjoint paths.

Intelligent path selection can be used to enhance the performance of multipath routing. For instance, a certain subset of paths may be selected for use based on a variety of criteria such as characteristics of the paths and interactions with the link layer. From a fault tolerance perspective, more reliable paths should be selected to reduce the chance of routes failures. Path selection also plays an important role for QoS routing. In QoS routing, only a subset of paths that together satisfies the QoS requirement is selected.

After a source begins sending data along multiple routes, some or all of the routes may break due to node mobility and/or link and node failures. As in unipath routing, route maintenance must be performed in the presence of route failures. As discussed previously for unipath routing, route discovery can be triggered upon failure of the route. In the case of multipath routing, route discovery can be triggered each time one of the routes fails or only after all the routes fail. Waiting for all the routes to fail before performing a route discovery would result in a delay before new routes are available. This may degrade the QoS of the application. However, initiating route discovery every time one of the routes fails may incur high overheads. Performing route discovery when N routes fail, where N is less than the number of paths available, may be a compromise between the two options.

Split Multipath Routing. Split Multipath Routing (SMR) proposed in [12] is an on-demand multipath source routing protocol. SMR is similar to DSR, and is used to construct maximally disjoint paths. Unlike DSR, intermediate nodes do not keep a route cache, and therefore, do not reply to RREQs. This is to allow the destination to receive all the routes so that it can select the maximally disjoint paths. Maximally disjoint paths have as few links or nodes in common as possible. Duplicate RREQs are not necessarily discarded. Instead, intermediate nodes forward RREQs that are received through a different incoming link, and whose hop count is not larger than the previously received RREQs. The proposed route selection algorithm only selects two routes. However, the algorithm can be extended to select more than two routes. In the algorithm, the destination sends an RREP for the first RREQ it receives, which represents the shortest delay path. The destination then waits to receive more RREQs. From the received RREQs, the path that is maximally disjoint from the shortest delay path is selected. If more than one maximally disjoint path exists, the shortest hop path is selected. If more than one shortest hop path exists, the path whose RREQ was received first is selected. The destination then sends an RREP for the selected RREQ.

AOMDV. AOMDV [13] is an extension to the AODV protocol for computing multiple loop-free and link-disjoint paths. To keep track of multiple routes, the routing entries for each destination contain a list of the next-hops along with the corresponding hop counts. All the next hops have the same sequence number. For each destination, a node maintains the advertised hop count, which is defined as the maximum hop count for all the paths. This is the hop count used for sending route advertisements of the destination. Each duplicate route advertisement received by a node defines an alternate path to the destination. To ensure loop freedom, a node only accepts an alternate path to the destination if it has a lower hop count than the advertised hop count for that destination. Because the maximum hop count is used, the advertised hop count therefore does not change for the same sequence number. When a route advertisement is received for a destination with a greater sequence number, the next-hop list and advertised hop count are reinitialized.

AOMDV can be used to find node-disjoint or link-disjoint routes. To find node-disjoint routes, each node does not immediately reject duplicate RREQs. Each RREQ arriving via a different neighbor of the source defines a node-disjoint path. This is because nodes cannot broadcast duplicate RREQs, so any two RREQs arriving at an intermediate node via a different neighbor of the source could not have traversed the same node. In an attempt to get multiple link-disjoint routes, the destination replies to duplicate RREQs regardless of their first hop. To ensure link-disjointness in the first hop of the RREP, the destination only replies to RREQs arriving via unique neighbors. After the first hop, the RREPs follow the reverse paths, which are node-disjoint and thus link-disjoint. The trajectories of each RREP may intersect at an intermediate node, but each takes a different reverse path to the source to ensure link-disjointness.

AODVM. AODVM [11] is an extension to AODV for finding multiple node-disjoint paths. Intermediate nodes are not allowed to send a route reply directly to the source. Also, duplicate RREQ packets are not discarded by intermediate nodes. Instead, all received RREQ packets are recorded in an RREQ table at the intermediate nodes. The destination sends an RREP for all the received RREQ packets. An intermediate node forwards a received RREP packet to the neighbor in the RREQ table that is along the shortest path to the source. To ensure that nodes do not participate in more than one route, whenever a node overhears one of its neighbors broadcasting an RREP packet, it deletes that neighbor from its RREQ table. Because a node cannot participate in more than one route, the discovered routes must be node-disjoint.

Comparison of Unipath and Mulitpath Routing. The main advantage of DSR over AODV is its simplicity. In DSR, while nodes do maintain route caches, they do not need to maintain routing tables with forwarding information, as in AODV. However, with DSR, more overhead is incurred in routing data packets, since the entire route must be specified in the packet header. The multipath extensions to DSR and AODV inherit these advantages and disadvantages from their parent protocols. Therefore, the main advantage of SMR, and any multipath extension to DSR, is simplicity.

Both AODV and DSR allow for intermediate nodes to respond to RREQs, which can reduce the time for route discoveries. However, both SMR and AODVM do not allow intermediate nodes to reply to route discoveries, in order to ensure that the destination can select disjoint paths. The primary advantage of AOMDV is that it allows intermediate nodes to reply to RREQs, while still selecting disjoint paths. All three multipath protocols do not reject duplicate RREQs at intermediate nodes. This allows for the discovery of more paths. However, it also results in more message overhead during route discovery due to increased flooding. Additionally, in the multipath protocols the destination replies to multiple RREQs, which results in more overhead. However, in SMR and AOMDV, the destination only replies to a subset of the received RREQs.

The primary disadvantages of multipath routing protocols compared to unipath protocols are complexity and overhead. In the case of mulipath extensions to AODV, maintaining multiple routes to a destination results in larger routing tables at intermediate nodes. In multipath routing, as discussed in the next section, the method by which packets are allocated to the multiple routes must be taken into account. Multipath routing can result in packet reordering. In unipath routing, traffic allocation is not an issue, since only one route is used.

Traffic Allocation. Once the source node has selected a set of paths to the destination, it can begin sending data to the destination along the paths. The traffic allocation strategy used deals with how the data is distributed amongst the paths. The choice of allocation granularity is important in traffic allocation. The allocation granularity specifies the smallest unit of information allocated to each path. For instance, a per-connection granularity would allocate all traffic for one connection to a single path. A per-packet granularity would distribute the packets from multiple connections amongst the paths. A per-packet granularity results in the best performance [14]. This is because it allows for finer control over the network resources. It is difficult to evenly distribute traffic amongst the paths in the per-connection case, because all the connections experience different traffic rates. If a round-robin traffic allocation approach is used, however, a per-packet granularity may result in packets arriving out-of-order at the destination. Packet reordering is an issue that needs to be dealt with in multipath routing, possibly at the transport layer.

4.3 Link Layer Issues

Nodes in an ad hoc network communicate through the wireless medium. If a shared channel is used, neighboring nodes must contend for the channel. When the channel is in use by a transmitting node, neighboring nodes hear the transmission and are blocked from receiving from other sources. Furthermore, depending on the link layer protocol, neighboring nodes may have to defer transmission until the channel is free. Even when multiple channels are used, the quality of neighboring transmissions may be degraded due to interference. Nodes within transmission range of each other are said to be in the same collision domain.

We now consider the situation where multipath routing is used and the multiple paths are used simultaneously to route data. Even if the multiple paths are node-disjoint, transmissions along the routes may interfere if some nodes among the routes are in the same collision domain. While node-disjointedness may ensure failure independence, it does not ensure transmission independence. When choosing multiple paths, it is important to choose paths that are as independent as possible to ensure the least interference between the paths.

Multiple metrics can be used to calculate the relative degree of independence among a set of paths, namely correlation [15] and coupling [16]. The correlation factor between two node-disjoint paths is defined as the total number of links connecting the paths [15]. Note that the correlation factor only applies to node-disjoint paths. The coupling between two paths is calculated as the average number of nodes that are blocked from receiving data along one of the paths when a node in the other path is transmitting [16]. The advantage of using coupling as a metric is that it can be used for both disjoint and non-disjoint routes. Non-disjoint routes are considered highly coupled. Choosing paths that have low coupling or correlation can improve the performance of multipath routing.

5   Applications of Multipath Routing

Multipath routing can be used to support a variety of applications in MANETs. In this section, we present proposed multipath routing methods that support reliability (fault tolerance), energy conservation, minimization of end-to-end delay, and satisfying bandwidth requirements. Finally, we discuss an example architecture which uses source coding that utilizes multipath route discovery and maintenance, along with traffic allocation to support video over MANETs.

5.1   Supporting Reliability (Fault Tolerance)

Satisfying Reliability Requirements. MP-DSR is a multipath QoS-aware extension to DSR proposed in [17]. The protocol attempts to provide end-to-end reliability as the QoS metric. End-to-end reliability is defined as the probability of sending data successfully within a time window. The end-to-end reliability is calculated from the reliabilities of the paths used for routing. The path reliability is calculated from the link availabilities. Link availability is defined as the probability that a link is available from time $t_0 + t$, given that it is an active link at time t_0 [18]. Path reliability is the product of the link availabilities along the path, assuming the link availabilities are independent. The end-to-end reliability is $1 - \prod_{k \epsilon K}(1 - k)$, where k is the path reliability of a path, and K is the set of all paths. Essentially, the end-to-end reliability is the probability that at least one path does not fail within the given time window. Figure 6 shows an example of how to compute the end-to-end reliability given a multipath route and the link availabilities of each path. Notice that the end-to-end reliability is higher than any of the path reliabilities.

The operation of MP-DSR is as follows. The application first supplies MP-DSR with a path reliability requirement. MP-DSR then determines the number of paths needed and lowest path reliability requirement each path must provide. The source then sends RREQ messages intended for the destination node to its neighbors containing this requirement and the end-to-end requirement. The RREQ message also contains the traversed path, and the accumulated path reliability. The intermediate nodes check to see if the RREQ message still meets the reliability requirements. If so, the node updates the accumulated path reliability based on the availability of the link just traversed, and forwards the message to its neighbors. If the RREQ message no longer meets the requirements, the message is discarded. When the destination receives all the RREQ messages, it sorts the messages according to the path reliabilities, and selects a set of disjoint paths that together satisfy the end-to-end reliability requirement. An RREP message is sent along each path back to the source. When the source receives the RREPs, it can begin using the multiple paths to route data.

Route maintenance can be performed when all routes fail or when the timer window expires. If all routes fail, the route discovery process is simply reinitiated. When the timer window expires, the source sends a route check messages along the paths to collect the path reliabilities. The destination replies to the route check messages. The source collects all the replies, and checks to see if the paths still meet the reliability requirement within a certain tolerance level. If validation is unsuccessful, then route discovery is triggered. One advantage of MP-DSR is that QoS characteristics are collected using local information available at intermediate nodes. Therefore, global knowledge is not required.

Hybrid Network for Enhanced Reliability. As discussed before, based on the results in [11], it may be difficult to find a suitable number of node disjoint paths between two nodes to provide the necessary fault tolerance and reliability. However, some ad hoc networks may contain heterogeneous nodes, where some

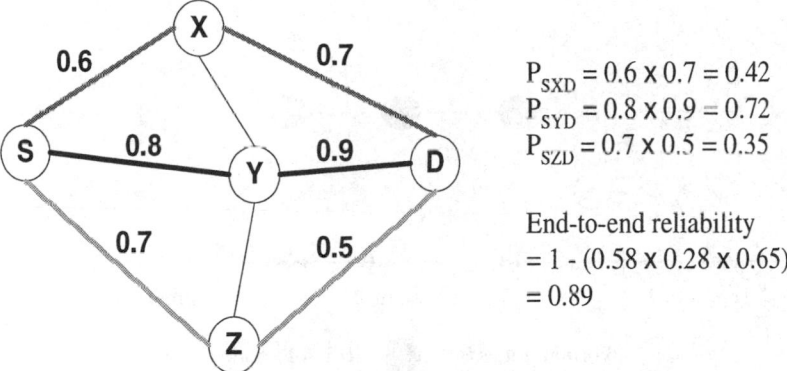

$P_{SXD} = 0.6 \times 0.7 = 0.42$
$P_{SYD} = 0.8 \times 0.9 = 0.72$
$P_{SZD} = 0.7 \times 0.5 = 0.35$

End-to-end reliability
$= 1 - (0.58 \times 0.28 \times 0.65)$
$= 0.89$

Fig. 6. Example network with multipath route consisting of three paths from S to D. The calculated path reliabilities are given to the right. The end-to-end reliability is calculated to be 0.89.

nodes are more reliable than other nodes. For instance, in a battlefield, low-powered sensors or handhelds may be deployed in the field, with high-powered, reliable, and secure nodes located in tanks or large vehicles. A scheme is proposed in [11] that takes advantage of reliable nodes to construct a reliable path between two nodes. In this scheme, a reliable path essentially consists of reliable nodes or reliable nodes connected by multiple node disjoint paths. Specifically, a reliable path is formed by concatenating reliable segments. A segment is deemed reliable if it consists totally of reliable nodes, or if there are a certain number of disjoint paths connecting the two reliable nodes. Refer to Figure 7 for an illustration of this scheme. AODVM is used to find node-disjoint paths between the nodes.

In order to see performance gains, the reliable nodes should not be deployed randomly, especially when the number of reliable nodes is small. It is difficult to find multiple disjoint paths between two nodes because there may exist sparse areas between the nodes that act as a "bottleneck" area, as in Figure 7. Therefore, reliable nodes are deployed in these areas to provide reliable paths. The objective is to position reliable nodes such that the probability of establishing a reliable path between any two arbitrary nodes is maximized. The reliable nodes gather topology information from surrounding nodes to determine where to position themselves. Depending on the mobility of the nodes, the reliable nodes may have to reposition themselves. The reliable nodes should be faster than the other nodes, such that they can adapt to node mobility in a timely fashion. Since the reliable nodes are more powerful than other nodes, it is not unreasonable to believe that they are faster. For instance, reliable nodes may be located in large vehicles while the other nodes are carried by pedestrians. Results in [11] show that the deployment of only a few reliable nodes (10% of total nodes in the network) can result in an increase in the probability of establishing reliable paths between arbitrary nodes.

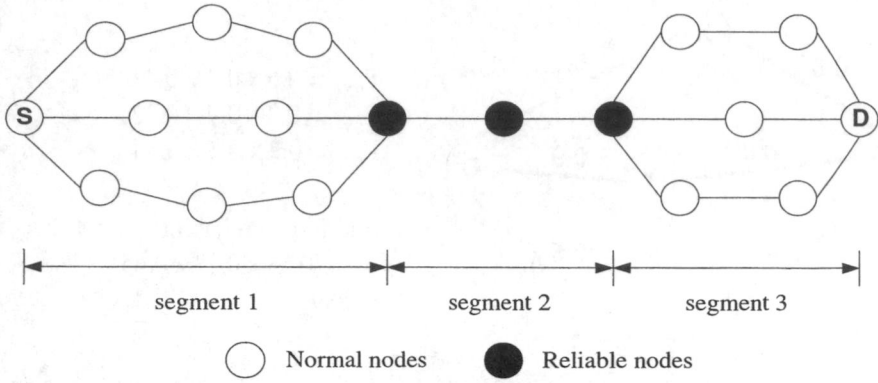

Fig. 7. Example of a reliable path between S and D [11]. The reliable path is formed
by the concatenation of three reliable segments. Segment 1 and segment 3 are reliable
because there are a suitable number of disjoint paths between the endpoints. In this
example there must be three disjoint paths between the endpoints for the segment to
be deemed reliable. Segment 2 is reliable because it consists entirely of reliable nodes.

Packet Salvaging for Fault Tolerance. When using multiple paths for fault
tolerance, if a source cannot send a packet due to a route failure, the packet
can be routed along an alternate route. However, if only the source maintains
multiple routes to the destination, when a route fails, a RERR message must
propagate from an intermediate node all the way to the source before the packet
can be routed along an alternate route. With packet salvaging [19], intermediate
nodes maintain multiple routes to the destination, and a RERR message propagates upstream only until an intermediate node can forward the packet along
an alternate route. This obviously reduces the packet delay for recovering from
a route failure.

A multipath extension to DSR that employs packet salvaging is proposed in
[20]. In this scheme, the first RREQ received by the destination is selected as
the primary route. The destination node then supplies each intermediate node
in the primary route with a link disjoint route to the destination. To do this, the
destination sends the RREP for a received RREQ directly to the intermediate
node the RREQ traversed, instead of to the source. It is possible that not all
intermediate nodes will receive an alternate route to the destination. Figure 8
depicts this scheme. Whenever an intermediate node along the primary route
fails to forward a packet to the next node due to a link failure, it switches to
its alternate route. It does this by replacing the unused portion of the route
in the packet header with the new alternate route. For instance, referring to
Figure 8, if link L_i fails, node n_i will replace $L_i - L_k$ in the packet header
with P_i. When a link along P_i breaks, a RERR message is sent upstream until
it reaches n_{i-1}, which then switches to path P_{i-1}. The intermediate node is
subsequently responsible for modifying all packet headers to use the alternate

route. When the source receives a RERR message, it means all alternate routes have failed, and therefore, a new route discovery must be performed.

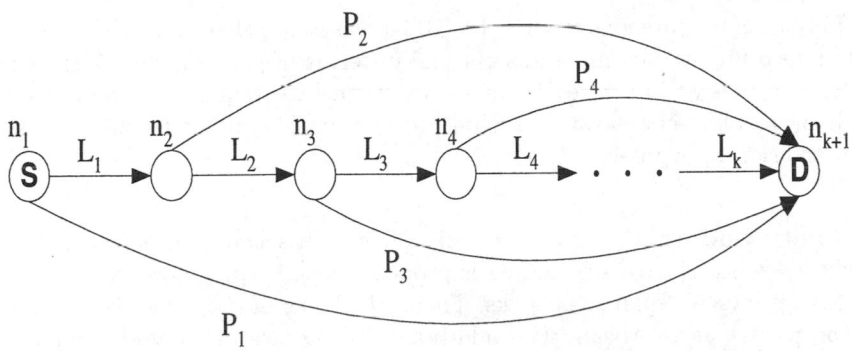

Fig. 8. An example of multipath protocol proposed in [20]. Each intermediate node n_i along the primary path has an alternate route P_i to the destination D.

The previous scheme only uses one route at a time. That is, alternate routes are only used when the current route fails. This may result in unused routes becoming stale. A similar multipath protocol called CHAMP (Caching and Multipath Routing) [19] uses round-robin traffic allocation to keep routes fresh. It also employs cooperative packet caching to improve fault tolerance. In CHAMP, the source broadcasts an RREQ when it doesn't have a route to the destination. Intermediate nodes forward the RREQ until it reaches the destination. While receiving RREQs, an intermediate node keeps track of the neighbors that are on an equal shortest path back to the source. The destination sends an RREP if the route in the RREQ is less than or equal to the shortest route received. A node only accepts the shortest routes to the destination, and the routes must be of equal length. When routing a packet to a destination, a node sends the packet along the least used route, thereby spreading data packets over all the available routes. Routes are of equal length in order to help reduce out-of-order packets arriving at the destination. CHAMP allows for non-disjoint paths.

CHAMP takes advantage of temporal locality in routing, where a dropped packet is a recently sent packet. Each node keeps a cache of packets it recently forwarded. If a node is unable to forward a packet to the next hop neighbor along a route, the route is removed from the route cache. If an alternate route to the destination is available, the node forwards the packet along this route. If no alternate routes are available, the node broadcasts a RERR message with the packet information. When a node receives a RERR message for a particular packet, it removes the corresponding route from its route cache. It then checks to see if it has the packet in its data cache. If it has the packet cached and an alternate route to the destination is available, the packet is forwarded along this route. Otherwise, the node re-broadcasts the RERR message, and the process continues. If the source node of a packet receives a RERR message for that packet and no alternate routes to the destination are available, the source initiates a route discovery.

In both of the previous studies [19] [20], it was found that maintaining just one alternate route at each node was optimal in terms of performance. Maintaining multiple routes at intermediate nodes results in less frequent route discoveries. Both protocols demonstrate how link disjoint paths work adequately well for fault tolerance purposes.

Diversity Coding. A multipath traffic allocation scheme for ad hoc networks that uses M-for-N diversity coding is proposed in [21]. In this scheme, a packet is split up into N equal size blocks. Then, M blocks of the same size are added to the packet as overhead. The additional blocks are calculated from the N blocks, and provide redundant information. The $N + M$ total blocks are then allocated to the multiple available paths. Based on the M-for-N coding scheme, if no more than M blocks are lost, the original packet can be reconstructed from the received blocks.

The main idea behind this scheme is to allocate the blocks amongst the routes such that the probability of losing no more than M blocks is maximized. Therefore, it is reasonable to believe that more blocks should be assigned to more reliable routes. In this scheme, each path is assigned a probability of success. The paths are assumed to be total erasure channels, in which either all the data or no data sent along the channel is received at the destination. Therefore, if a path fails, all the blocks sent along that path are lost. Otherwise, all the blocks are received. An equation is developed in [21] to compute the probability of success, or the probability that no more than M blocks are lost, based on the probabilities assigned to each path and how the blocks are allocated. The equation is maximized to determine which routes to use and how the blocks should be allocated to each route. An observation in [21] was made that in the limit as the number of paths becomes large, the probability of success is 100%.

This scheme requires the existence of a mechanism to assign path success probabilities. Mechanisms for calculating path reliability do exist [17] [18]. The coding scheme used should also be relatively fast. The amount of overhead added to each packet obviously should be less than the actual size of the packet in order to see performance gains.

5.2 Minimizing End-to-End Delay

A traffic allocation scheme is proposed in [22] that uses weighted round robin packet distribution to improve the delay and throughput. In this scheme, the source node maintains a set of multiple paths to the destination, and distributes the load amongst the paths according to the RTT of each path. Paths that have a smaller RTT are allocated more consecutive packets. It is assumed in this scheme that the RTT of each path is inversely proportional to the available bandwidth. To justify this, it is assumed that ad hoc networks have less heterogeneity than WAN, and the bandwidth-delay product is a constant. Therefore, traffic is distributed amongst the paths proportional to the available bandwidth of the paths. In addition to improving the end-to-end delay, the scheme can also improve network congestion. A multipath extension to DSR called MSR is used by the source node to find multiple paths to the destination. Disjoint paths are preferred to ensure path independence.

The scheme proposed in [23] attempts to minimize the end-to-end delay for sending a volume of data from source to destination by using multipath routing and intelligent traffic allocation. In this scheme, data is routed along multiple paths in sequential blocks. Since routes may break before all the data is transmitted, new paths must be discovered. The scheme divides the total data into chunks and uses a new set of multiple paths to route each chunk. Pre-emptive route discoveries are done to find new paths before route errors occur. In order to determine how to distribute data amongst the paths, the scheme uses a mechanism to predict the lifetime of a path. The lifetime of a path is determined by the link affinity, which is a prediction of the life span of a link. The affinity is calculated based on the transmission range of the nodes, distance between nodes, and the average velocity of the nodes. A path's stability is defined as the smallest link affinity along the path.

The route discovery process is essentially an extension to DSR. One extension to DSR is that intermediate nodes append the link affinity value to route reply packets. Therefore, when the source receives the route replies, it can determine the stability of each path. Also, the source node records the time it sends a route request and the times each route reply is received, so that it has the total time of route discovery. This time indicates the delay of the path due to congestion, packet transmission, and hop count. In the scheme, an optimization problem is defined that minimizes the average network delay for sending the data, based on the constraints that data is distributed among multiple paths and the distributed data must reach the destination within the lifetime of those paths. Given the set of discovered stable paths with their corresponding delay and hop counts, the optimization problem selects a subset of the paths and determines how data will be distributed amongst the paths. A chunk of the total data is then distributed among the selected paths. Before transmission of the data chunk is complete, a new route discovery process is performed to find new routes. The route discovery is performed at a time such that the new routes are available just before transmission of the chunk is complete. The optimization problem is then calculated again for the new routes, and another data chunk is transmitted. The process continues until the total data volume is transmitted to the destination. This is a good example of a comprehensive scheme that employs mechanisms for route discovery, traffic allocation, and route maintenance to improve the QoS.

5.3   Satisfying Bandwidth Requirements

A bandwidth reservation scheme for ad hoc networks that uses multipath routing is proposed in [24]. The protocol attempts to find multiple paths that collectively satisfy the bandwidth requirements. The original bandwidth requirement is essentially split into multiple sub-bandwidth requirements. Each sub-path is then responsible for one sub-bandwidth requirement. This protocol is on-demand and uses the local bandwidth information available at each node for discovering routes. A ticket-based approach is used to search for multiple paths. In this approach a number of probes are sent out from the source, each carrying a ticket. Each probe is responsible for searching one path. The number of tickets sent controls the amount of flooding that is done. Each probe travels along a path that contains the necessary bandwidth.

The source initially sends a certain number of tickets each containing the total bandwidth requirement. The tickets are sent along links that contain sufficient bandwidth to meet the requirement. When an intermediate node receives a ticket, it checks to see which links have enough bandwidth to meet the requirement. If it finds some, it then chooses a link, reserves the bandwidth, and forwards the ticket on the link. If no links have the required bandwidth, the node reserves bandwidth along multiple links such that the sum of the reserved bandwidths equals the original requirement. In this way, the bandwidth requirement is split into sub-bandwidth requirements equaling the bandwidths reserved along each of the links. The original ticket is split into sub-tickets, with each sub-ticket being forwarded along one of the links. Each sub-ticket is then responsible for finding a multi-path satisfying the sub-bandwidth requirement. If no links can be found to satisfy the bandwidth requirements, the intermediate node drops the ticket. Links with more available bandwidth are preferred. An example of the route discovery process is given in Figure 9. The destination eventually receives multiple tickets or sub-tickets comprising a whole ticket. The destination chooses a single ticket, which represents a uni-path, or a group of sub-tickets comprising a whole ticket, which represents a multipath, and sends a route reply. The route replies traverse the reverse paths taken by the sub-tickets, confirming the bandwidth reservations along the way.

The above ticket-based approach does not specify how the available link bandwidth is determined. Specifically, the approach does not deal with the radio interference problem. To address this issue, a multipath protocol that uses the CDMA-over-TDMA model is proposed in [25]. In this model, bandwidth is calculated and reserved based on the free time slots of links. In this approach, the network is flooded during route discovery to search for paths. Each RREQ packet accumulates the link-state of the path it traverses. The link-state corresponds to the free time slot list for each of the traversed links. For each RREQ packet the destination receives, the maximal bandwidth available on the path is determined based on the free time slots. The destination keeps track of the accumulated bandwidth for the paths it has discovered. Once the destination receives enough paths to satisfy the bandwidth requirement, it can send the route replies and reserve the time slots along the paths.

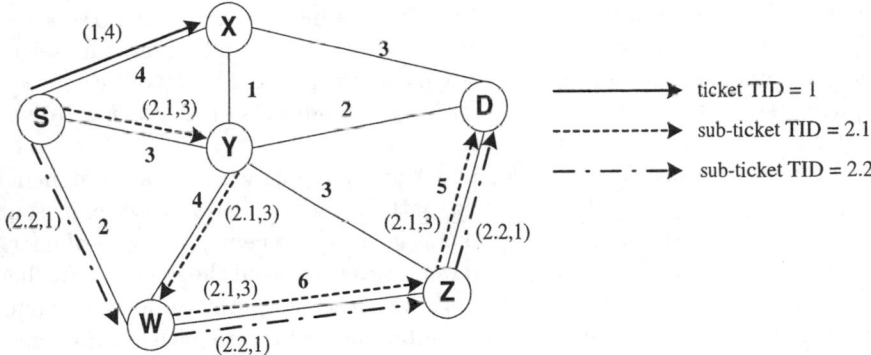

Fig. 9. Example of the route discovery process for bandwidth reservation scheme proposed in [24]. The links are labeled with their available bandwidth. Each message is labeled with the tuple (ticket ID, bandwidth requirement). Ticket IDs are assigned in a hierarchical fashion (Ticket 2.1 is a sub-ticket of ticket 2). Node S initially distributes two tickets. Node X drops ticket 1, because not enough bandwidth is available to proceed. Node S divides ticket 2 into sub-tickets 2.1 and 2.2, with sub-requirements 3 and 1 respectively. The sub-tickets merge at node W and take the same path to node D. D receives 2 paths to the destination: SYWZD and SWZD.

In the time-slotted scheme, intermediate nodes do not drop the RREQ packets, as in the ticket-based scheme. The time-slotted approach accumulates the link-state of the entire path, and the destination can construct a partial network topology with corresponding link state in an on-demand fashion. The ticket-based approach makes hop-by-hop decisions based on only local link-state, and intermediate nodes drop RREQ packets when not enough bandwidth is available on the local links. Because intermediate nodes do not drop RREQs, the time-slotted approach can potentially find more paths. The bandwidth allocation along the paths can also be maximized, since the link-state is available for the entire path. However, these advantages come at the expense of higher message overhead, because the network is flooded and RREQs must store the entire path link-state. Both approaches allow for non-disjoint paths.

5.4 Minimizing Energy Consumption

Since a MANET may consist of nodes which are not able to be re-charged in an expected time period, energy conservation is crucial to maintaining the life-time of such a node. In networks consisting of these nodes, where it is impossible to replenish the nodes' power, techniques for energy-efficient routing as well as efficient data dissemination between nodes is crucial. An energy-efficient mechanism for unipath routing in sensor networks called directed diffusion has been proposed in [26]. Directed diffusion is an on-demand routing approach. In directed diffusion, a (sensing) node which has data to send periodically broadcasts it.

When nodes receive data, they send a reinforcement message to a pre-selected neighbor which indicates that it desires to receive more data from this selected neighbor. As these reinforcement messages are propagated back to the source, an implicit data path is set up; each intermediate node sets up state that forwards similar data towards the previous hop.

Directed diffusion is an on-demand routing approach; it was designed for energy efficiency so it only sets up a path if there is data between a source and a sink. However, the major disadvantage of the scheme, in terms of energy-efficiency, is the periodic flooding of data. In order to avoid the flooding overhead, [27] proposes the setup and maintenance of alternate paths in advance using a localized path setup technique based upon the notion of path reinforcement. The goal of a localized reinforcement-based mechanism is for individual nodes to measure short term traffic characteristics and choose a primary path as well as a number of alternate paths based upon their empirical measurements. An alternate path is intended to be used when the primary fails. "Keep-alive" data is sent through the alternate paths even when the primary path is in use. Because of this continuous "keep-alive" data stream, nodes can rapidly switch to an alternate path without going through a potentially energy-depleting discovery process for a new alternate path.

A multipath routing technique which uses braided multipaths is also proposed in [27]. Braided multipaths relax the requirement for node disjointness. Multiple paths in a braid are only partially disjoint from each other and are not completely node-disjoint. These paths are usually shorter than node-disjoint multipaths and thus consume less energy resources; alternate paths should consume an amount of energy comparable to the primary path. A simple localized technique for constructing braids is as follows. A source sends out a primary path reinforcement to its (primary path) neighbor as well as alternate path reinforcements to its (alternate path) neighbors. Each node in the network performs this same neighbor and path selection process. The evaluation of the performance of the proposed energy constrained algorithms is a function of the overall goal of minimizing energy resources. It was found that energy-efficient multipath routing using the braided multipath approach expends only 33% of the energy of disjoint paths for alternate path maintenance in some cases, and have a 50% higher resilience to isolated failures.

5.5   Example Application: Source Coding for Video Support

A system for transporting video over ad hoc networks using multipath routing and source coding is proposed in [28]. The system considers two types of source coding: *multiple-description coding* and *layered coding*. Both types of coding create multiple sub-streams out of a single video stream. In multiple-description coding, each sub-stream has equal importance, and therefore, each sub-stream contributes equally to the quality level. In layered coding, a base layer stream is created along with multiple enhancement layer streams. The base layer stream provides a basic level of quality, while each enhancement layer stream if correctly received adds to the quality. Therefore, in layered coding, base layer packets should be sent along more reliable paths to ensure that they are received at the destination.

In the system, the source maintains multiple paths to the destination and reserves bandwidth along the paths such that the total bandwidth falls within an acceptable range. The total number of paths is not necessarily equal to the number of streams. Therefore, a path may carry packets from different streams. Similarly, packets from one stream may be allocated to different paths. The task of the source is to allocate the packets from each stream among the paths such that a minimum level of quality can be observed at the receiver. Depending on the path conditions and application requirements, the source chooses to use multiple-description coding or layered coding. The source coder also must adjust the rate allocation to each stream depending on the available bandwidth. Using intelligent path selection and traffic allocation along with adaptive source coding, the system can adapt well to fluctuating network conditions caused by path failures or changes in available bandwidth. Refer to Figure 10 for a depiction of the system architecture.

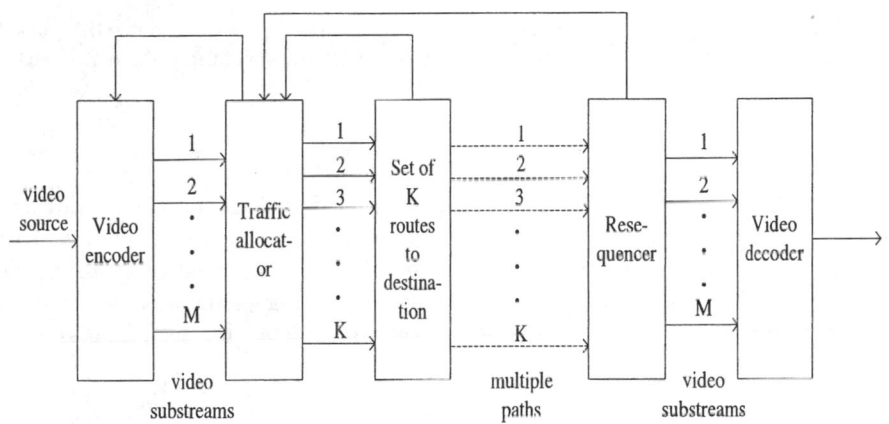

Fig. 10. Architecture for multipath source coding system proposed in [28].

A scheme to provide reliable transport for video specifically using layered coding along with multipath routing is proposed in [29]. In the proposed scheme, the video data is encoded into two layers: the base layer and one enhancement layer. The source uses two disjoint paths to the destination to route data. Base layer packets are sent along one path, while enhancement layer packets are sent along the other path. Base layer packets are protected using Automatic Repeat Request (ARQ). When a base layer packet is lost, the destination sends an ARQ request to the source. When an ARQ request is received, the source retransmits the base layer packet along the enhancement layer path to ensure timely arrival of the base layer packet, and the enhancement layer packet being transmitted at that time instance is discarded. The necessary bandwidth for the base layer and enhancement layer paths are reserved using a signaling protocol. This scheme is depicted in Figure 11.

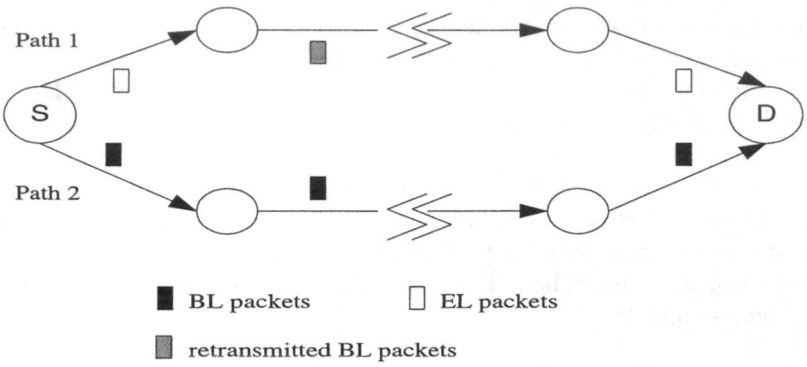

■ BL packets   ☐ EL packets

▨ retransmitted BL packets

Fig. 11. Multipath layered coding scheme proposed in [29]. Base layer (BL) packets are sent along path 2, while enhancement layer (EL) packers and retransmitted BL packets are sent along Path 1.

As packet drops increase due to congestion or degraded network conditions, enhancement layer packets are dropped at the source in favor of retransmitted base layer packets. This attempts to ensure that a basic level of quality is always achieved at the destination. Using layered coding along with ARQ would work well when using lossy paths and if the extra delay for retransmissions is acceptable.

6   Future Work

In this paper, we have presented multipath routing in ad hoc networks mostly in terms of the network layer. We have not made mention of the interaction of multipath routing with the transport layer, in particular, TCP. The main issue that must be dealt with at the transport layer is the arrival of out-of-order packets when multiple paths are used in a round-robin fashion. In TCP, out-of-order packets are assumed to signal congestion in the network, at which point TCP reduces its window size. This can have a detrimental effect on the overall throughput seen by TCP connections. Therefore, the implementation of a TCP-friendly multipath protocol is necessary. We have discussed CHAMP, which uses equal length paths to reduce out-of-order packets. However, finding only equal length paths puts a restriction on the number of paths you can find. If unequal paths are chosen, there could also be ways to perform intelligent traffic allocation depending on path lengths and path delays such that out-of-order packets are minimized. For instance, sending later packets over shorter paths and earlier packets over longer paths may result in reduced out-of-order packets at the receiver. This implies intelligently sending packets out-of-order such that they arrive in-order at the receiver.

In our discussion of using multipath routing to support QoS, most of the protocols proposed only provide QoS in terms of specific metrics, such as bandwidth, delay, or reliability. However, it may be necessary to develop mechanisms to support QoS in terms of multiple metrics. For instance, when searching for multiple paths that have the required bandwidth, it is desirable to find reliable paths. Given the faulty nature of MANETs, constructing a multipath route that meets the bandwidth requirements while also meeting certain reliability requirements would result in better performance. Also, the mechanisms proposed for supporting QoS in terms of delay only attempt to minimize or improve on the delay. It would be desirable to develop a multipath protocol that can provide delay bounds or guarantees, which are required by some real-time applications. Using multipath routing to provide adaptive QoS using source coding is also a promising technique that can be expanded upon for applications other than video.

References

1. Hedrick, C.: Routing Information Protocol. Internet Request For Comments 1058 (1988)
2. Moy, J.: OSPF Version 2. Internet Request For Comments 1247 (1991)
3. Perkins, C.E., Royer, E.M.: Ad-hoc On-Demand Distance Vector Routing. Proceedings of the 2nd IEEE Workshop on Mobile Computing Systems and Applications (1999)
4. Johnson, D.B., Maltz, D.A.: Dynamic Source Routing in Ad Hoc Wireless Networks. Mobile Computing. (1996) 153-181
5. Ash, G.H., Kafker, A.H., Krishnan, K.R.: Servicing and Real-Time Control of Networks with Dynamic Routing. Bell System Technical Journal, Vol. 60, No. 8 (1981)
6. Gibbons, R.J., Kelley, F.P., Key, P.B.: Dynamic Alternative Routing - Modelling and Behavior. Proceedings of the 12th International Teletraffic Conference (1988)
7. ATM Forum. ATM user-network interface specification, Version 3.0.
8. Murthy, S., Garcia-Luna-Aceves, J.J.: An Efficient Routing Protocol for Wireless Networks, Mobile Networks and Applications, Vol. 1, No. 2 (1996)
9. Perkins, C.E., Bhagwat, P.: Highly Dynamic Destination-Sequenced Distance-Vector Routing (DSDV) for Mobile Computers. ACM SIGCOMM (1994) 234-244
10. Pham, P.P., Perreau, S.: Performance Analysis of Reactive Shortest Path and Multi-path Routing Mechanism With Load Balance. IEEE INFOCOM (2003)
11. Ye, Z., Krishnamurthy, S.V., Tripathi, S.K.: A Framework for Reliable Routing in Mobile Ad Hoc Networks. IEEE INFOCOM (2003)
12. Lee, S.-J., Gerla, M.: Split Multipath Routing with Maximally Disjoint Paths in Ad Hoc Networks. IEEE International Conference on Communications, Vol. 10 (2001)

13. Marina, M.K., Das, S.R.: On-demand Multipath Distance Vector Routing in Ad Hoc Networks. Proceedings of the International Conference for Network Procotols (2001)
14. Krishnan, R., Silvester, J.A.: Choice of Allocation Granularity in Multipath Source Routing Schemes. IEEE INFOCOM (1993)
15. Wu, K., Harms, J.: Performance Study of a Multipath Routing Method for Wireless Mobile Ad Hoc Networks. Proceedings of Symposium on Modeling, Analysis and Simulation on Computer and Telecommunication Systems (2001) 99-107
16. Pearlman, M.R., Haas, Z.J., Sholander, P., Tabrizi, S.S.: On the Impact of Alternate Path Routing for Load Balancing in Mobile Ad Hoc Networks. Proceedings of the ACM MobiHoc, (2000) 3-10
17. Leung, R., Liu, J., Poon, E., Chan, A.-L.C., Li, B.: MP-DSR: A QoS-aware Multipath Dynamic Source Routing Protocol for Wireless Ad-hoc Networks. Proceedings of the 26th IEEE Annual Conference on Local Computer Networks (2001) 132-141
18. Jiang, S., He, D., Rao, J.: A Prediction-based Link Availability Estimation for Mobile Ad Hoc Networks. IEEE INFOCOM (2001)
19. Valera, A., Seah, W.K.G, Rao, S.V.: Cooperative Packet Caching and Shortest Multipath Routing in Mobile Ad Hoc Networks. IEEE INFOCOM (2003)
20. Nasipuri, A., Das, S.R.: On-Demand Multipath Routing for Mobile Ad Hoc Networks. Proceedings of the 8th International Conference on Computer Communications and Networks (1999)
21. Tsirigos, A., Haas, Z.J.: Multipath Routing in the Presence of Frequent Topological Changes. IEEE Communications Magazine, Vol. 39, No. 11 (2001)
22. Wang, L., Shu, Y., Dong, M., Zhang, L., Yang, O.W.W.: Adaptive Multipath Source Routing in Ad Hoc Networks. IEEE International Conference on Communications, Vol. 3 (2001)
23. Das, S.K., Mukherjee, A., Bandyopadhyay, S., Paul, K., Saha, D.: Improving Quality-of-Service in Ad hoc Wireless Networks with Adaptive Multi-path Routing. IEEE Global Telecommunications Conference (2000)
24. Liao, W.-H., Tseng, Y.-C., Wang, S.-L., Sheu, J.-P.: A Multi-path QoS Routing Protocol in a Wireless Mobile Ad Hoc Network. IEEE International Conference On Networking (2001)
25. Chen, Y.-S., Tseng, Y.-C., Shue, J.-P., Kuo, P.-H.: On-Demand, Link-State, Multi-Path QoS Routing in a Wireless Mobile Ad-Hoc Network. Proceedings of European Wireless (2002)
26. Intanagonwiwat, C., Govindan, R., Estrin, D.: Directed Diffusion: A Scalable and Robust Communication Paradigm for Sensor Networks. Proceedings of ACM Mobicom (2000)
27. Ganesan, D., Govindan, R., Shenker, S., Estrin, D.: Highly Resilient Energy Efficient Multipath Routing in Wireless Sensor Networks. Mobile Computing and Communications Review, Vol. 1, No. 2 (2002)
28. Wang, Y., Panwar, S., Lin, S., Mao, S.: Wireless Video Transport Using Path Diversity: Multiple Description vs. Layered Coding. Proceedings of the IEEE International Conference on Image Processing (2002)
29. Mao, S., Lin, S., Panwar, S.S., Wang, Y.: Reliable Transmission of Video Using Automatic Repeat Request and Multi-path Transport. IEEE Vehicular Technology Conference (2001)

Algorithmic Design for Communication in Mobile Ad Hoc Networks*

Azzedine Boukerche[1] and Sotiris Nikoletseas[2]

[1] School of Information Technology and Engineering (SITE)
University of Ottawa, Canada
boukerch@site.uottawa.ca
[2] Department of Computer Engineering and Informatics
University of Patras, and Computer Technology Institute (CTI), Greece
nikole@cti.gr

Abstract. As a result of recent significant technological advances, a new computing and communication environment, Mobile Ad Hoc Networks (MANET), is about to enter the mainstream. A multitude of critical aspects, including mobility, severe limitations and limited reliability, create a new set of crucial issues and trade-offs that must be carefully taken into account in the design of robust and efficient algorithms for these environments. The communication among mobile hosts is one among the many issues that need to be resolved efficiently before MANET becomes a commodity.

In this paper, we propose to discuss the communication problem in MANET as well as present some characteristic techniques for the design, the analysis and the performance evaluation of distributed communication protocols for mobile ad hoc networks. More specifically, we propose to review two different design techniques. While the first type of protocols tries to create and maintain routing paths among the hosts, the second set of protocols uses a randomly moving subset of the hosts that acts as an intermediate pool for receiving and delivering messages. We discuss the main design choices for each approach, along with performance analysis of selected protocols.

1 Introduction

With the recent dramatic progress in computer platforms and wireless communications, mobile ad hoc networks (MANET) are about to enter the mainstream without however having created yet a solid algorithmic basis for coping with a new set of crucial issues and inherent trade-offs. Critical features, including mobility, severe limitations in power and communication capabilities and limited reliability, have to be carefully taken into account towards designing efficient

* Dr. A. Boukerche's work was partially supported by the Canada Research Chair (CRC) Program, the Canada Foundation for Innovation Grant and Ottawa Distinguished Researcher Award (OIT/ODRA #2021722), and Dr. S. Nikoletseas's work was partially supported by the IST/FET Program of the European Union under contract numbers IST-1999-14186 (ALCOM-FT) and IST-2001-33116 (FLAGS).

M.C. Calzarossa and E. Gelenbe (Eds.): MASCOTS 2003, LNCS 2965, pp. 235–253, 2004.
© Springer-Verlag Berlin Heidelberg 2004

and robust algorithms for such environments. The communication among mobile hosts is one among the many important issues that need to be resolved efficiently before MANET becomes a commodity.

The problem of communication among remote mobile hosts is one of the most fundamental problems in mobile ad-hoc networks and is at the core of many distributed algorithms, such as counting the number of hosts, leader election, data processing just to mention a few. In mobile ad-hoc networks, a node MH_s that wants to communicate with MH_r might not be within communication range, however it could communicate if other hosts lying "in between" them are willing to forward the packets on their behalves.

The simplest way to establish communication in such networks is to perform the basic flooding paradigm. By flooding, we mean a distributed process of broadcasting a packet to all nodes within the transmission radius of the initiating sender and so on, until all nodes in the sender's "connected component" receive the packet. It is evident that such an approach will waste a large portion of the available bandwidth and therefore it is inefficient to use such schemes.

In order to avoid the flooding problem, several protocols have been proposed in the related literature. A possible categorization of these protocols is based on the approach taken to establish a reliable communication. One approach tries to construct new paths and update these paths dynamically between the mobile hosts. The second approach, instead avoids path creation and maintenance by taking advantage of the hosts movement and accidental meetings in the network area.

1.1   The Path Construction and Maintenance Approach

In an attempt to overcome the inefficiency of the flooding protocols, some algorithms have been proposed that try to construct a a dynamic-graph, e.g., connectivity related dynamic data structure that represents the topology of the network by a series of message passing. By taking advantage of such a data structure, messages can be transmitted over a limited number of intermediate hosts that interconnect MH_s with MH_r, also known as *routing paths* or *routes*. Indeed, this approach is common in ad-hoc mobile networks that either cover a relatively small space (i.e., the temporary network has a small diameter in terms of the transmission range), or it is dense (i.e., comprised of thousands of wireless hosts). In such a case, since almost all "intermediate" locations are occupied by some hosts, communication is possible and broadcasting can be efficiently accomplished as outlined in [1].

In [9,39], the authors propose the Dynamic Source Routing (DSR) protocol, which uses *on-demand route discovery*. There exists several DSR protocol variations. Ecah of the variations tries to optimize the route discovery overhead. Global positioning system information has been used to reduce the message overhead mainly due to the flooding paradigm. In [6,7], Boukerche et. al. propose two modifications to the DSR ad-hoc routing protocol: (1) Use geographic location information for propagating route requests, and (2) Randomly choosing only a subset of neighbors to which to forward a route request message. The results obtained in [7] are quite encouraging.

In [40], the authors present the Ad Hoc On Demand Distance Vector routing (or AODV) protocol that also uses a demand-driven route establishment procedure. More recently two mechanisms have been proposed to deal with the congestion control and communication problems within the AODV routing protocol. First, the authors in [3] have investigated a randomized approach of the AODV protocol. Then they have presented a preemptive ad hoc on-demand distance vector routing protocol for mobile ad hoc networks [4,5]. They have reported a set of simulation experiments to evaluate their performance when compared to the original AODV on-demand routing protocol. The results they have obtained were quite encouraging.

In [34,35], the authors have presented a Temporally-Ordered Routing Algorithm (TORA), designed to minimize reaction to topological changes by localizing routing-related messages to a small set of nodes near the change. [36,23] attempt to combine proactive and reactive approaches in the Zone Routing Protocol (ZRP), by initiating route discovery phase on-demand, but limit the scope of the proactive procedure only to the initiator's local neighborhood. Geographic location information using global positing system (GPS) has also been investigated for the ZRP protocol in [8] using a similar approach as in [7]. In [28], the authors have proposed the Location-Aided Routing (LAR) protocol that uses a global positioning system to provide location information to improve the performance of the protocol by limiting the search for a new route in a smaller "request zone". Several other contributions in this volume study algorithms and implementations for MANETs, in particular [20] which discusses a new routing protocol that allows for Quality-of-Service based routing including power awareness, and [27,33].

1.2   The "Support" Approach

In contrast to all such methods, in [15,11,14] a new framework is introduced for protocols designed for highly dynamic movement of the mobile users. The idea of the approach is to take advantage of the mobile hosts natural movement by exchanging information whenever mobile hosts meet incidentally. Specifically, the protocol framework proposes the idea of forcing a small subset of the deployed hosts to move as per the needs of the protocol. This set of hosts is called the "support" of the network (denoted by Σ). Assuming the availability of such hosts, the designer can use them suitably *by specifying their motion* in certain times that the algorithm dictates.

This approach may be particularly useful in cases of rapid deployment of a number of mobile hosts, where it is possible to have a small team of fast moving and versatile vehicles, to implement the support. These vehicles can be cars, jeeps, motorcycles or helicopters. This small team of fast moving vehicles can also be a collection of independently controlled mobile modules, i.e. robots. This specific approach is similar to [42] that studies the problem of motion coordination in distributed systems consisting of such robots, which can connect, disconnect and move around. Having the support moving in a coordinated way, *i.e. as a chain of nodes*, has some similarities to [42].

Definition 1. *The subset of the mobile hosts of an ad-hoc mobile network whose motion is determined by a network protocol P is called the* support Σ *of* $\mathcal{P}$. *The part of P which indicates the way that members of Σ move and communicate is called the* support management subprotocol M_Σ *of* $\mathcal{P}$.

The scheme follows the general design principle of mobile networks (with a fixed subnetwork however) called the "two tier principle" [26], stated for mobile networks with a fixed subnetwork however, which says that any protocol should try to move communication and computation to the fixed part of the network. The assumed set of hosts that are coordinated by the protocol *simulates* such a (skeleton) network; the difference is that the simulation actually constructs a coordinated *moving* set.

In a recent paper ([29]), Q.Li and D.Rus present a model which has some similarities to the support approach. The authors give an interesting, yet different, protocol to send messages, which forces all the mobile hosts to slightly deviate (for a short period of time) from their predefined, deterministic routes, in order to propagate the messages. Their protocol is, thus, *compulsory* for any host and it works only for deterministic host routes. It considers the propagation of only one message (end to end) each time, in order to be correct. In contrast, the support scheme allows for simultaneous processing of many communication pairs. In their setting, [29] shows optimality of message transmission times.

2   The AODV Protocol

The Ad-Hoc On-Demand Distance Vector (AODV) routing protocol described in [38] builds on the DSDV algorithm ([37,39]). AODV is an improvement on DSDV because it typically minimizes the number of required broadcasts by creating routes on a demand basis, as opposed to maintaining a complete list of routes as in the DSDV algorithm. AODV is a pure on-demand route acquisition system, since nodes that are not on a selected path do not maintain routing information or participate in routing table exchanges ([38]).

When a source node desires to send a message to some destination node and does not already have a valid route to that destination, it initiates a path discovery process to locate the other node. It broadcasts a route request (RREQ) packet to its neighbors, which then forward the request to their neighbors, and so on, until either the destination or an intermediate node with a "fresh enough" route to the destination is located. AODV utilizes destination sequence numbers to ensure that all routes are loop-free and contain the most recent route information. Each node maintains its own sequence number, as well as a broadcast ID. The broadcast ID is incremented for every RREQ the node initiates, and together with the node's IP address, uniquely identifies a RREQ. Along with its own sequence number and the broadcast ID, the source node includes in the RREQ the most recent sequence number it has for the destination. Intermediate nodes can reply to the RREQ only if they have a route to the destination whose corresponding destination sequence number is greater than or equal to that contained in the RREQ.

During the process of forwarding the RREQ, intermediate nodes record in their route tables the address of the neighbor from which the first copy of the broadcast packet is received, thereby establishing a reverse path. If additional copies of the same RREQ are later received, these packets are discarded. Once the RREQ reaches the destination or an intermediate node with a fresh enough route, the destination/intermediate node responds by unicasting a route reply (RREP) packet back to the neighbor from which it first received the RREQ. As the RREP is routed back along the reverse path, nodes along this path set up forward route entries in their route tables which point to the node from which the RREP came. These forward route entries indicate the active forward route. Associated with each route entry is a route timer which will cause the deletion of the entry if it is not used within the specified lifetime. Because the RREP is forwarded along the path established by the RREQ, AODV only supports the use of symmetric links.

Routes are maintained as follows. If a source node moves, it is able to reinitiate the route discovery protocol to find a new route to the destination. If a node along the route moves, its upstream neighbor notices the move and propagates a link failure notification message (an RREP with infinite metric) to each of its active upstream neighbors to inform them of the erasure of that part of the route [39]. These nodes in turn propagate the link failure notification to their upstream neighbors, and so on until the source node is reached. The source node may then choose to reinitiate route discovery for that destination if a route is still desired.

3   The Randomized AODV Protocol

The AODV protocol is extended with a drop factor that induces a randomness feature to result in Randomized Ad-Hoc On-Demand Routing (R-$AODV$) protocol [3]. During the route discovery process, every intermediary or router node between the source and the destination nodes makes a decision to either broadcast/forward the RREQ packet further towards the destination or drop it. Before forwarding a RREQ packet, every node computes the drop factor which is a function of the inverse of the number of hop counts traversed by the RREQ packet. This drop factor lies in the range of 0 to 1. Also, the node generates a random number from 0 to 1. If this random number is higher than the drop factor, the node forwards the RREQ packet. Otherwise, the RREQ packet is dropped. Dropping of RREQ packets does not necessarily result in a new route discovery process by the source node. This is due to the fact that the original broadcast by the source node results in multiple RREQ packets via the neighbors and this diffusing wave results quickly in a large number of RREQ packets traversing the network in search of the destination. A major proportion of these packets are redundant due to the fact that in the ideal case, a single RREQ packet can find the best route. Also, a number of these packets diffusing in directions away from the destination shall eventually timeout. Hence, in $R\text{-}AODV$, the aim is to minimize on these redundant RREQ packets, or alternatively, drop as much as possible of these redundant RREQ packets. The drop policy is conservative and its value becomes lesser with higher number of hops. As RREQ packets get near the destination node, the chances of survival of RREQ packets is higher. Hence, the first phase of the route discovering process, that is, finding the destination node, is completed as soon as possible and a RREP packet can be transmitted from the destination node back to the source node.

In $R\text{-}AODV$ [3], the dropping of redundant RREQ packets reduces a proportion of RREQ packets that shall never reach the destination node, resulting in a decrease of network congestion. Hence, the ratio of the number of packets received by the nodes to the number of packets sent by the nodes, namely, the throughput, should be higher in $R\text{-}AODV$ compared to AODV.

The following algorithm has been used in the decision making process of whether to drop the RREQ packets by the intermediary or routing nodes (see [3] for more information).

```
Step 1: Calculate drop_factor
        drop_factor = (1/(Hop_count_of_RREQ_packet + 1))
Step 2: Calculate a random value in the range of 0 to 1.
Step 3: If (random_value > drop_factor)
        then broadcast/forward RREQ_packet
        else drop RREQ_packet
```

4   Adaptive Preemptive AODV Protocol

In this section, we review the preemptive protocol first, as described in [4,5], then we discuss how this preemptive protocol is added into the original AODV.

The preemptive protocol initiates a route rediscovery before the existing path breaks. It overlaps the route discovery routine and the use of the current active path, thereby reducing the average delay per packet. During the development of $P_r AODV$, we have investigated several preemptive mechanisms. In [4,5], the following two approaches have been investigated:

(i) *Schedule a rediscovery in advance:* In this approach, when a reply packet returns to the source through an intermediate node, it collects the information of the links. Therefore, when the packet arrives at the source, the information about the condition of all links will be known, including the minimum value of the lifetime of the links. Hence, we can schedule a rediscovery $T_{rediscovery}$ time before the path breaks.

(ii) *Warn the source before the path breaks:* Some mechanisms are needed to take care of finding which path is likely to break. We can monitor the signal power of the arrived packets as follows: when the signal power is below a threshold value, we begin the ping-pong process between this node and its immediate neighbor nodes. This node sends to its neighbors in the upstream a hello packet called *ping*, and the neighboring nodes will respond with a hello packet called *pong*. Such *ping-pong* messages should be monitored carefully. In [4], the following approach was used, when bad packets were received (or we timeout on ping packets) a warning message should be sent back to the source during the monitoring period. Upon receiving a warning message, a path rediscovery routine is invoked.

Warning the source: Recall that a preemptive protocol should monitor the *signal power* of the receiving packet. In [4], the authors' approach is to monitor the transfer time for the hello packets. According to [21], $P_n = \frac{P_0}{r^n}$, where P_0 is a constant for each link and n is also a constant, we can see that the distance, which connects to the transfer time of the packet via each link, has a one-to-one relationship to the signal power. Thus, by monitoring the transfer time of packets via each link, we can identify when the link is going to break.

In this protocol, when a node (destination) receives a real packet other than AODV packets, it sends out a pong packet to its neighbor (source), i.e., where the packet comes from. Note that there is no need to send this pong packet if we can include a field in the real packet to store the current time the packet is sent. However, in the original packet, it doesn't have such a field, hence, we need an additional *pong* packet to get such information. In both ping and pong packets, we include a field containing the current time that the packet is sent. When the source monitoring the transfer time is greater than some specific value c_1, it begins the *ping-pong* process. If more than two pong packets arrived with the transfer time greater than a constant c_2, the node will warn the source of the real packet.

Upon receiving the warning message, the source checks if the path is still active. This can easily be done, since each route entry for a destination associates a value, *rt_last_send*, which indicates if the path is in use or not. If the path is still in use, then a rediscovery routine is invoked.

Theoretically, these two cases can take care of finding new paths before current paths break. The first case is quite helpful if the data transmission rate is slow. If only one data packet is sent out per time-expired period, then even the current route is still needed. Note that this entry may not have packets to go through it during the next time period. This can happen if the interval between two continuous packets is greater that the timeout period of the current route. The second case can be used to find a new path whenever a warning message is generated. However, those two cases are also based on some assumptions, such as the span of a link which can be estimated by the time out of ping packets and the values of those constants are accurate. Once we break those assumptions, the test results will be unexpected.

5   The SUPPORT Protocols

5.1   The Motion Graph Model

In [15] the authors propose an *explicit* model of motions in the light of the fact that the motions of the hosts are the cause of the fragility of the virtual links. Thus they distinguish explicitly between (a) the *fixed* (for any algorithm) *space* of possible motions of the mobile hosts and (b) the kind of motions that the hosts perform inside this space.

They model the space of motions only combinatorially, i.e. as a graph. Future research is expected to complement this effort by introducing geometry details into the model.

In particular, they abstract the environment where the stations move (in three-dimensional space with possible obstacles) by a *motion-graph* (i.e. neglecting the detailed geometric characteristics of the motion. They first assume that each mobile host has a transmission range represented by a sphere tr centered by itself. This means that any other host inside tr can receive any message broadcast by this host. They approximate this sphere by a cube tc with volume $\mathcal{V}(tc)$, where $\mathcal{V}(tc) \simeq \mathcal{V}(tr)$. Given that the mobile hosts are moving in the space $\mathcal{S}$, $\mathcal{S}$ is divided into consecutive cubes of volume $\mathcal{V}(tc)$.

Definition 2. *The motion graph $G(V, E)$, ($|V| = n$, $|E| = m$), which corresponds to a quantization of $\mathcal{S}$ is constructed in the following way: a vertex $u \in G$ represents a cube of volume $\mathcal{V}(tc)$. An edge $(u, v) \in G$ if the corresponding cubes are adjacent.*

The number of vertices n, actually approximates the ratio between the volume $\mathcal{V}(\mathcal{S})$ of space $\mathcal{S}$, and the space occupied by the transmission range of a mobile host $\mathcal{V}(tr)$. Given the transmission range tr, n depends linearly on the volume of space $\mathcal{S}$ regardless of the choice of tc, and $n = O\left(\frac{V(\mathcal{S})}{V(tr)}\right)$.

Since the edges of G represent neighboring polyhedra each node is connected with a constant number of neighbors, which yields that $m = \Theta(n)$. In the sequel, tc is taken to be a cube, thus G has maximum degree of six and $m \leq 6n$.

Thus *motion graph G* is (usually) a *bounded degree graph* as it is derived from a regular graph of small degree by deleting parts of it corresponding to motion or communication obstacles.

5.2   The Basic Idea of the Protocols

In simple terms, the support protocols work as follows: The nodes of the support move fast enough so that they visit (in sufficiently short time) the entire motion graph. Their motion is accomplished in a distributed way via a *support motion subprotocol P_1*. When some node of the support is within communication range of a sender, an underlying *sensor subprotocol P_2* notifies the sender that it may send its message(s).

The messages are then stored "somewhere within the support structure". When a receiver comes within communication range of a node of the support, the receiver is notified that a message is "waiting" for him and the message is then forwarded to the receiver.

The messages received by the support are propagated within the structure when two or more members of the support meet on the same site (or are within communication range). A synchronization subprotocol P_3 is used to dictate the way that the members of the support exchange information.

In a way, the support Σ plays the role of a (moving) skeleton subnetwork (whose structure is defined by the motion subprotocol P_1), through which all communication is routed. From the above description, it is clear that the size, k, and the shape of the support may affect performance.

Note that the proposed scheme does not require the propagation of messages through hosts that are not part of Σ, thus its security relies on the support's security and is not compromised by the participation in message communication of other mobile users. For a discussion of intrusion detection mechanisms for ad-hoc mobile networks see [43].

5.3   The SNAKE Protocol

The main idea of the protocol proposed in [15,11,14] is as follows. There is a set-up phase of the ad-hoc network, during which a predefined set, k, of hosts, become the nodes of the support. The members of the support perform a leader election by running a randomized breaking symmetry protocol in anonymous networks (see e.g. [24]). This is run once and imposes only an initial communication cost. The elected leader, denoted by MS_0, is used to co-ordinate the support topology and movement. Additionally, the leader assigns local names to the rest of the support members MS_1, MS_2, ..., MS_{k-1}.

The nodes of the support move in a coordinated way, always remaining pairwise adjacent (i.e., forming a chain of k nodes), so that they sweep (given some time) the entire motion graph. This encapsulates the *support motion subprotocol* P_1^S. Essentially the motion subprotocol P_1^S enforces the support to move as a "snake", with the head (the elected leader MS_0) doing a random walk on the motion graph and each of the other nodes MS_i executing the simple protocol "move where MS_{i-1} was before". More formally, the movement of Σ is defined as follows:

Initially, MS_i, $\forall i \in \{0, 1, \ldots, k-1\}$, start from the same area-node of the motion graph. The direction of movement of the leader MS_0 is given by a memoryless operation that chooses *randomly the direction* of the next move. Before leaving the current area-node, MS_0 sends a message to MS_1 that states the new direction of movement. MS_1 will change its direction as per instructions of MS_0 and will propagate the message to MS_2. In analogy, MS_i will follow the orders of MS_{i-1} after transmitting the new directions to MS_{i+1}. Movement orders received by MS_i are positioned in a queue Q_i for sequential processing. The very first move of MS_i, $\forall i \in \{1, 2, \ldots, k-1\}$ is delayed by a δ period of time.

[15] assumes that the mobile support hosts move with a common speed. Note that the above described motion subprotocol P_1^S enforces the support to move as a "snake", with the head (the elected leader MS_0) *doing a random walk on the motion graph G* and each of the other nodes MS_i executing the simple protocol "move where MS_{i-1} was before". This can be easily implemented because MS_i will move following the edge from which it received the message from MS_{i-1} and therefore this protocol does not require common sense of orientation.

The purpose of the random walk of the head is to ensure a *cover* (within some finite time) of the whole motion graph, without memory (other than local) of topology details. Note that this memoryless motion also ensures fairness. The value of the *random walk principle* of the motion of the support is further justified in the correctness and the efficiency analyses of the protocol in [15], where the basic communication protocol should meet some performance requirements *regardless of the motion of the hosts not in Σ*.

A modification of M_Σ^S proposed in [15] is that the head does a random walk on *a spanning subgraph* of G (e.g. a spanning tree). This modified M_Σ (call it M_Σ') is more efficient in our setting since "edges" of G just represent adjacent locations and "nodes" are really possible host places.

A simple approach to implement the support *synchronization subprotocol* P_3^S presented in [15] is to use a forward mechanism that transmits incoming messages (in-transit) to neighboring members of the support (due to P_1^S at least one support host will be close enough to communicate). In this way, all incoming messages are eventually copied and stored in every node of the support. This is not the most efficient storage scheme and can be refined in various ways in order to reduce memory requirements.

5.4   The RUNNERS Protocol

A different approach to implement M_Σ is to allow each member of Σ not to move in a snake-like fashion, but to perform an *independent* random walk on the motion graph G, i.e., the members of Σ can be viewed as "runners" running on G. In other words, instead of maintaining at all times pair wise adjacency between members of Σ, all hosts sweep the area by moving independently of each other. When two runners meet, they exchange any information given to them by senders encountered using a new synchronization subprotocol P_3^R. As in the snake case, the same underlying sensor sub-protocol P_2 is used to notify the sender that it may send its message(s) when being within communication range of a node of the support.

As presented in [15], the runners protocol does not use the idea of a (moving) backbone subnetwork as no motion subprotocol P_1^R is used. However, all communication is still routed through the support Σ and it is expected that the size k of the support (i.e., the number of runners) will affect performance in a more efficient way than that of the snake approach. This expectation stems from the fact that each host will meet each other in parallel, accelerating the spread of information (i.e., messages to be delivered). This superiority of the RUNNERS protocol over the SNAKE protocol has been indeed verified (using large scale simulations) in [12]. Also, the recent work of Dimitriou, Nikoletseas and Spirakis [19] on concurrent random walks of interacting particles on graphs provides an analytical tool for the estimation of communication times in the RUNNERS protocol.

A member of the support needs to store all undelivered messages, and maintain a list of receipts to be given to the originating senders. For simplicity, we can assume a generic storage scheme where all undelivered messages are members of a set S_1 and the list of receipts is stored on another set S_2. In fact, the unique ID of a message and its sender ID is all that is needed to be stored in S_2.

When two runners meet at the same vertex of the motion graph G, the synchronization subprotocol P_3^R is activated. This subprotocol imposes that when runners meet on the same vertex, their sets S_1 and S_2 are synchronized. In this way, a message delivered by some runner will be removed from the set S_1 of the rest of runners encountered, and similarly delivery receipts already given will be discarded from the set S_2 of the rest of runners. The synchronization subprotocol P_3^R is partially based on the *two-phase commit* algorithm as presented in [30] and works as follows.

Let the members of Σ residing on the same site (i.e., vertex) u of G be $MS_1^u, \ldots, MS_j^u$. Let also $S_1(i)$ (resp. $S_2(i)$) denote the S_1 (resp. S_2) set of runner MS_i^u, $1 \leq i \leq j$. The algorithm assumes that the underlying sensor sub-protocol P_2 informs all hosts about the runner with the lowest ID, i.e., the runner MS_1^u. P_3^R consists of two rounds.

Round 1: All $MS_1^u, \ldots, MS_j^u$ residing on vertex u of G, send their S_1 and S_2 to runner MS_1^u. Runner MS_1^u collects all the sets and combines them with its own to compute its new sets S_1 and S_2: $S_2(1) = \bigcup_{1 \leq l \leq j} S_2(l)$ and $S_1(1) = \bigcup_{1 \leq l \leq j} S_1(l) - S_2(1)$.

Round 2: Runner MS_1^u broadcasts its decision to all the other support member hosts. All hosts that received the broadcast apply the same rules, as MS_1^u did, to join their S_1 and S_2 with the values received. Any host that receives a message at Round 2 and which has not participated in Round 1, accepts the value received in that message as if it had participated in Round 1.

This simple algorithm guarantees that mobile hosts which remain connected (i.e., are able to exchange messages) for two continuous rounds, will manage to synchronize their S_1 and S_2. Furthermore, on the event that a new host arrives or another disconnects during Round 2, the execution of the protocol will not be affected. In the case where runner MS_1^u fails to broadcast in Round 2 (either because of an internal failure or because it has left the site), then the protocol is simply re-executed among the remaining runners.

Remark that the algorithm described above does not offer a mechanism to remove message receipts from S_2; eventually the memory of the hosts will be exhausted. A simple approach to solve this problem and effectively reduce the memory usage is to construct an order list of IDs contained in S_2, for each receiver. This ordered sequence of IDs will have gaps - some messages will still have to be confirmed, and thus are not in S_2. In this list, we can identify the maximum ID before the first gap. It is now possible to remove from S_2 all message receipts with smaller ID than this ID.

Recent research towards an analysis of the RUNNERS protocol. In [19], Dimitriou, Nikoletseas and Spirakis study the problem of information propagation between concurrent random walks. The obtained results can be used in the analysis of the expected communication time of the "RUNNERS" protocol.

In particular, they study the following problem: Consider k particles, 1 red and $k-1$ white, chasing each other on the nodes of a graph G. If the red one catches one of the white, it "infects" it with its color. The newly red particles are now available to infect more white ones. When is it the case that all white will become red?

The authors model this problem by k concurrent random walks, one corresponding to the red particle and $k - 1$ to the white ones. The *infection time* T_k of infecting *all* the white particles with red color is then a random variable that depends on k, the initial position of the particles, the number of nodes and edges of the graph, as well as on the structure of the graph.

They develop a set of probabilistic tools that are used to obtain upper bounds on the expected value of T_k for general graphs and important special cases (including cliques and expander graphs). They also evaluate and validate their analytical results by large scale experiments.

A recent AODV - RUNNERS **comparison.** In [10], the authors focus on the effect of mobility and users' density on the performance of routing protocols. In particular, they choose to investigate the impact on AODV (following the path maintenance approach) and RUNNERS (following the support approach). They have implemented the two protocols and performed a large scale and detailed simulation study of their performance. The main findings are: the AODV protocol behaves well in networks of high user density and low mobility rate, while its performance drops for sparse networks of highly mobile users. On the other hand, the RUNNERS protocol seems to tolerate well (and in fact benefit from) high mobility rates and low densities.

5.5   Special Variations of the SUPPORT Protocol

In [11,16,17] the authors introduce two practical models of ad-hoc mobile networks and accordingly variate the "RUNNERS" protocol: a) In *hierarchical* ad-hoc networks, comprised of dense subnetworks of mobile users interconnected by a very fast yet limited backbone infrastructure, a "runners" support is used in each lower level of the hierarchy (i.e. in each dense subnetwork). This hierarchical implementation of the support approach is shown to significantly outperform a simple implementation of it in hierarchical ad-hoc networks. b) In *highly changing* ad-hoc networks, where the deployment area changes in a highly dynamic way and is unknown to the protocol, an adaptive runners protocol is introduced. This protocol instead of using a fixed sized support for the whole duration of the protocol, employs a support of some initial (small) size which adapts to the actual levels of traffic and the (unknown and possibly rapidly changing) network area, by changing its size in order to converge to an optimal size, thus satisfying certain Quality of Service criteria.

6   Analysis of the SNAKE Protocol

Time-efficiency of semi-compulsory protocols (i.e. protocols affecting the motion of *some* of the hosts) for ad-hoc networks is not possible to estimate without a scenario for the motion of the mobile users not in the support (i.e. the non-compulsory part). In a way similar to [13,24], in [15] the authors propose an "on-the-average" analysis by assuming that the movement of each mobile user *is a random walk on the corresponding motion graph G*. This kind of analysis is proposed as a necessary and interesting first step in the analysis of efficiency of any semi-compulsory or even non-compulsory protocol for ad-hoc mobile networks. In fact, the assumption that the mobile users are moving randomly (according to uniformly distributed changes in their directions and velocities, or according to the random waypoint mobility model, by picking random destinations) has been used in [25], [22].

In the light of the above, [15] assumes that any host not belonging in the support, conducts a random walk on the motion graph, independently of the other hosts.

The head of the snake does a *continuous time random walk* on the motion graph $G(V, E)$. This is without loss of generality (since, if it is a discrete time random walk, all results transfer easily, see [2]).

Let the random walk of a host on G that induces a continuous time Markov chain M_G as follows: The states of M_G are the vertices of G and they are finite. Let s_t denote the state of M_G at time t. Given that $s_t = u$, $u \in V$, the probability that $s_{t+dt} = v$, $v \in V$, is $p(u, v) \cdot dt$ where

$$p(u, v) = \begin{cases} \frac{1}{d(u)} & \text{if } (u, v) \in E \\ 0 & \text{otherwise} \end{cases}$$

and $d(u)$ is the degree of vertex u.

The authors in [15] assume that all random walks are *concurrent* and that there is a global time t, not necessarily known to the hosts.

Note 3. Since the motion graph G is finite and connected, the continuous Markov chain abstracting the random walk on it is automatically time-reversible.

Definition 4. *For a vertex j, let $T_j = \min\{t \succeq 0 : s_t = j\}$ be the first hitting time of the walk onto that vertex and let $E_i T_j$ be its expected value, given that the walk started at vertex i of G.*

Definition 5. *For the random walk of any particular host, let $\pi()$ be the stationary distribution of its position after a sufficiently long time.*

Let $E_\pi[\ ]$ denote the expectation for the chain started at the stationary distribution π.

The communication time of M_Σ^S is the total time needed for a message to arrive from a sending node u to a receiving node v. Since the communication time, T_{total}, between MH_S, MH_R is bounded above by $X + Y + Z$, where X is the time for MH_S to meet Σ, Y is the time for MH_R to meet Σ (after X) and Z is the message propagation time within Σ, in order to estimate the expected values of X and Y (these are random variables), the authors in [15] work as follows:

(a) Note first that X, Y are, statistically, of the same distribution, under the assumption that u, v are randomly located (at the start) in G. Thus $E(X) = E(Y)$.

(b) Replace the meeting time of u and Σ by a hitting time, using the following thought experiment:
(b1) Fix the support Σ in an "average" place inside G.
(b2) Then collapse Σ to a single node (by collapsing its nodes to one but keeping the incident edges). Let H be the resulting graph, σ the resulting node and $d(\sigma)$ its degree.
(b3) Then estimate the hitting time of u to σ assuming u is somewhere in G, according to the *stationary distribution*, π of its walk on H. Let the expected value of this hitting time be $E_\pi T_\sigma^H$.

Thus, now $E(T_{total}) = 2E_\pi T_\sigma^H + \mathcal{O}(k)$.

Note that the equation above amortizes over meeting times of senders (or receivers) because it uses the stationary distribution of their walks for their position when they decide to send a message.

Now, proceeding as in [2], they get the following lemma.

Lemma 6. *[15] For any node σ of any graph H in a continuous-time random walk*

$$E_\pi T_\sigma^H \leq \frac{\tau_2(1 - \pi(\sigma))}{\pi(\sigma)}$$

where $\pi(\sigma)$ is the (stationary) probability of the walk at node (state) σ and τ_2 is the relaxation time of the walk.

Note that in the above bound, $\tau_2 = \frac{1}{\lambda_2}$ where λ_2 is the second eigenvalue of the (symmetric) matrix $S = \{s_{i,j}\}$ where $s_{i,j} = \sqrt{\pi(i)}\, p_{i,j}\, (\sqrt{\pi(i)})^{-1}$ and $P = \{p_{i,j}\}$ is the transition matrix of the Markov Chain corresponding to the walk. Since S is symmetric, it is diagonalizable and the spectral theorem gives the following representation: $S = U\Lambda U^{-1}$ where U is orthonormal and Λ is a diagonal real matrix of diagonal entries $0 = \lambda_1 < \lambda_2 \leq \cdots \leq \lambda_{n'}$, $n' = n - k + 1$ (where $n' = |V_H|$). These λ's are the eigenvalues of both P and S. In fact, λ_2 is an indication of the *expansion* of H and of the asymptotic rate of convergence to the stationary distribution, while relaxation time τ_2 is the corresponding interpretation measuring time.

It is a well-known fact (see e.g. [32]) that $\forall v \in V_H$, $\pi(v) = \frac{d(v)}{2m'}$ where $m' = |E_H|$ is the number of the edges of H and $d(v)$ is the degree of v in H. Thus $\pi(\sigma) = \frac{d(\sigma)}{2m'}$.

By estimating $d(\sigma)$ and m' and remarking that the operation of locally collapsing a graph does not reduce its expansion capability and hence $\lambda_2^H \geq \lambda_2^G$ (see *Lemma 1.15* [18], ch. 1, p.13), they get the following:

Theorem 7. *[15]*

$$E(X) \;=\; E(Y) \;\leq\; \frac{1}{\lambda_2(G)}\,\Theta\!\left(\frac{n}{k}\right)$$

Theorem 8. *[15] The expected communication time of the* SNAKE *protocol is bounded above by the formula*

$$E(T_{total}) \;\leq\; \frac{2}{\lambda_2(G)}\,\Theta\!\left(\frac{n}{k}\right) \;+\; \Theta(k)$$

Note that the above upper bound is minimized when $k = \sqrt{\frac{2n}{\lambda_2(G)}}$, a fact also verified by simulations (see [13] and also [15]).

The authors in [15] remark that this upper bound is tight in the sense that by [2], Chapter 3,

$$E_\pi T_\sigma^H \;\geq\; \frac{\left(1 - \pi(\sigma)\right)^2}{q_\sigma \pi(\sigma)}$$

where $q_\sigma = \sum_{j \neq \sigma} p_{oj}$ in H_j is the total exit rate from σ in H. By using martingale type arguments [15] show sharp concentration around the mean degree of σ.

6.1   Robustness

Now, we examine the robustness of the support management subprotocol M_Σ^S and M_Σ^R under single stop-faults. We present two Theorems proved in [15]

Theorem 9. *[15] The support management subprotocol M_Σ^S is 1-fault tolerant.*

Proof. If a single host of Σ fails, then the next host in the support becomes the head of the rest of the snake. We thus have two, independent, random walks in G (of the two snakes) which, however, will meet in finite expected time and re-organize as a single snake via a very simple re-organization protocol which is the following:

When the head of the second snake Σ_2 meets a host h of the first snake Σ_1 then the head of Σ_2 follows the host which is "in front" of h in Σ_1, and all the part of Σ_1 after and including h waits, to follow Σ_2's tail. $\square$

Note that in the case that more than one faults occur, the procedure for merging snakes described above may lead to deadlock (see Figure 1).

Theorem 10. *[15] The support management subprotocol M_Σ^R is resilient to t faults, for any $0 \leq t < k$.*

Proof. This is achieved using redundancy: whenever two runners meet, they create copies of all messages in transit. In the worst-case, there are at most $k - t$ copies of each message. Note, however, that messages may have to be re-transmitted in the case that only one copy of them exists when some fault occurs.

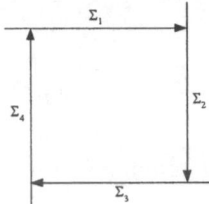

Fig. 1. Deadlock situation when four "snakes" are about to be merged.

To overcome this limitation, the sender will continue to transmit a message for which delivery has not been confirmed, to any other members of the support encountered. This guarantees that more than one copy of the message will be present within the support structure.   □

7   Conclusions

In this paper, we focus on the fundamental problem of communication in mobile ad hoc networks. Towards demonstrating alternative choices in the design of communication algorithms, we reviewed two characteristic approaches in the relevant literature: one that tries to create and maintain routes between mobile hosts, and another that, instead, uses a randomly moving subset of the hosts as "helpers" for receiving and delivering messages. We presented several protocols for each approach, along with a discussion of performance aspects of these protocols.

References

1. M. Adler and C. Scheideler, "Efficient communication strategies for ad-hoc wireless networks," *10th Annual Symposium on Parallel Algorithms and Architectures, ACM*, 1998, pp. 259–268.
2. D. Aldous and J. Fill, "Reversible markov chains and random walks on graphs," http://stat-www.berkeley.edu/users/aldous/book.html, 1999.
3. A. Boukerche, "Congestion control mechanisms for on-demand vector ad hoc routing protocols," *Europar'2003 Conference*.
4. A. Boukerche and L. Zhang "Preemptive on-demand routing protocol for wireless and mobile ad Hoc networks," *IEEE/ACM/SCS 36th Annual Simulation Symposium*, pp. 73-80, 2003.
5. A. Boukerche and L. Zhang, "A performance evaluation of an efficient preemptive on-demand routing protocol for mobile ad hoc networks," *Journal of Wireless and Mobile Communication* (To appear).
6. A. Boukerche, J. Linus, Agrawal Saurabah, "A performance study of dynamic source routing protocols for mobile and wireless ad hoc networks", *EuroPar'2002 Conference*, pp. 957-964, LNCS 2400, 2002.

7. A. Boukerche, S. Vaidya, "Route Discovery Optimization Scheme Using GPS Systems", *SCS/IEEE Annual Simulation Symposium* pp. 20-26, 2002.
8. A. Boukerche, S. Roger, "GPS Quey Optimization in Wireless and Mobile Ad Hoc Networks", *6th IEEE Symposium on Computer and Communications* pp. 198-2003, 2001.
9. J. Broch, D.B. Johnson, and D.A. Maltz, "The dynamic source routing protocol for mobile ad-hoc networks," *Tech. report, IETF*, Internet Draft, December 1998, draft-ietf-manet-dsr-01.txt.
10. I. Chatzigiannakis, E. Kaltsa, and S. Nikoletseas, "On the effect of user mobility and density on the performance of routing protocols for ad-hoc mobile networks," *Tech. report, CTI*, 2003.
11. I. Chatzigiannakis and S. Nikoletseas, "Design and analysis of an efficient communication strategy for hierarchical and highly changing ad-hoc mobile networks," *ACM/Baltzer Journal of Mobile Networks and Applications (MONET)*, (To appear).
12. I. Chatzigiannakis, S. Nikoletseas, N. Paspalis, P. Spirakis, and C. Zaroliagis, "An experimental study of basic communication protocols in ad-hoc mobile networks," *5th International Workshop on Algorithmic Engineering (WAE 2001)*, Springer-Verlag, 2001, Lecture Notes in Computer Science, LNCS 2141, pp. 159–171.
13. I. Chatzigiannakis, S. Nikoletseas, and P. Spirakis, "Analysis and experimental evaluation of an innovative and efficient routing approach for ad-hoc mobile networks," *4th International Workshop on Algorithmic Engineering (WAE 2000)*, Springer-Verlag, 2000, Lecture Notes in Computer Science, LNCS 1982, pp. 99–110.
14. I. Chatzigiannakis, S. Nikoletseas, and P. Spirakis, "An efficient communication strategy for ad-hoc mobile networks", *15th International Workshop on Distributed Algorithms (DISC 2001)*, Springer-Verlag, 2001, Lecture Notes in Computer Science, LNCS 2180, pp. 285–299.
15. I. Chatzigiannakis, S. Nikoletseas, and P. Spirakis, "Distributed communication algorithms for ad hoc mobile networks," *Journal of Parallel and Distributed Computing (JPDC)* **63** (2003), no. 1, 58–74, Special issue on Wireless and Mobile Ad-hoc Networking and Computing, edited by A. Boukerche.
16. I. Chatzigiannakis, S. Nikoletseas, and P. Spirakis, "An Efficient Routing Protocol for Hierarchical Ad-Hoc Mobile Networks," in *Proc. 1st International Workshop on Parallel and Distributed Computing Issues in Wireless Networks and Mobile Computing*, satellite workshop of *15th Annual International Parallel Distributed Processing Symposium* – IPDPS'01.
17. I. Chatzigiannakis and S. Nikoletseas, "An Adaptive Compulsory Protocol for Basic Communication in Highly Changing Ad-hoc Mobile Networks," in *Proc. 2nd International Workshop on Parallel and Distributed Computing Issues in Wireless Networks and Mobile Computing*, satellite workshop of *16th Annual International Parallel Distributed Processing Symposium* – IPDPS'02.
18. F.R.K. Chung, "Spectral graph theory," *Regional conference series in mathematics* **92** (1994), *CBMS Conference on Recent Advances in Spectral Graph Theory*.
19. T. Dimitriou, S. Nikoletseas, and P. Spirakis, "The infection time of graphs," *Tech. report, CTI*, 2003.
20. E. Gelenbe, R. Lent, M. Gellman, P. Liu, and P. Su, "CPN and QoS Driven Smart Routing in Wired and Wireless Networks", In M. Calzarossa and E. Gelenbe, editors, *Performance Tools and Applications to Networked Systems*, volume 2965 of *Lecture Notes in Computer Science*, Springer, 2004.

21. T. Goff, N. Abu-Ghazaleh, D. Phatak and R. Kahvecioglu, "Premeptive Routing in Ad Hoc networks", *Journal of Parallel and Distributed Computing*, pp. 123-140, Vol. 63, No. 2, 2003.

22. Z.J. Haas and M.R. Pearlman, "The performance of a new routing protocol for the reconfigurable wireless networks," *International Conference on Communications, IEEE*, 1998.

23. Z.J. Haas and M.R. Pearlman, "The zone routing protocol (ZRP) for ad-hoc networks," *Tech. report, IETF*, Internet Draft, June 1999, draft-zone-routing-protocol-02.txt.

24. K.P. Hatzis, G.P. Pentaris, P.G. Spirakis, V.T. Tampakas, and R.B. Tan, "Fundamental control algorithms in mobile networks," *11th Annual Symposium on Parallel Algorithms and Architectures, ACM*, 1999, pp. 251–260.

25. G. Holland and N. Vaidya, "Analysis of TCP performance over mobile ad hoc networks," *5th Annual ACM/IEEE International Conference on Mobile Computing (MOBICOM 1999), IEEE/ACM*, 1999, pp. 219–230.

26. T. Imielinski and H.F. Korth, "Mobile computing," Kluwer Academic Publishers, 1996.

27. A. Klemm, C. Lindemann, and O.P. Waldhorst, "Peer–to–Peer Computing in Mobile Ad Hoc Networks", In M. Calzarossa and E. Gelenbe, editors, *Performance Tools and Applications to Networked Systems*, volume 2965 of *Lecture Notes in Computer Science*, Springer, 2004.

28. Y. Ko and N.H. Vaidya, "Location-aided routing (LAR) in mobile ad-hoc networks," *4th Annual ACM/IEEE International Conference on Mobile Computing (MOBICOM 1998), IEEE/ACM*, 1998, pp. 66–75.

29. Q. Li and D. Rus, "Sending messages to mobile users in disconnected ad-hoc wireless networks," *6th Annual ACM/IEEE International Conference on Mobile Computing (MOBICOM 2000), IEEE/ACM*, 2000, pp. 44–55.

30. N.A. Lynch, "Distributed algorithms," Morgan Kaufmann Publishers Inc., 1996.

31. K. Mehlhorn and S. Näher, "LEDA: A platform for combinatorial and geometric computing," Cambridge University Press, 1999.

32. R. Motwani and P. Raghavan, "Randomized algorithms," Cambridge University Press, 1995.

33. S. Mueller, R.P. Tsang, and D. Ghosal, "'Multipath Routing in Mobile Ad Hoc Networks: Issues and Challenges,", In M. Calzarossa and E. Gelenbe, editors, *Performance Tools and Applications to Networked Systems*, volume 2965 of *Lecture Notes in Computer Science*, Springer, 2004.

34. V. Park and M. Corson, "A highly adaptive distributed routing algorithm for mobile wireless networks," *16th Annual IEEE International Conference on Computer Communications and Networking (INFOCOM 1997), IEEE*, 1997, pp. 1405–1413.

35. V.D. Park and M.S. Corson, "Temporally-ordered routing algorithms (TORA) version 1 functional specification," *Tech. report, IETF*, Internet Draft, October 1999, draft-ietf-manet-tora-spec-02.txt.

36. M. Pearlman and Z. Haas, "Determining the optimal configuration for the zone routing protocol," *Journal on Selected Areas in Communications* **17** (2003), no. 8, 1395–1414.

37. C. Perkins, "Highly dynamic destination-sequenced distance-vector routing (DSDV) for mobile computers," *18th Conference on Communications Architectures, Protocols and Applications (SIGCOMM 1994), ACM*, 1994, pp. 234–244.

38. C. Perkins and E.M. Royer, "Ad-hoc on demand distance vector (AODV) routing," *2nd Annual Workshop on Mobile Computing Systems and Applications, IEEE*, 1999, pp. 90–100.

39. C.E. Perkins, "Ad hoc networking," Addison-Wesley, Boston, 2001.
40. C.E. Perkins and E.M. Royer, "Ad-hoc on demand distance vector (AODV) routing," *Tech. report, IETF*, Internet Draft, September 1999, draft-ietf-manet-aodv-04.txt.
41. C.E. Perkins, E.M. Royer, and S.R. Das, "Ad-hoc on demand distance vector (AODV) routing," *Tech. report, IETF*, Internet Draft, June 2002, draft-ietf-manet-aodv-11.txt.
42. J.E. Walter, J.L. Welch, and N.M. Amato, "Distributed reconfiguration of metamorphic robot chains," *19th ACM Annual Symposium on Principles of Distributed Computing (PODC 2000), ACM*, 2000, pp. 171–180.
43. Y. Zhang and W. Lee, "Intrusion detection in wireless ad-hoc networks," *6th Annual ACM/IEEE International Conference on Mobile Computing (MOBICOM 2000), IEEE/ACM*, 2000, pp. 275–283.

Performance Management in Dynamic Computing Environments

Gabriele Kotsis

Abteilung Telekooperation, Johannes Kepler University Linz
Altenberger Strasse 69, A-4040 Austria
gabriele.kotsis@jku.ac.at,
http://www.tk.uni-linz.ac.at

Abstract. In this paper we argue that the traditional approach to performance analysis needs to be revisited to be applicable in modern computing environments. The major reasons are the highly dynamic and changing QoS requirements, which require fast and proactive performance tuning activities and the interdependencies between user behaviour and system performance, which must be adequately covered in the workload modelling process. We argue for a paradigm shift towards performance management and will discuss the impacts on workload modelling and on performance analysis. Some initial results towards a performance management approach will be presented.

1 Motivation

In the past decade of computing, we have observed several trends which significantly changed the ways in which we interact with computers. One of the most significant developments was the evolution of the world wide web. Starting as a network for exchange of research results, the WWW is now being used as a platform for any possible application one can imaging, including personal communication, commercial transactions, e-learning, or distributed computing to name a few.

New services and applications are offered to the user characterized by a high degree of interactivity, extensive use of multimedia data (audio, video), or wireless access to the Web. The devices to access information, computing, and communication resources offered by the world wide web are getting smaller, thus mobile, but at the same time more powerful allowing for innovative ways of interaction.

Ubiquitous or pervasive computing are new paradigms, going beyond the concept of any time any place accessibility towards ambient, intelligent, context aware embedding of communication and computation power into objects of our daily lifes.

Such computing environments are characterized by a user transparent mapping of requests to services. The users no longer issue explicitly commands to interact with the system, rather components of the system try to sense and identify the users' intentions and to offer the appropriate services. The users do not even need to be aware of the servers in the system.

M.C. Calzarossa and E. Gelenbe (Eds.): MASCOTS 2003, LNCS 2965, pp. 254–264, 2004.
© Springer-Verlag Berlin Heidelberg 2004

This different interaction behaviour has a significant influence on the techniques and tools used to evaluate the performance of such systems. In this paper, we are going to address a selection of problems that arise in this context. In section 2 we will briefly describe the characteristics of pervasive computing environments and identify the basic principles of operation. Based on that description, we will discuss the concept of quality of service (QoS) in such dynamically changing computing environments and elaborate QoS demands. In section 3 we will discuss the implications on workload modelling followed by a presentation of the performance management approach as a method to guarantee QoS in modern computing environments 4. We conclude with an outlook on future work.

2   QoS Requirements in Modern Computing Environments

Following the notation introduced in [1], we can identify persons, things, and places as the main entities in a pervasive computing environment. Those entities communicate and interact with each other, typically without the need for a centralised coordination, but in a dezentralised, autonomous, and distributed way. The entities have sensors which enable them to collect information on the environment they are in and on other entities. The type of sensors range from positioning systems used to identify the location, time sensors, bio-medical sensors, movement sensors, etc. The sensed information must be processed in order to aggregate the sensed data into information and in oder to transform this information into knowledge about the envrionment. The entitites thus become context or environment aware. The ultimate goals is to use this context knowledge in order to provide intelligent, adaptive services to other entities.

Taking up the user's point of view, we can identify three major success factors for pervasive computing environments: first of all, such services must be easy to use. It should not be necessary to perform time consuming installations or adaptations in order to use services. Services should support the work of humans, not hinder it. We are all very well aware of examples, where so called "intelligent support" for our work tasks is rather annoying then being helpful. Just think for example on some helpers in your text processing software asking you whether you want to write a letter and want some support in writing a letter ... just because you typed something that could look like an address. Second, the concept of sensing and collecting information about other entities, including humans, immediately causes concerns about security and privacy. How much information am I willing to provide to whom? How can integrity be ensured? Finally, the performance and reliability of such services is crucial. Information must be delivered in time, for multi media data, continous transmission rates are crucial, error rates must be minimised, etc.

We can summarise all those issues under the term quality of service. The notion of quality of service is well known in computer networks, where it refers to the concept that performance characteristics, such as transmission delays or error rates, can be measured, improved, and, to some extent, guaranteed in advance for various types of services.

In pervasive computing, the notion of QoS should be extended to the capturing of characteristics related to security, usability, and performability as the three major requirements in a pervasive computing environment.

The success of ubiquitous or pervasive computing environments will depend on the quality of service (QoS) level at which they can be operated. Mechanisms are needed to guarantee QoS in the above defined wider sense. The mechanisms described in the previuous section used for sensing data, transforming data into information, for transforming information into context knowledge and finally for putting this knowledge into action must be QoS driven.

We will propose in the following section a methodology for performance management, which shows how a pervasive computing environment can be made QoS aware. We will focus on performance, but the concepts can be extended to ensure usability as well as security.

3   Workload Modelling Techniques

According to trends and changes in the computing environment, both, the type and the volume of the load in the system is changing. Understanding the interaction between users and applications and services and being able to study the effects of this user behaviour will be crucial in the design and development of applications and services, both from a performance as well as an user interface design point of view, in capacity planning of existing and future networking infrastructures (including the planning of link capacity of the performance tuning of web servers), but also in introducing new services from a marketing point of view.

Being able to adequately capture and model the load in pervasive computing environments is therefore a crucial factor in performance management. In this section, we will therefore discuss the impacts on workload modelling.

We will review and present three major trends in workload modelling, which are highly relevant for performance management: user behaviour oriented workload modelling, hierarchical workload modelling and adaptive workload modelling.

3.1   Adaptive Workload Modelling

From a load modelling point of view, the difference in using computing resource has changed the type of model for workload characterization. While in the early days of computing (70s) the typical systems were used in batch or interactive mode, static workload models [2] could adequately represent the user behaviour. In the 80s, dynamic workload models were introduced, which were able to represent variabilities in user behavior ([3,4,5]). Within the last years, generative workload models [6,7,8] have been proposed as a suitable method capturing the dynamics and changes in the system.

The next step is the development of adaptive or learning workload models. This class of models is able to capture the load history of the system under study and to use this history to predict future load in the system.

First attempts towards this kind of model are reported for example in [9].

3.2 User Behaviour Oriented Workload Modelling

Given the tight interaction between the users and the systems in pervasive computing environments, we believe that it is crucial to represent the user behaviour in the workload model.

In the past performance evaluation of network systems has been focused on the resource level (see e.g. [10] for an overview of approaches in traffic modelling). With the growing popularity of high interactive systems like the WWW it is necessary to analyze this additional considerations and create more global models. On the application level user orientation can help to optimize interaction between applications and network protocols or overall network efficiency considerations such as file caching. Further log file analysis at this level gives design and usability suggestions for WWW pages, sites and browsers as reported by [11] and [12].

In literature, several approaches have been proposed for considering explicitely user behaviour in the workload models. Those approaches include Workload Patterns [13], User Behaviour Graphs[14], Trace-Based Characterisation [15], and probabilistic attributed context free grammars (PACFGs) [16].

3.3 Hierarchical Workload Modelling

The motivation for a hierarchical approach are twofold. On the one hand, a layered approach allows for a description of the traffic at various levels from the modelling point of view. In particular, we identify a top-level, user behaviour oriented point of view, the workload generator level, which translates down to the lowest level, the service level, at which we can characterise the actual type of traffic from a resource-oriented point of view. On the other hand, aggregation techniques can be applied to make the models more tractable from an evaluation point of view.

Hierarchical descriptions of workloads have already proven their usability for representing workloads of interactive [17], distributed [18,19], or parallel [20] systems. This modelling concept has more recently been applied in the area of load modelling in web traffic characterisation [21].

3.4 Towards a General Workload Modelling Framework

At the moment there does not exist a single model being able to adequately represent the various types of load observed in modern computing environments. The existing workload models do not allow the analyst to explicitly characterise the user behaviour. To overcome these problems a user behaviour model framework was proposed in [22] and implemented in [23]. The authors propose a user behaviour model framework that is constructed in a top down manner, consisting of various layers starting from a user-oriented view of the workload at the top level and a network traffic oriented view at the lowest level. Layers offer services to the next higher layer and require services from the layers below. The framework is meant to enable the modeler to plug in his own models at the layers of his choice, thus choosing the right balance between the simulation complexity and creating representative results. In this thesis, a workload generator will be specified (see Chapter 4) following this modelling framework. To understand the structure of the workload generator it is therefore necessary to briefly describe the modelling framework.

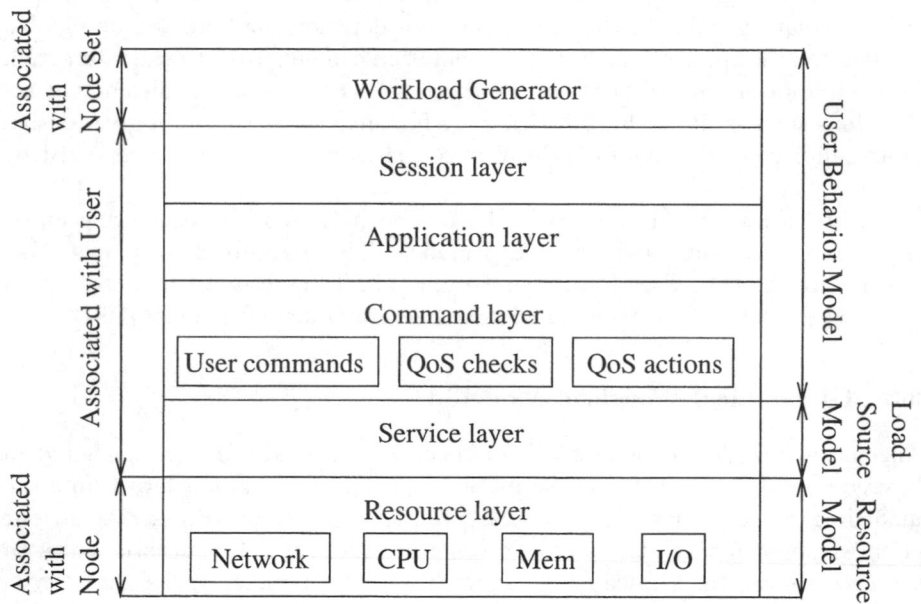

Fig. 1. Hierarchical Framework for User Behaviour Modelling

The goal of the modeling framework is to find sets of layered models as shown in Fig. 1 that can be plugged onto each other to represent some typical user behaviour that has been either observed or that is projected to be seen in the future. The models will be created top down, starting always at the highest layer and then going down to the chosen level of detail, each layer adding another level of detail to the model.

Once a model has been identified at a certain layer, it can be re-used in other modelling studies. Thus, this modelling framework should evolve into a library of workload models, allowing also the non-expert to use this kind of modelling approach. For example, an internet service provider could use the model library to construct a workload mix of web users. All he or she would have to do is to define the appropriate number of users for certain pre-defined web classes at the top level, the user layer. The underlying modelling methodology would automatically translate this top level workload view into the resource-oriented view needed for example in a capacity planning study. In the BISANTE project [24], such re-usable models have been defined for selected types of user behaviours (e.g. modelling web user behaviour of students in a computer Lab, or studying the effects of user behaviour in an ISP rubnetwork to name a few).

4   Performance Management Approach

It follows quite natural (and has been observed in the history of performance evaluation, see for example the report given in [25]) that changes in the computing environment (parallel and distributed computing, network computing, mobile computing, pervasive computing, etc.) and new system features (security and reliability mechanisms, agent-based systems, intelligent networks, adaptive systems, etc.) pose new challenges to performance evaluation by raising the need for new analysis methodologies.

Based on a schematic presentation of the traditional performance evaluation methodology, we will propose a performance management approach and give some application examples.

4.1   What Is Performance Management?

The traditional approach to evaluate the performance of computer systems and networks is an "off-line" performance analysis as depicted in Figure 2 (a) (see for example [26,27,28]). Starting with a characterisation of the system under study and a characterisation of the load, a performance model is built and performance results are obtained by applying performance evaluation techniques (including analytical, numerical, or simulaton techniques). Alternatively, performance measurements of the real system can replace the modelling and evaluation step. In any case, an interpretation of results follows which can trigger either a refinement of the model (or new measurements) if the results do not provide the required insight, or which will lead to performance tuning activities (system side or load side) if performance deficiencies are detected.

But the many aspects of emerging computing environments create a variety of performance influencing factors which cannot be adequately represented in a model. Applications in such environments are typically characterised by a complex, irregular, data-dependent execution behavior which is highly dynamic and has time varying resource demands. The performance delivered by the execution platform is hard to predict as it is usually constituted of heterogeneous components.

In addition, a demand is observed for immediate, embedded performance tuning actions replacing the traditional off-line approach of performance analysis. The human expert driven process of constructing a model, evaluating it,

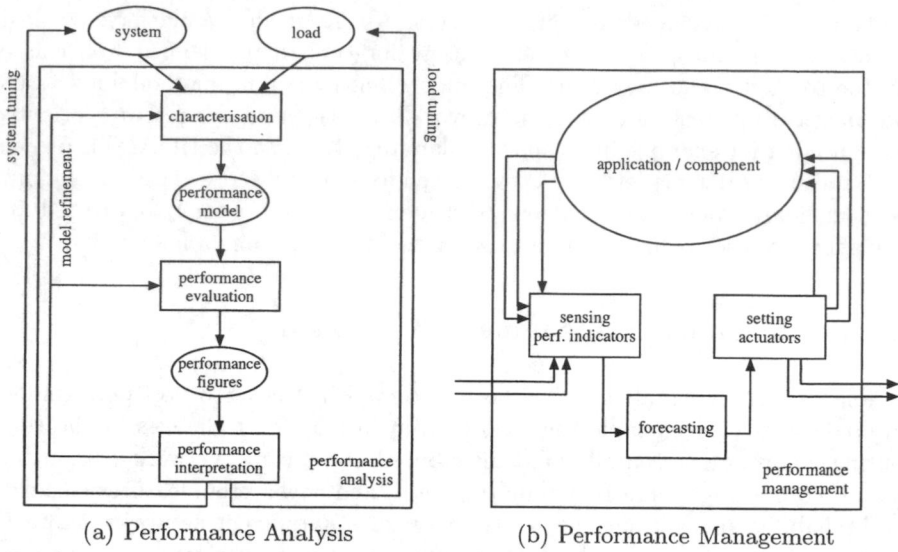

(a) Performance Analysis   (b) Performance Management

Fig. 2. From Performance Analysis to Performance Management

validating and interpretating the results and finally putting performance tuning activities into effect is no longer adequate if real-time responsiveness of the system is needed.

Therefore we propose an on-line, dynamic performance management approach as outlined in Figure 2 (b).

The term performance management was chosen because of the similarities to management in the economic sense, which in its functional dimension involves creation and maintenance of a (socio-economic) system and steering activities, i.e. the definition of goals and the scheduling and control of actions to reach those goals [29].

Performance Management as proposed in this paper focuses on the steering component, namely the definition of performance criteria that have to be met (corresponding to the goals) and the scheduling and control of performance tuning activities.

The distinctive features of this new approach are given as follows.

– Performance management must be an **automated** process which tries to take the human "out of the loop". The human expert is still needed to set-up the system and to define the performance goals, but mechanisms for sensing the current state of the system, comparing the obtained values against the defined goals, deciding on tuning activitites and finally putting these activitites into effect are fully automated and embedded within the system.
– Performance management must be **environment aware** sensing the current state (in terms of currently delivered performance) of the execution platform.

This is achieved by mechanisms (indicators) are provided to sense the state of the environment.

- Performance management follows a **distributed** approach to avoid performance hazard of a centralized coordinator in that it **locally** monitors system performance and collects local state information.Malony
- Performance management must incorporate a mechanism that **autonomously** controls the system behaviour.
- Performance management should be **pro-active** because of the (real) time gap between state information collection and launch of control actions. A forcast mechanism predicts the next system state based on which a control decision is made.
- The control loop consisting of sensing the state, predicting the future state, and scheduling of control actions ensures that performance management is **self-adaptive** with respect to performance fluctuations in the environment resulting from sharing of communication resources, background processes, several layers of software, etc.

4.2   Applications of Performance Management

In [30] we have proposed an agent-based mechanism for the realisation of the performance management architecture sketched Figure 2 (b). The area of application was balancing resource usage in a distributed computing environment wherre users submit different types of request to the system in a transparent way.

In paper [31] the performance management approach has been applied on a distributed application. The concept here was to "wrap" the application with a pro-active performance management module which is responsible for collecting system state information via sensors, predicting the future system state, and finally setting actuators to control/steer the performance.

While in paper [30] agents had the possibility to communicate with each other to obtain a performance management decision, the approach presented in [31] is fully autonomous and decisions are made locally in a distributed way.

5   Conclusions and Future Research

In this paper, we have presented (the vision of) an performance management approach for guaranteeing QoS in terms of performance in pervasive computing environments. Similar to the basic operation principles of pervasive computing environments, which are environment aware, the performance management approach is "performance aware", in that mechanisms are included into pervasive computing entities and services which are able to capture the current state of the system with respect to performance. Based on the data about the current state, the system is able to interpret and predict future states and to automatically and autonomously steer and control the performance. We put emphasis on the workload moedlling part as a central aspect in performance evaluation.

According to the architecture of performance management as outlined in this paper, future research will be oriented along four directions:

More work is needed in identifying appropriate sensors and sensing mechanisms. Challenges are similar to those in real-time monitoring, namely to provide the values in real-time and to maintain a low degree of intrusion, i.e. not to disturb the system behaviour by measurements.

With respect to the forecasting mechanism, different techniques can be applied, including for example neural networks, genetic algorithms or other learning mechanisms. The major problem here is the tradeoff between the computation costs (amount of time needed to come up with a result) and the prediction accuracy.

Different application domains will be investigated, including pervasive computing environments in the home, in the office, or in electronic commerce.

Last but not least, the performance management approach will be extended to a QoS management approach, following the wider definition of QoS as given in the paper. This requires the definition of mechanisms for modelling and capturing usability and security. Appropriate "sensors" as well as "actuators" for ensuring usability and security have to be identified.

References

1. Kindberg, T., Barton, J., Morgan, J., Becker, G., Caswell, D., Debaty, P., Gopal, G., Frid, M., Krishnan, V., Morris, H., Schettino, J., Serra, B., Spasojevic, M.: People, places, things: Web presence for the real world. In: Proceedings of the Workshop on Mobile Computing Systems and Application. (2000)
2. Ferrari, D.: Workload Characterization and Selection in Computer Performance Measurement. Computer **5** (1972) 18–24
3. Calzarossa, M., Italiani, M., Serazzi, G.: A Workload Model Representative of Static and Dynamic Characteristics. Acta Informatica **23** (1986) 255–266
4. Ferrari, D., Serazzi, G., Zeigner, A.: Measurement and Tuning of Computer Systems. Prentice-Hall (1983)
5. Haring, G.: On Stochastic Models of Interactive Workloads. In Agrawala, A., Tripathi, S., eds.: PERFORMANCE '83, North-Holland (1983) 133–152
6. Dulz, Q., Hofman, S.: Grammar-based workload modeling of communication systems. In: Proc. of Intl. Conference on Modeling Techniques and Tools For Computer Performance Evaluation, Tunis (1991)
7. Raghavan, S., Vasukiammaiyar, D., Haring, G.: Generative models for networkload in a single server environment. Technical report, University of Maryland, College Park (1993)
8. Barford, P., Crovella, M.: Generating representative web workloads for network and server performance evaluation. In: Measurement and Modeling of Computer Systems. (1998) 151–160
9. Jagannathan, S., Swaminathan, N., Raghavan, S.: Predicting behavior patterns using adaptive workload models. In: Proceedings of the 7th Intl Conference on Modelling and Simulation of Communication and Telecommunication Systems (1999)

10. Hlavacs, H., Kotsis, G., Steinkellner, C.: Traffic source modeling. Technical Report TR-99101, Institute of Applied Computer Science and Information Systems, University of Vienna (1999)

11. Catledge, L.D., Pitkow, J.E.: Characterizing browsing strategies in the www. In: Proceedings of Third International WWW Conference. (1994)

12. Claffy, K.C., Braun, H.: Web traffic characterization: an assessment of the impact of caching documents from ncsa's web server. http://www.ncsa.uiuc.edu/SDG/IT94/Proceedings/DDay/claffy/main.html (1994)

13. Calzarossa, M., Serazzi, G.: A Characterization of the Variation in Time of Workload Arrival Patterns. IEEE Trans. on Computers C-34 (1985) 156–162

14. Calzarossa, M., Marie, R., Trivedi, K.: System Performance with User Behavior Graphs. Performance Evaluation 11 (1990) 155–164

15. Cunha, C.R., Bestavros, A., Crovella, M.E.: Characteristics of www client-based traces. Technical Report BU-CS-95-010, Computer Science Department Boston University (1995)

16. Kotsis, G., Krithivasan, K., Raghavan, S.: A workload characterization methodology for www applications. In et.al., H., ed.: Performance and Management of Complex Communication Networks, Chapman and Hall (1998) 153–173

17. Haring, G.: Fundamental Principles of Hierarchical Workload Description. In Serazzi, G., ed.: Workload Characterization of Computer Systems and Computer Networks, North Holland (1986) 101–110

18. Calzarossa, M., Haring, G., Serazzi, G.: Workload Modelling for Computer Networks. In U. Kastens and F. J. Rammig, ed.: Architektur und Betrieb von Rechnersystemen. 10. GI/ITG-Fachtagung, Springer Verlag, Informatik Fachberichte Nr. 168 (1988) 324–339

19. Raghavan, S.V., Vasukiammaiyar, D., Haring, G.: Hierarchical approach to building generative network load models. Computer Networks and ISDN (1994) (to appear).

20. Calzarossa, M., Haring, G., Kotsis, G., Merlo, A., Tessera, D.: A hierarchical approach to workload characterization for parallel systems. In: High Performance Computing and Networking, LNCS vol. 919, Springer (1995) 102–109

21. Kotsis, G., Krithivasan, K., Raghavan, S.: Generative workload models of internet traffic. In: Proceedings of the ICICS Conference, Singapore, IEEE Singapore (1997) 152–156

22. Hlavacs, H., Kotsis, G.: Modeling user behaviour - a layered approach. In: Proceedings of the Seventh International Symposium on Modeling, Analysis and Simulation of Computer and Telecommunication Systems, MASCOTS'99, IEEE Computer Society Press (1999)

23. Hlavacs, H., Hotop, E., Kotsis, G.: Workload generation in opnet by user behaviour modelling. In: Proceedings of OPNETWORK 2000, August 28 - September 1, Washington, USA, OPNET / Mil3 (2000) Awarded with the Distinguished Paper Award.

24. Kotsis, G.: Bisante — multi media cd-rom on project results from the ec funded project bisante (broadband integrated satellite network traffic evaluation) (2000) University of Vienna.

25. Haring, G., Lindemann, C., Reiser, M., eds.: Performance Evaluation of Computer Systems and Communication Networks. Dagstuhl-Seminar-Report, No 189 (1997)

26. Ferrari, D.: Computer Systems Performance Evaluation. Prentice-Hall, Englewood Cliffs, New Jersey (1978)

27. Menasce, D.A., Almeida, V.A.E., Dowdy, L.W., eds.: Capacity Planning and Performance Modeling - From Mainframes to Client-Server Systems. Prentice Hall (1994)
28. Drakopoulos, E.: Enterprise network planning and design: Methodology and application. Computer Communications **22** (1999) 340–352
29. Woll, A., ed.: Wirtschaftslexikon. R. Oldenbourg Verlag (1990)
30. Haring, G., Kotsis, G., Puliafito, A., Tomarchio, O.: A transparent architecture for agent based resource management. In: Proceedings of the International Conference on Telecommunications (ICT'98), IEE, IEEE (1998) 338–342
31. Ferscha, A., Johnson, J., Kotsis, G., Anglano, C.: Pro-active performance management of distributed applications. In Boukerche, A., Das, S., Wilsey, P., Williamson, C., eds.: Proceedings of the Sixth International Symposium on Modeling, Analysis and Simulation of Computer and Telecommunication Systems, MASCOTS'98, IEEE Computer Society Press (1998) 146–152

Software Performance Modeling Using UML and Petri Nets*

José Merseguer and Javier Campos

Dpto. de Informática e Ingeniería de Sistemas
Universidad de Zaragoza
C/María de Luna, 1, 50018 Zaragoza, Spain
{jmerse,jcampos}@unizar.es

Abstract. Software systems are today one of the most complex artifacts, they are simultaneously used by hundred-thousand of people sometimes in risk real time operations, such as auctions or electronic commerce. Nevertheless, it is a common practice to deploy them without the
expected performance. Software Performance Engineering has emerged
as a discipline to complement Software Engineering research in order
to address this kind of problems. In this work, we survey some recent
contributions in the field of Software Performance Engineering. The approach surveyed has as main features that it uses the UML diagrams to
specify the functional and performance requeriments of the system and
the stochastic Petri nets formalism to analyse it.

1 Introduction

The measurement of any system or device developed in the framework of any
engineering field should be cleary identified as a stage in the life cycle of the
product. Traditionally, three methods have been proposed, sometimes in complementary ways, to reveal how a system performs: direct measurement, simulation
and analythical techniques. Several reasons have pointed the last two methods
as the most significant, among others, for engineers because they allow to test
the device before its development and for researchers since mathematical tools
can be applied to model and experiment such devices at a lower cost. Both methods share the goal of creating a performance model of the system/device that
accurately describes its load, routing rates and activities duration. Performance
models are often described in some stochastic formalism that provides the required analysis and simulation capabilities, e.g. queuing network models [20],
stochastic Petri nets [29] or stochastic process algebra [17]. On the other hand,
the success of these methods and performance models has been mostly attached
to the available tools to apply and assist them.

In contrast with these common engineering practices, software systems are
sometimes deployed without the performance expected by clients, being usual

---

* This work has been developed within the projects P084/2001 of the Aragonese Government and TIC2002-04334-C03-02 and TIC2003-05226 of the Spanish Ministry
of Science and Technology.

M.C. Calzarossa and E. Gelenbe (Eds.): MASCOTS 2003, LNCS 2965, pp. 265–289, 2004.
© Springer-Verlag Berlin Heidelberg 2004

among developers to test the performance of their systems only when they have been implemented, following the well-known "fix-it-later" approach. It has been common to believe that powerful hardware at minor extra cost will solve performance problems when the software is deployed. The inadequacy of these actitudes has been justified by the youth of the software engineering discipline.

In the last years, software performance engineering [40] (SPE) has grown as a field to propose methods and tools to effectively overcome these problems attached to software development. The emergence of such a discipline becames a necessity when thinking that today complex distributed systems, most of them operating in the Internet, are avid of high quality requirements of security, dependability, performance or responsiveness.

The state of the research proposals of the SPE field suggests that they cannot be considered today to be applied in the industry. It is true that some of these advances have been applied and a variety of examples can be found about their effective application, but in our opinion they are not stablished as a systematic approach. The software industry, being concerned about the problems involving software quality, has actively participated throughout the OMG [30] consortium in the development of a standard to channel research advances into its industrial application. We refer to the UML Profile for Schedulability, Performance and Time Specification [31], the Profile later on.

The election of the Unified Modeling Language [32] (UML) to build the Profile on it, is not a coincidence, since the SPE community decided in its first international workshop [44] to consider the UML as the reference notation to describe software systems and their performance requirements. Among other goals, the Profile tries to:

- Enable the construction of models that could be used to make quantitative predictions regarding these characteristics.
- Facilitate communication of design intent between developers in a standard way.
- Enable inter operability between various analysis and design tools.

This paper surveys some recent contributions in the field of SPE, also it briefly describes the state of our current work and the lines that will be followed in a near future. In these last years, we have been working in the development of a SPE proposal, actually a method and a tool, to evaluate performance estimates of software systems in the early stages of the software life cycle.

Despite the fact that we started the development of the approach surveyed here previously to the definition of the Profile, it follows the most of the directions stated by the Profile. Moreover in some of the directions that the approach diverges from the Profile, we have tried to rechannel them to follow it.

Among the changes promoted in this approach to follow the Profile, it can be noticed that the tagged valued language and the metamodel to describe performance requirements is now followed instead of the previous *pa-UML* [27,28].

In other aspects the proposal differs significantly from the Profile. In [25] a SPE role for the use case and statechart diagrams is defined in the context of object oriented approach to describe the life of system classes, that clearly differs from the Profile, that is based on an scenario-based approach using activity and sequence diagrams.

In the following the big picture of the approach is given. The use case diagram is proposed to calculate, among other estimates, the system usage by actors. The statechart diagram is used to model the life of each class in the system and to describe its performance requirements. The actions in the statechart are described at a lower level by means of the activity diagram. Statecharts and activity diagrams are translated into stochastic Petri nets to conform a performance model representing the whole system. The sequence diagram, representing patterns of interactions among objects, is translated togheter with the statechart of the corresponding objects into a stochastic Petri net representing a performance model of a concrete execution of the system. Other diagrams such as the deployment or classes are now being studied to be coupled with the others.

At the same time we have been working in the development of a tool, at a prototype level, to support this proposal. Starting from the ArgoUML [36] CASE tool, it can be enriched with the capabilities to describe performance requeriments according to the Profile. Moreover, a module that translates the UML diagrams into the corresponding Petri nets has been developed and attached to the tool. The obtained Petri net files are sufficient to feed the GreatSPN [16] tool that can simulate or analyze them in order to obtain the desired estimates.

As a new work, some steps are announced in this paper towards the integration of the results in SPE surveyed here with previous results in the efficient performance analysis of stochastic Petri nets [7,8,9,33]. We consider interesting to merge both fields since it is expected to reach performance analysis techniques for the models obtained from the translation that contribute to palliate the state explosion problem inherent to the Petri net formalism.

The rest of the article is structured as follows. In section 2, an introduction to the notations and formalisms related with the proposal is given, assuming the reader is familiar with both UML and Petri nets terminologies and concepts. Section 3 depicts our proposal of a method for SPE and travels around each UML diagram in the proposal surveying its performance role under our interpretation and in some cases its translation into the Petri net formalism. Section 4 is devoted to envisage the future steps of our work in the field of the efficient analysis of the obtained performance models. Related work is explored in section 5. Finally conclusions of the work are obtained in section 6.

2   Context of the Work

The work surveyed in this article relies on a number of fields and formalisms, they are addressed in this section. We start by recalling the main proposals of the SPE. After, the UML notation is introduced and then we study the part of the Profile of interest for this work. Finally, the stochastic formulation of the Petri net formalism used in the resulting performance models is described, as well as some issues concerned with their analysis.

2.1   Software Performance Engineering

The term software performance engineering (SPE) was first introduced by C.U. Smith in 1981 [41]. Several complementary definitions have been given in the literature to describe the aim of the SPE. Among them it can be remarked the following:

- In [40], SPE is proposed as a method (a systematic and quantitative approach) for constructing software systems to meet performance objectives, taking into account that SPE augments other software engineering methodologies but it does not replace them.
- SPE is defined in [38] as a collection of methods for the support of the performance-oriented software development of application systems throughout the entire software development process to assure an appropriate performance related product quality.
- Finally, in [42] new perspectives for SPE are devised, then proposing that SPE must provide principles, patterns [26,15] and antipatterns [43] for creating responsive software, the data required for evaluation, procedures for obtaining performance specifications and guidelines for the types of evaluation to be conducted at each development stage.

It is important to remark that the previous definitions emphasize that SPE cannot be placed outside the context of software engineering. This fact contrasts with other engineering fields, such as telecommunication, where performance practices have been applied successfully in "isolation", i.e. not explicitly while developing the engines. Moreover SPE, as pointed out in [38], reuses and enlarges concepts and methods from many other disciplines such as: Performance management, performance modeling, software engineering, capacity planning, performance tunning and software quality assurance.

The SPE process proposed in [40] still remains as a reference for a very general proposal to establish the basic steps that a SPE process should consider. Firstly, the goals or quantitative values must be defined, obviously changing from one stage of the software life cycle to other. Data *gathering* is accomplished by defining the proper scenarios interpreting how the system will be typically used and its possible deviations (defining upper and lower bounds when uncertainties are present). The construction and evaluation of the performance model associated with the system is one of the fundamentals in SPE. Validation and verification of the performance model are on-going activities of the SPE process.

The common paradigms of stochastic models used in SPE are the queuing network models [20], stochastic process algebras [17] and stochastic Petri nets [29].

2.2   The Unified Modeling Language

The Unified Modeling Language [32] (UML) is a semi formal language developed by the OMG to specify, visualize and document models of software systems and non-software systems too. UML defines twelve types of diagrams, divided into three categories: static diagrams, behavioral diagrams and diagrams to organize and manage application modules. Behavioral diagrams (sequence, collaboration, use case, statechart and activity) constitute a major aim in this work since the most performance issues of systems can be represented by means of them.

The semantics of the behavioral UML diagrams is specified in the *Behavioral Elements* package, which is decomposed into the following subpackages: Common behavior, collaborations, use cases, state machines and activity graphs. The last four subpackages give semantics to each one of the UML behavioural diagrams.

In this work a translation of each UML behavioural diagram into a stochastic Petri net is proposed. The translation is based on the UML metamodel defined for each subpackage. Figure 1 shows the UML metamodel for the state machines subpackage, the rest of the metamodels can be found in [32].

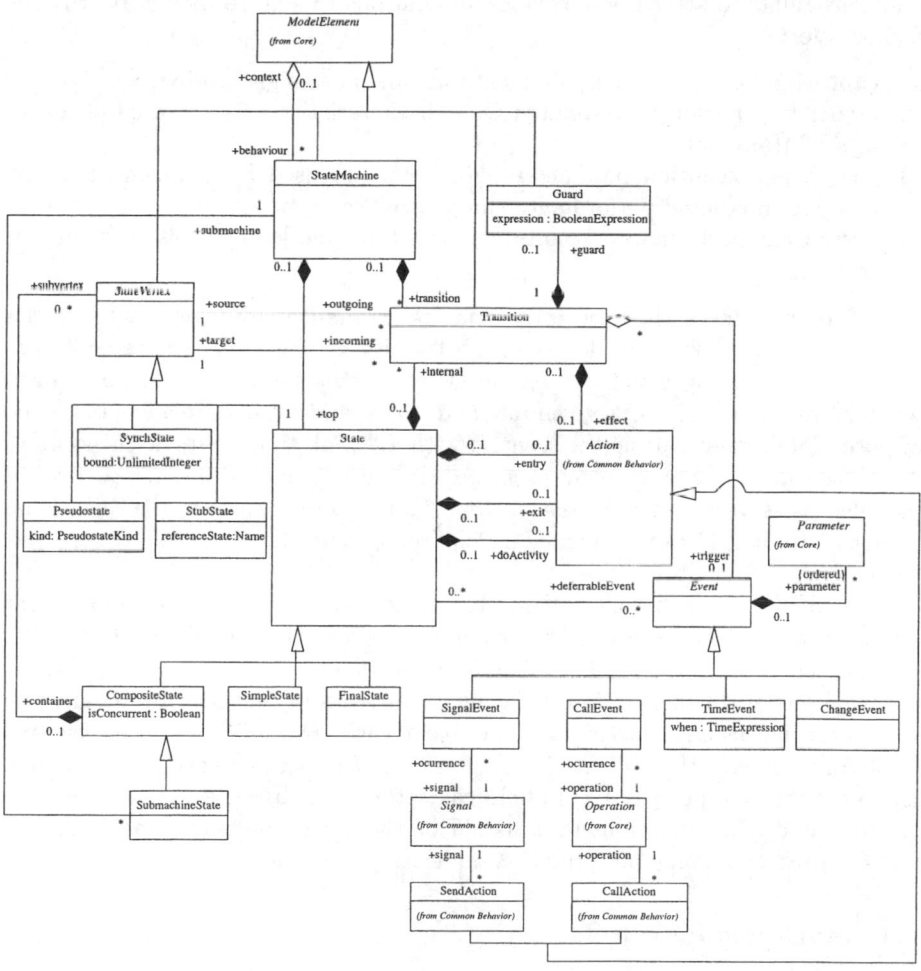

Fig. 1. UML state machines metamodel.

2.3   UML Profile for Schedulability, Performance, and Time Specification

The adoption of this Profile [31] by the OMG in March 2002 was of major importance for the SPE community. Its main purposes are:

- to encompass different real-time modeling techniques,
- to annotate UML real-time models to predict timeliness, performance and schedulability characteristics based on the analysis of these software models.

Although the profile is oriented to real time systems, the annotations proposed in the *performance sub-profile* remain valid for more general purposes, such as the specification of distributed software systems in non real time environments, which are the ones of interest in this work.

The sub-profile extends the UML metamodel with stereotypes, tagged values and constraints to attach performance annotations to a UML model. It provides facilities for:

1. capturing performance requirements within the design context,
2. associating performance-related QoS characteristics with selected elements of a UML model,
3. specifying execution parameters which can be used by modeling tools to compute predicted performance characteristics,
4. presenting performance results computed by modeling tools or found in testing.

In order to meet these objectives the performance sub-profile extends the UML metamodel with the following abstractions. The QoS requirements are placed on *scenarios*, which are executed by *workloads*. The workload is *open* when its requests arrive at a given rate and *closed* when there are a fixed number of potential users executing the scenario with a "think time" outside the system. The scenarios are composed by *steps*, i.e. elementary operations. *Resources* are modeled as servers and have *service time*. *Performance measures* (utilizations, response times, ...) can be defined as required, assumed, estimated or measured values.

The SPE approach surveyed in this work proposes a method and a tool following the objectives given by the sub-profile and the guidelines of the SPE summarized in section 2.1. The main difference with the sub-profile is that it proposes an scenario-based approach (using collaborations and activity diagrams), while this approach tries to identify the role of each UML diagram (use cases, interactions, statecharts and activity diagrams) in the performance process under an object-oriented perspective. In both cases, the final objective is to obtain a set of annotated UML diagrams that should be the input to create a performance model in terms of some performance modeling paradigm.

2.4   Stochastic Petri Nets

A Petri net [39] is a mathematical tool aimed to model a wide-range of concurrent systems. Formally:

Definition 1. *A Petri net is a 5th-tuple* $\mathcal{N} = \langle P, T, \mathbf{Pre}, \mathbf{Post}, \mathbf{m_0} \rangle$ *such that,*
P is a set of places,
T is a set of transitions,
$P \cap T = \emptyset$,
$\mathbf{Pre} : P \times T \rightarrow I\!N$ *is the input function,*
$\mathbf{Post} : T \times P \rightarrow I\!N$ *is the output function,*
$\mathbf{m_0} : P \rightarrow I\!N$ *is the initial marking.*

Stochastic Petri nets (SPNs) were proposed [29] as a non deterministic model, then associating with each transition a random firing time. In this work an extension of the SPNs is considered, the class of Generalized Stochastic Petri Nets (GSPNs) [1]:

Definition 2. *A GSPN system is a 8th-tuple* $\langle P, T, \Pi, \mathbf{Pre}, \mathbf{Post}, H, W, \mathbf{m_0} \rangle$
where,
$P, T, \mathbf{Pre}, \mathbf{Post}, \mathbf{m_0}$ *is a PN as in Def. 1,*
$\Pi : T \rightarrow I\!N$ *is the priority function that maps transitions onto priority levels,*
$H : P \times T \rightarrow I\!N$ *is the inhibition function,*
$W : T \rightarrow I\!R$ *is the weight function that assigns rates (of negative exponential distribution) to timed transitions and weights to immediate transitions.*

The GSPN have been extended to label places and transitions:

Definition 3. *A labeled GSPN system (LGSPN) is a triplet* $\mathcal{LS} = (S, \psi, \lambda)$
where,
S is a GSPN system as in Def. 2,
$\psi : P \rightarrow L^P \cup \tau$ *is a labeling function for places,*
$\lambda : T \rightarrow L^T \cup \tau$ *is a labeling function for transitions,*
L^T, L^P *and* τ *are sets of labels,*
τ-*labeled net objects are considered to be "internal", not visible from the other components.*

The LGSPN formalism introduces an operator to *compose* ordinary GSPNs models (i.e., $\mathbf{Pre}(p,t) \in \{0,1\}$ and $\mathbf{Post}(t,p) \in \{0,1\}$), that is of great importance in this work:

Notation 1 (Place and transition superposition of two ordinary LGSPNs).
Given two LGSPN ordinary systems $\mathcal{LS}_1 = (S_1, \psi_1, \lambda_1)$ *and* $\mathcal{LS}_2 = (S_2, \psi_2, \lambda_2)$, *the LGSPN ordinary system* $\mathcal{LS} = (S, \psi, \lambda)$:

$$\mathcal{LS} = \mathcal{LS}_1 \underset{L_T, L_P}{||} \mathcal{LS}_2$$

is the composition over the sets of (no τ) *labels* $L_T \subseteq L^T$ *and* $L_P \subseteq L^P$.

The definition of this operator can be found in [4]. Figure 2 depicts an informal representation of this operator. This operator is used in this work, among others, to compose, using the labels of some places and transitions, LGSPNs that represent system modules in order to obtain an LGSPN that represents the whole system.

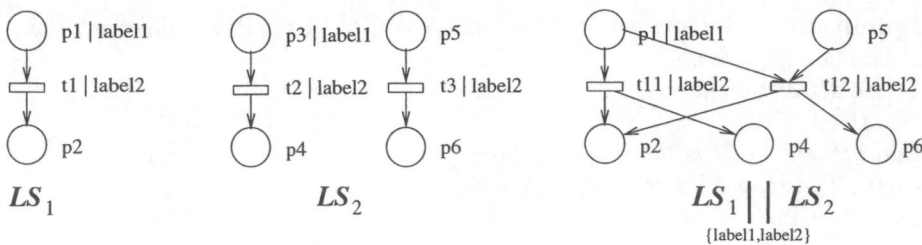

Fig. 2. Superposition of places and transitions

2.5 Analysis of Stochastic Models

The performance models obtained by the application of the SPE method proposed in this work can be used to estimate *performance measures*, therefore they should be simulated or analyzed, but only the second approach has been applied so far.

Traditionally, techniques for the analysis (computation of performance measures or validation of logical properties) of Petri nets are classified in three complementary groups, enumeration, transformation, and structural analysis:

- *Enumeration* methods are based on the construction of the reachability graph (coverability graph in case of unbounded models), but they are often difficult to apply due to their computational complexity, the well-know *state explosion problem*.
- *Transformation* methods obtain a Petri net from the original one belonging to a subclass easier to analyze but preserving the properties under study, see [5].
- *Structural analysis* techniques are based on the net structure and its initial marking, they can be divided into two subgroups: *Linear programming* techniques, based on the state equation and *graph based* techniques, based on "ad hoc" reasoning, frequently derived from the firing rule.

A complementary classification based on the quality of the obtained results is: exact, approximation and bounding techniques.

- *Exact* techniques are mainly based on algorithms for the automatic construction of the infinitesimal generator of the isomorphic Continuous Time Markov Chain (CTMC). ç Refer to [1] for numerical solutions of GSPN systems.
- *Approximation* techniques do not obtain the exact solution but an approximation. Some of them substitute the computation of the isomorphic CTMC by the solution of smaller components [33].
- Finally, *bounds* [7,9] are techniques that offer the further results from the reality. Nevertheless, they can be useful in the early phases of the software life-cycle, in which many parameters are not known accurately.

3   Towards a SPE Method

"Large software systems are the most complex artifacts of human civilization" [6]. It is widely accepted that complex systems need formal models to be properly modeled and analyzed, also when the kind of analysis to accomplish is with performance evaluation purposes.

Today still remains the gap among the classical performance evaluation techniques and the classical proposals for software development [37,19]. Neither of these proposals for software development deal with performance analysis, at most they propose to describe some kind of performance requirements. So, it could be argued that there does not exist an accepted *process* to model and study system performance in the software development process. The Profile [31] should be the starting point to bridge this gap, at least by providing the concepts and notations for the vocabulary of performance requeriments and the guides in the development of performance tools. Then a process should propose the steps to follow to obtain a performance model and it should give trends to analyze this model.

The process presented in this section follows the directions pointed by SPE and the Profile, recalled in sections 2.1 and 2.3, respectively. It emphasizes the previous matters: formal modeling and the need of connecting software and performance well-know practices.

The process to evaluate software performance should be "more or less transparent" for the software designer. By "transparent" is meant that the software designer should be concerned as less as possible to learn new processes since the tasks of analysis and design already imply the use of a process. Therefore, ideally the process to evaluate performance of software systems should not exist, it should be integrated in the common practices of the software engineer. It is our intention that the proposal presented here can be used together with any software life-cycle process, and that the performance model can be semi-automatically obtained as a *by-product* of the software life-cycle. Another important issue for the process is that the performance model should be obtained in the early stages of the software life-cycle. In this way proper actions to solve performance problems take less effort and less economical impact.

Figure 3 gives the big picture of the steps followed by this process, that are the following ones:

1. From the problem domain the software requirements should be modeled using the desired software life cycle paradigm and the UML notation. The meaning of the statecharts will be the description of the behavior of the active classes of the system. The activity diagrams will specify the refinement of the activities in the statecharts. The sequence diagrams will be used to model concrete executions of interest in the context of the system. While developing the models, performance requirements will be modeled by means of the Profile. The role of each diagram in this step and the proposed annotations are described later in this section.
2. In this step functions that translate each UML diagram into an LGSPN should be applied. From these models and applying the composition rules given in section 3.5 two different performance models can be obtained in

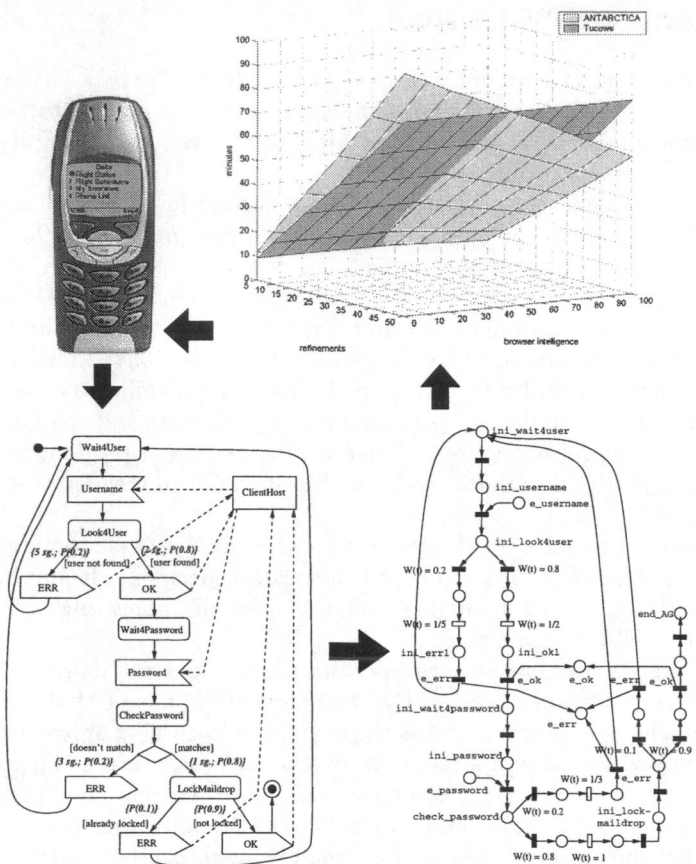

Fig. 3. SPE process.

terms of LGSPNs: a performance model representing the whole system or representing a concrete execution of the system. Translation and composition rules have been embedded in an augmented CASE tool [22].

3. Finally, the parameters to be computed and the experiments to test the system should be defined. Also, analytical or simulation techniques to solve the formal model should be applied. This phase can be made easier by integrating the augmented CASE tool with stochastic Petri net analysers.

The following sections (from 3.1 to 3.4) address the first and second steps of the proposal. Therefore, they revise the role that each UML diagram should play (under the prespective of the approach) in the SPE process, as well as the kind of performance requirements it is able to gather. Concerning the functions that translate the diagrams into LGSPN, due to space limitations, just the flavor for those of the statechart is given. The third step of the approach is partially addressed in section 4 by revising efficient techniques to analyze the model.

3.1   Use Case Diagram

In UML a use case diagram shows actors and use cases together with their relationships. The relationships are associations between the actors and the use cases, generalizations between the actors, and generalizations, extends and includes among the use cases [32].

A use case represents a coherent unit of functionality provided by a system, a subsystem or a class as manifested by sequences of messages exchanged among the system (subsystem, class) and one or more actors together with actions performed by the system (subsystem, class). The use cases may optionally be enclosed by a rectangle that represents the boundary of the containing system or classifier [32].

In the use case diagram in Figure 4 we can see: two actors, three use cases and four associations relationships between actors and use cases, like that represented by the link between the actor1 and the UseCase1.

Role of the use case diagram concerning performance. The use case diagram allows to model the usage of the system for each actor. We propose the use case diagram with performance evaluation purposes to show the use cases of interest to obtain performance figures. Among the use cases in the diagram a subset of them will be of interest and therefore marked to be considered in a performance evaluation process.

The role of the use case diagram is to show the use cases that represent executions of interest in the system. Then, a performance model can be obtained for each execution (use case) of interest, that should be detailed by means of the sequence diagram [4].

The existence of a use case diagram is not mandatory to obtain a performance model. In [24] it was shown how a performance model for the whole system can be obtained from the statecharts that describe it.

It is important to recall the proposal in [11], that consists in the assignment of a probability to every edge that links a type of actor to a use case, i.e. the probability of the actor to execute the use case. The assignment induces the same probability to the execution of the corresponding set of sequence diagrams that describes it. Since we propose to describe the use case by means of only one sequence diagram, we can express formally our case as follows.

Let us suppose to have a use case diagram with m users and n use cases. Let $p_i(i = 1, \ldots, m)$ be the i-th user frequency of usage of the software system and let P_{ij} be the probability that the i-th user makes use of the use case $j(j = 1, \ldots, n)$. Assuming that $\sum_{i=1}^{m} p_i = 1$ and $\sum_{j=1}^{n} P_{ij} = 1$, the probability of a sequence diagram corresponding to the use case x to be executed is:

$$P(x) = \sum_{i=1}^{m} p_i \cdot P_{ix} \tag{1}$$

The previous formula, taken from [11], is important because it allows to assign a "weight" to each particular execution of the system. As an example, see in Figure 4 the annotations attached to UseCase3.

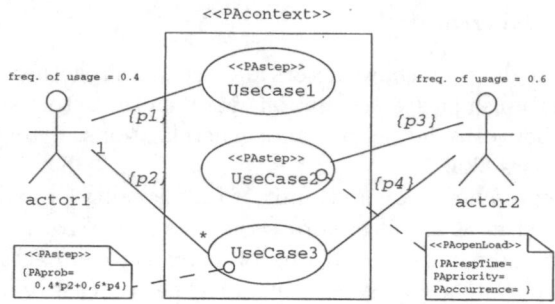

Fig. 4. Use case diagram with performance annotations.

The relationships between the actors themselves, and between the use cases themselves are not considered with performance evaluation purposes.

Performance annotations. The use case diagram should represent a *Performance Context*, since it specifies one or more scenarios that are used to explore various dynamic situations involving a specific set of resources. Then, it is stereotyped as ≪PAcontext≫. Since there is not a class or package that represents a use case diagram (just the ≪useCaseModel≫ stereotype) the base classes for ≪PAcontext≫ are not incremented.

Each use case used with performance evaluation purposes could represent a *step* (no predecessor neither successor relationship is considered among them). Then, they are stereotyped as ≪PAstep≫, therefore the base classes for this stereotype should be incremented with the class *UseCase*. A load (≪PAclosedLoad≫ or ≪PAopenLoad≫) can be attached to them. Obviously each one of these *steps* should be refined by other *Performance Context*, i.e. a *Collaboration*.

The probabilities attached to each association between an actor and a use case, although not consistently specified, represent the frequencies of usage of the system for each actor (see p1, p2, p3 and p4 in Figure 4). They are useful to calculate the probability for each *Step*, i.e. use case, using equation (1).

3.2   Statechart Diagram

A UML statechart diagram can be used to describe the behavior of a model element such as an object or an interaction. Specifically, it describes possible sequences of states and actions through which the element can proceed during its lifetime as a result of reacting to discrete events. A statechart maps into a UML state machine that differs from classical Harel state machines in a number of points that can be found in [32]. Recent studies of their semantics can be found in [24,13].

A state in a statechart diagram is a condition during the life of an object or an interaction during which it satisfies some condition, performs some action, or waits for some event. A simple transition is a relationship between two states indicating that an object in the first state will enter the second state. An event is a noteworthy occurrence that may trigger a state transition [32].

A composite state is decomposed into two or more concurrent substates (regions) or into mutually exclusive disjoint substates [32].

Role of the statechart concerning performance. The *profile* proposes determining system's performance characteristics using scenarios, described by collaborations or activity graphs. By contrast, we have explored an alternative that consists in determining those characteristics from an object's life viewpoint. In order to take a complete view of the system behavior, it is necessary to understand the life of the objects involved in it, being the statechart diagram the adequate tool to model these issues. Then, it is proposed to capture performance requirements at this level of modeling: for each class with relevant dynamic behavior a statechart will specify its routing rates and system usage and load.

The performance requeriments gathered by modeling the statecharts for the system are sufficient enough to obtain a performance model [24]. In this case, all the statecharts togheter represent a *Performance Context* where to explore all the dynamic situations in the system. A particular (and strange) situation arises when only one statechart describes all the system behaviour, then it becomes a *Performance Context*.

Moreover, the statecharts that describe the system (or a subset of them) togheter with a sequence diagram constitute a *Performance Context* that can be used to study parameters associated to concrete executions [4].

In a statechart diagram the useful model elements from the performance evaluation viewpoint are the activities, the guards and the events.

Activities represent tasks performed by an object in a given state. Such activities consume computation time that must be measured and annotated. Activity graphs are adecuate to refine this level of the statechart.

Guards show conditions in a transition that must hold in order to fire the corresponding event. Then they can be considered as system's routing rates.

Events labeling transitions correspond to events in a sequence diagram showing the server or the client side of the object. Objects can reside in the same machine or in different machines for the case of distributed systems. In the first case, it can be assumed that the time spent to send the message is not significant in the scope of the modeled system. Of course, the actions taken as a response of the message can spend computation time, that should be modeled. For the second case, for those messages that travel through the net, we consider that they spend time, then they represent a load for the system that should be modeled.

Performance annotations. As proposed in [31] for the activity-based approach, the open or closed workload (induced by the object in our case) is associated with the first step in the diagram, in this case the transition stereotyped ≪create≫, see Figure 5.

In pa-UML [23] the annotations for the duration of the activities show the time needed to perform them. If it is necessary, a minimum and a maximum values could be annotated. If different durations must be tested for a concrete activity then a variable can be used. Examples of these labels are {1sec},

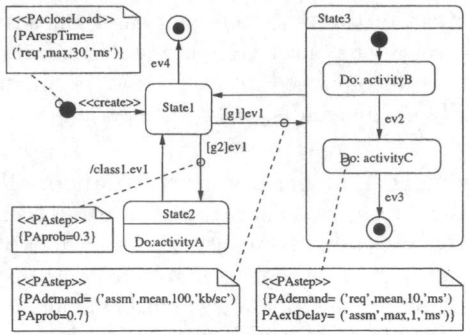

Fig. 5. Statechart with performance annotations.

{0.5sec..50sec} or {time1}. Using the *profile*, an activity in a state will be stereotyped ≪PAstep≫, then the expressivity is enriched by allowing to model not only response time but also its demand, repetition, intervals, operations or delay, see Figure 5. The successor/predecessor relationship inherent to the ≪PAstep≫ stereotype is not stablised in this case (causing that the probability atribute is not used), firstly because in a state at most one activity can appear [32], but also because it is not of interest to set order among all the activities in the diagram.

By stereotyping the transitions as ≪PAstep≫, it is possible:

A *To consider the guards as routing rates.* The probability of event success represents routing rates in pa-UML by annotating such probability in the guard. Using the *profile*, the attribute probability could be used also to avoid non-determinism (i.e. transitions labeled with the same event and outgoing from the same state), be aware that this attribute does not provoke a complete order among the succesor *steps* (transitions).
  As an example, see Figure 5. The succesor steps of the ≪create≫ transition will be transitions ev4, [g1]ev1 and [g2]ev2, while the predecessor *steps* of transition [g1]ev1 are transitions ev3, ≪create≫ and /class1.ev1. Therefore, it is only necessary to assign probabilities to transitions [g1]ev1 and [g2]ev2.

B *To model the network delays caused by the load of the events (messages) that label them.* The annotations for the load of the messages in pa-UML are attached to the transitions (outgoing or internal) by giving the size of the message (i.e. {1K..100K} or {1K}). Using the *profile*, the size of the message can be specified as a *step* and the network delay as a demand, see transition [g1]ev1 in Figure 5.

Translation into LGSPN. The approach taken to translate a statechart into an LGSPN consists in the translation of each element in the metamodel of the state machines (see Figure 1) into an LGSPN subsystem. Just to take the flavor of the translation, Figure 6 depicts the LGSPNs subsystems obtained by the translation of a simple state. The translation of the other elements in the metamodel can be found in [23,24].

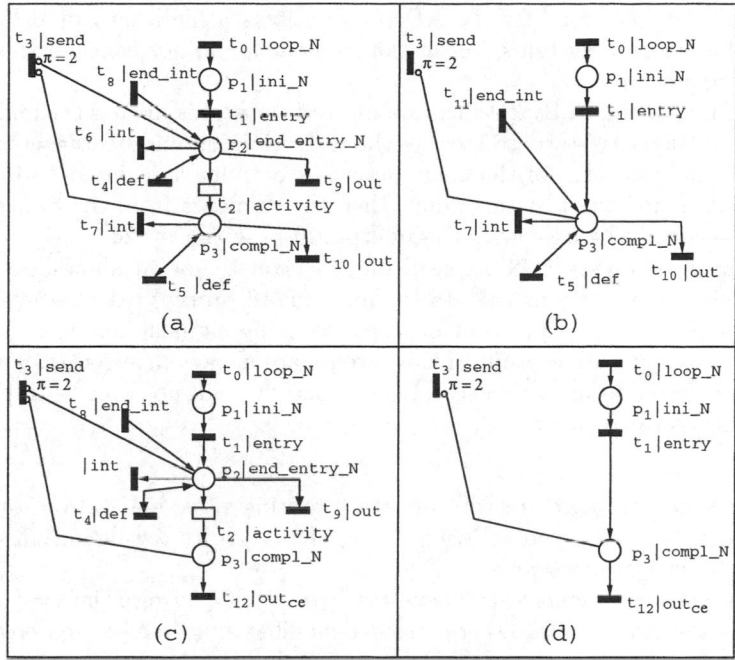

Fig. 6. Different labelled "basic" systems for a simple state N: (a) with activity and no immediate outgoing transition; (b) no activity and no immediate outgoing transition; (c) with activity and immediate outgoing transition; (d) no activity and with immediate outgoing transition.

In order to obtain an LGSPN ($\mathcal{LS}_{sc}$) for a given statechart sc, the subsystems that represent its metamodel elements are composed using the operator in Notation 1. The details of the composition method can be found also in [23, 24].

3.3 Activity Diagram

Activity diagrams (ADs) represent UML activity graphs and are just a variant of UML state machines (see [32]), in fact, a UML activity graph is a specialization of a UML state machine. The main goal of ADs is to stress the internal control flow of a process in contrast to statecharts, which are often driven by external events.

Role of the activity diagram concerning performance. Considering the fact that ADs are suitable for internal flow process modeling, they are relevant to describe activities performed by the system, usually expressed in the statechart as *doActivities* in the states. The AD will be annotated with the information to model routing rates and the duration of the basic actions.

Other interpretations for the AD propose it as a high level modeling tool, that of the workflow systems, but at the moment we do not consider this role in our SPE approach.

According to the UML specification most of the states in an AD should be an action or subactivity state, so most of the transitions should be triggered by the ending of the execution of the entry action or activity associated to the state. Since UML is not strict at this point, then the elements from the SMs package could occasionally be used with the interpretation given in [24].

As far as this issue is concerned, our decision is not to allow other states than action, subactivity or call states, and thus to process external events just by means of call states and control icons involving signals, i.e. signal sendings and signal receipts. As a result of this, events are always deferred (as any event is always deferred in an action state), so an activity will not ever be interrupted when it is described by an AD.

Performance annotations. To annotate routing rates and action durations, we consider the stereotype ≪PAstep≫ together with its tag definitions PAprob and PArespTime, respectively.

PArespTime = (<source-modifier>,'max',(n,'s.')), or PArespTime = (<source-modifier>,'dist',(n-m,'s.')). Where <source-modifier>::= 'req'|'assm'|'pred'|'msr'; 'dist' is assumed to be an exponential negative distribution and n-m expresses a range of time.

Annotations will be attached to transitions in order to allow the assignment of different action durations depending on the decision. It implies that the class *Transition* should be included as a base class for the stereotype ≪PAstep≫.

Translation into LGSPN. The performance model obtained from an activity diagram in terms of LGSPNs as proposed in [22] can be used with performance evaluation purposes with two goals: A) just to obtain some performance measures of the model element they describe or B) to compose this performance model with the performance models of the statecharts that use the activity modeled in order to obtain a final performance model of the system described by the referred statecharts. The translation process can be found in [22].

3.4   Sequence Diagram

A sequence diagram describes a communication pattern performed by instances playing the roles to accomplish a specific purpose, i.e. an interaction. The semantics of the sequence diagram is provided by the collaboration package [32].

Role of the sequence diagram concerning performance. For our purposes a sequence diagram should detail the functionality expressed by a use case in the use case diagram, by focusing in the interactions among its participants.

We consider the following relevant elements and constructions of the sequence diagram from the performance point of view to model the load of the system.

Objects can reside in the same machine or in different machines in the case of distributed systems. In the first case it can be assumed that the time spent to send the message is not significant in the scope of the modeled system. Of course the actions taken as a response of the message can spend computation time, but it will be modeled in the statechart diagram. For the second case, those messages that *travel through the net*, it is considered that they spend time, then representing a load for the system that should be modeled in this diagram.

A *condition* can be attached to each message in the diagram, representing the possibility that the message could be dispatched. Even multiple messages can leave a single point each one labeled by a condition. From the performance point of view it can be considered that routing rates are attached to the messages.

A set of messages can be dispatched multiple times if they are enclosed and marked as an *iteration*. This construction also has its implications from the performance point of view.

Performance annotations. A deeper explanation of some of the annotations given in the following can be found in [3].

The sequence diagram should be considered as a *performance context* since it will be used to obtain a performance model in terms of LGSPN. Then it will be stereotyped as ≪PAcontext≫.

Each message will be considered as a ≪PAstep≫, it allows that:

- as in the case of statecharts and ADs the *conditions* of the messages are represented by probabilities, then with tagged value PAprob;
- *iterations* will be represented by the tagged value PArep;
- to represent the time spent by a message *travelling through the net*, its size (e.g. 160 Kbytes) and the speed of the net (e.g. 8 Kbytes/sec) should be considered. For this example, the delay of the message is 20 sec., it will be represented by the tagged value PArespTime as follows ('assm','mean',(20,'sec')).

Translation into LGSPN. In [4], the translation process of a given sequence diagram sd into its corresponding LGSPN $\mathcal{LS}_{sd}$ can be found.

The LGSPN $\mathcal{LS}_{sd}$ will be composed (using the operator in Notation 1) with the LGSPNs that represent the statecharts involved in the interaction. Therefore obtaining a performance model that represents a concret execution of the system, obviously that execution described by the sequence diagram.

3.5   LGSPN Performance Models

In this section we address the topic of how to obtain a performance model in terms of LGSPNs from a system description in terms of a set of UML diagrams.

Two different kind of performance models can be obtained using the translations surveyed in the previous subsections.

A Suppose a system described by a set of UML statecharts $\{sc_1, \ldots, sc_k\}$ and a set of activity diagrams refining some of their *doActivities*, $\{ad_1, \ldots, ad_l\}$.

Then $\{\mathcal{LS}_{sc_1}, \ldots, \mathcal{LS}_{sc_k}\}$ and $\{\mathcal{LS}_{ad_1}, \ldots, \mathcal{LS}_{ad_l}\}$ represent the LGSPNs of the corresponding diagrams.

$\mathcal{LS}_{sc_i-ad_j}$ will represent an LGSPN for the statechart sc_i and the activity diagram ad_j.

Then a performance model representing the whole system can be obtained by the following expression:

$$\mathcal{LS} = \overset{i=1,\ldots,k\,j=1,\ldots,l}{\underset{Labels}{||}} \mathcal{LS}_{sc_i-ad_j}$$

The works in [24,22,23] detail the composition method.

B If a concrete execution of the system $\mathcal{LS}$ in [A] is described by a sequence diagram sd, $\mathcal{LS}_{sd}$ represents its corresponding LGSPN.

Then a performance model representing this concrete execution of the system can be obtained by the following expression:

$$\mathcal{LS}_{execution} = \mathcal{LS} \underset{Labels}{||} \mathcal{LS}_{sd}$$

The works in [4,3] detail the composition method.

4   Analysis of the System

As stated in section 2.5, a fundamental question in the use of stochastic PN models for performance evaluation, even under Markovian stochastic interpretation, is the so called *state explosion problem*. A general approach to deal with (computational) complexity is to use a *divide and conquer* (D&C) strategy, what requires the definition of a decomposition method and the subsequent composition of partial results to get the full solution. On the other hand, the trade-off between computational cost and accuracy of the solution leads to the use of approximation or bounding techniques instead of the computation of exact values. In this context, a pragmatic compromise to be handled by the analyzer of a system concerns the definition of faithful models, that may be very complex to exactly analyse (what may lead to the use of approximation or just bounding techniques), or simplified models, for which exact analysis can be, eventually, accomplished. D&C strategies can be used with exact, approximate, or bounding techniques.

The D&C techniques for performance evaluation present in the literature consider either *implicit or explicit decomposition* of PN models.

In [7], an *implicit* decomposition into P-semiflows is used for computing throughput *bounds* in polynomial time on the net size for free choice PN's with arbitrary pdf of time durations. A generalization to arbitrary P/T models (and also to coloured models) is presented in [9]. These techniques that use an implicit decomposition are directly applicable to the PN models obtained from UML diagrams (section 3). The bounds are computed from the solution of proper linear programming problems, therefore they can be obtained in polynomial time on the size of the net model, and they depend only on the mean values of service time associated to the firing of transitions and the routing rates associated with transitions in conflict and not on the higher moments of the probability distribution functions of the random variables that describe the timing of the system. The idea is to compute vectors χ and $\overline{\mu}$ that maximize or minimize the throughput of a transition or the average marking of a place among those verifying several operational laws and other linear constraints that can be easily derived from the net structure. The detailed linear programming problem to solve can be found in [9] and it has been implemented in the *GreatSPN* software tool [16].

In [8], an *explicit* decomposition of a general P/T net into *modules* (connected through buffers) is defined by the analyzer or provided by model construction. The modules are complemented with an abstract view of their environment in the full model. A computational technique is also presented in [8] that uses directly the information provided by the modules to compute the *exact* global limit probability distribution vector without storing the infinitesimal generator matrix of the whole net system (by expressing it in terms of a tensor algebra formula of smaller matrices). In [33], the same decomposition used in [8] is applied to compute a throughput *approximation* of the model through an *iterative technique* looking for a fixed point.

In both [8] and [33], the model decomposition and, eventually, the solution composition process should be net-driven (i.e., derived at net level). Fortunately, the process that is surveyed in section 3 leads to a global PN system that consists on the composition of PN modules (those explained in section 3.5) through fusion of places, thus both the tensor algebra based technique [8] and the approximation technique [33] can be considered to deal with the state explosion problem.

We start by reviewing the decomposition technique for general PN systems that was proposed in [8] (more details can be found there).

An arbitrary PN system can always be observed as a set of *modules* (disjoint simpler PN systems) that asynchronously communicate by means of a set of *buffers* (places).

Definition 4 (SAM). [8] *A strongly connected PN system,* $\mathcal{S} = \langle P_1 \cup \ldots \cup P_K \cup B, T_1 \cup \ldots \cup T_K, \mathbf{Pre}, \mathbf{Post}, \mathbf{m_0} \rangle$, *is a* System of Asynchronously Communicating Modules, *or simply a SAM, if:*
1. $P_i \cap P_j = \emptyset$ *for all* $i, j \in \{1, \ldots, K\}$ *and* $i \neq j$;
2. $T_i \cap T_j = \emptyset$ *for all* $i, j \in \{1, \ldots, K\}$ *and* $i \neq j$;
3. $P_i \cap B = \emptyset$ *for all* $i \in \{1, \ldots, K\}$;
4. $T_i = P_i{}^\bullet \cup {}^\bullet P_i$ *for all* $i \in \{1, \ldots, K\}$.
The net systems $\langle \mathcal{N}_i, \mathbf{m_{0i}} \rangle = \langle P_i, T_i, \mathbf{Pre}_i, \mathbf{Post}_i, \mathbf{m_{0i}} \rangle$ *with* $i \in \{1, \ldots, K\}$ *are called* modules *of* $\mathcal{S}$ *(where* $\mathbf{Pre}_i, \mathbf{Post}_i$, *and* $\mathbf{m_{0i}}$ *are the restrictions of* $\mathbf{Pre}$, $\mathbf{Post}$, *and* $\mathbf{m_0}$ *to* P_i *and* T_i*). Places in* B *are called* buffers. *Transitions belonging to the set* $\mathrm{TI} = {}^\bullet B \cup B^\bullet$ *are called* interface transitions. *Remaining ones* $((T_1 \cup \ldots \cup T_K) \setminus \mathrm{TI})$ *are called* internal *transitions.*

A SAM with two modules (the subnets generated by nodes whose tag starts with P_i or T_i for $i = 1, 2$) and four buffers (B_1, B_2, B_3, B_4) is depicted in Fig. 7.a.

All the strongly connected PN systems belong to the SAM class with the only addition of a *structured view* of the model (either given by construction or

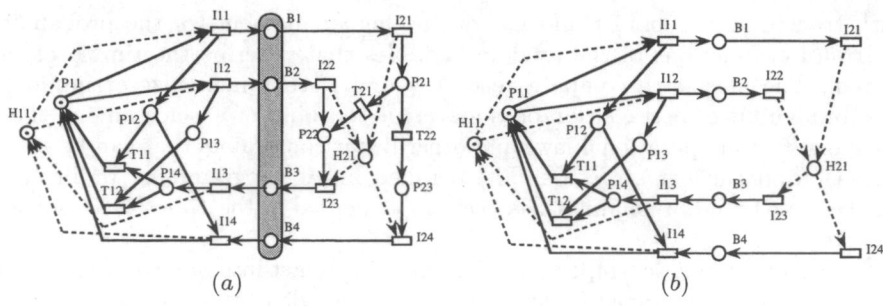

Fig. 7. (a) A SAM and (b) its LS1.

decided after observation of the model). In the case of PN's obtained from the SPE process introduced in section 3, the structured view is obtained from the net components that correspond to each superposed LGSPN.

In [8], a reduction rule has been introduced for the *internal behaviour* of modules of a SAM. Each module is decomposed into several pieces and each piece is substituted by a set H_i^j of new special places called *marking structurally implicit places* (MSIP's). Later, using that reduction, the original model can be decomposed into a collection of *low level systems* ($\mathcal{LS}_i$ with $i = 1, \ldots, K$) and a *basic skeleton* ($\mathcal{BS}$). In each $\mathcal{LS}_i$, only one module is kept while the internal behaviour of the others is reduced. In [8], the $\mathcal{LS}_i$ and the $\mathcal{BS}$ are used for a tensor algebra-based exact computation of the underlying CTMC.

An algorithm for the computation of the set H_i^j of MSIP's was proposed in [8]. A place is *implicit*, under interleaving semantics, if it can be deleted without changing the firing sequences. Each MSIP of H_i^j added to $\mathcal{N}$ needs an initial marking for making it implicit. In [10], an efficient method for computing such marking is presented.

The next step for the definition of the $\mathcal{LS}_i$ and the $\mathcal{BS}$ is to define an *extended system* ($\mathcal{ES}$). Consider, for instance, the SAM given in Fig. 7.a. The original net system $\mathcal{S}$ is the net without the places H_{11} and H_{21}. These places are the MSIP's computed to summarise the internal behaviour of the two modules. Place H_{11} summarises the module 1 (the subnet generated by nodes whose tags begin with P_1 or T_1), and place H_{21} summarises the module 2. The $\mathcal{ES}$ is the net system of Fig. 7.a (adding to $\mathcal{S}$ the places H_{11} and H_{21}).

From the $\mathcal{ES}$, we can build the $\mathcal{LS}_i$ and $\mathcal{BS}$. In each $\mathcal{LS}_i$ all the modules $\mathcal{N}_j$ with $j \neq i$, are reduced to their interface transitions and to the implicit places that were added in the $\mathcal{ES}$, while $\mathcal{N}_i$ is fully preserved. Systems $\mathcal{LS}_i$ represent different low level views of the original model. In the $\mathcal{BS}$ all the modules are reduced, and it constitutes a high level view of the system. In Fig. 7.b. the $\mathcal{LS}_1$ of Fig. 7.a is depicted. The $\mathcal{BS}$ is obtained by deleting from Fig. 7.b the nodes whose tags begin with P_1 and T_1.

Then, in [8] is presented how the rate matrix of a stochastic SAM can be expressed in terms of matrices derived from the rate matrix of the $\mathcal{LS}_i$ systems.

Let $\mathbf{Q}$ be the infinitesimal generator of a stochastic SAM. We can rewrite $\mathbf{Q}$ as $\mathbf{Q} = \mathbf{R} - \boldsymbol{\Delta}$, where $\boldsymbol{\Delta}$ is a diagonal matrix and $\boldsymbol{\Delta}[i,i] = \sum_{k \neq i} \mathbf{Q}[i,k]$. The same definition holds for the $\mathcal{LS}_i$ components: $\mathbf{Q}_i = \mathbf{R}_i - \boldsymbol{\Delta}_i$.

If states are ordered according to the high level state $\mathbf{z}$ (a state in the reachability set of $\mathcal{BS}$), then matrices $\mathbf{Q}$ and $\mathbf{R}$ (respectively, $\mathbf{Q}_i$ and $\mathbf{R}_i$) can be described in terms of blocks $(\mathbf{z}, \mathbf{z}')$. We shall indicate them with $\mathbf{Q}(\mathbf{z}, \mathbf{z}')$ and $\mathbf{R}(\mathbf{z}, \mathbf{z}')$ $(\mathbf{Q}_i(\mathbf{z}, \mathbf{z}')$ and $\mathbf{R}_i(\mathbf{z}, \mathbf{z}')$, respectively).

Diagonal blocks $\mathbf{R}_i(\mathbf{z}, \mathbf{z})$ have non null entries that are due *only* to the firing of transitions in $T_i \setminus \mathrm{TI}$ (internal behaviour), while blocks $\mathbf{R}_i(\mathbf{z}, \mathbf{z}')$ with $\mathbf{z} \neq \mathbf{z}'$ have non null entries due *only* to the firing of transitions in TI.

Let $\mathrm{TI}_{\mathbf{z}, \mathbf{z}'}$ with $\mathbf{z} \neq \mathbf{z}'$, be the set of transitions $t \in \mathrm{TI}$ such that $\mathbf{z} \overset{t}{\longrightarrow} \mathbf{z}'$ in the basic skeleton $\mathcal{BS}$. From a matrix $\mathbf{R}_i(\mathbf{z}, \mathbf{z}')$, with $\mathbf{z} \neq \mathbf{z}'$ we can build additional matrices $\mathbf{K}_i(t)(\mathbf{z}, \mathbf{z}')$, for each $t \in \mathrm{TI}_{\mathbf{z}, \mathbf{z}'}$, according to the following definition:

$$\mathbf{K}_i(t)(\mathbf{z}, \mathbf{z}')[\mathbf{m}, \mathbf{m}'] = \begin{cases} 1 & \text{if } \mathbf{m} \overset{t}{\longrightarrow} \mathbf{m}' \\ 0 & \text{otherwise} \end{cases}$$

where $\mathbf{m}$ and $\mathbf{m}'$ are two of the states with a high level view equal to $\mathbf{z}$ and $\mathbf{z}'$ respectively: $\mathbf{m}|_{H_1 \cup \ldots \cup H_K \cup B} = \mathbf{z}$, and $\mathbf{m}'|_{H_1 \cup \ldots \cup H_K \cup B} = \mathbf{z}'$.

Matrices $\mathbf{G}(\mathbf{z}, \mathbf{z}')$ can then be defined as:

$$\mathbf{G}(\mathbf{z}, \mathbf{z}) = \bigoplus_{i=1}^{K} \mathbf{R}_i(\mathbf{z}, \mathbf{z})$$

$$\mathbf{G}(\mathbf{z}, \mathbf{z}') = \sum_{t \in \mathrm{TI}_{\mathbf{z}, \mathbf{z}'}} w(t) \bigotimes_{i=1}^{K} \mathbf{K}_i(t)(\mathbf{z}, \mathbf{z}') \tag{2}$$

The steady-state distribution of a stochastic SAM can be computed using the $\mathbf{G}$ matrix given in equation (2).

Instead of using the above result to compute exact indices, the same decomposition technique for a SAM can be used for an approximate throughput computation. The method is, basically, a *response time preservation algorithm*. In each $\mathcal{LS}_i$ a unique module of S with all its places and transitions is kept. So, in $\mathcal{LS}_i$ interface transitions of module j (for $j \neq i$) approximate the response time of module j. The algorithm is the following [33]:

select the modules
derive the reachability graph of $\mathcal{LS}_i$ for $i = 1, \ldots, K$ and the one of $\mathcal{BS}$
give an initial service rate $\mu_i^{(0)}$ for $i = 2, \ldots, K$
$j := 0$ {counter for iteration steps}
repeat
    $j := j + 1$
    for $i := 1$ **to** K **do**
        solve $\mathcal{LS}_i$: In: $\mu_l^{(j)}$ for $l < i$ and $\mu_l^{(j-1)}$ for $i < l \leq K$
                Out: initial rates μ_i and   thr. $\chi_i^{(j)}$ of $\mathrm{TI} \cap T_i$
        solve $\mathcal{BS}$: In: $\mu_l^{(j)}$ for $l < i$ and $\mu_l^{(j-1)}$ for $i < l \leq K$
                μ_i and $\chi_i^{(j)}$ for transitions in $\mathrm{TI} \cap T_i$
                Out: actual rates $\mu_i^{(j)}$ $\mathrm{TI} \cap T_i$
until convergence of $\chi_1^{(j)}, \ldots, \chi_K^{(j)}$

The gain of state space decomposition techniques with respect to the classical exact solution algorithm (solution of the underlying CTMC of the original model) is in both memory and time requirements. With respect to space, the infinitesimal generator of the CTMC of the whole system is never stored. Instead of that, the generator matrices of smaller subsystems are stored. With respect to time complexity, in the case of the above iterative approximation algorithm we do not solve the CTMC isomorphous to the original system but those much smaller isomorphous to the derived subsystems.

5   Related Work

There exist several works in SPE that inspired the work presented in this article. They share similarities such as: SPE is driven by models, they were proposed before or in the first specific conference of the SPE [44], they began to devise UML as the design notation. Summarizing, they were the first attempts to position SPE as we understand it today. Concretely, we should mention: the Permabase project [45], the works developed at Carleton University [47,48] and the works by Pooley and King [21,35].

Another set of works can be considered as contemporary to the one presented here. They explicitly consider UML as the design language for the SPE process and they are more mature than those previously cited since they appeared later in time. Some of them were presented in [46,18] and others even in posterior relevant conferences or journals. Among them, the following stand out: the works of Cortellessa [11], the software architecture approaches [2], the work of De Miguel [12] and finally the research work at Carleton University [34].

6   Conclusion

In this paper we surveyed recent contributions in the field of SPE using UML and stochastic Petri nets. The UML behavioral diagrams were studied in order to:

1. find the possible roles that they can play in a SPE process,
2. describe the performance annotations that each one can support under these roles,
3. translate them into stochastic Petri nets.

The result of the proposed translation consists in a performance model that represents either the whole system (composition of the LGSPN representing each SC in the system) or a concrete execution of the system (composition of the LGSPN representing a SD togheter with the LGSPNs of the involved SCs). Since the performance model can suffer the state space explosion problem, we also surveyed here several works on efficient analysis techniques that can be used to solve the model.

References

1. M. Ajmone Marsan, G. Balbo, G. Conte, S. Donatelli, and G. Franceschinis, *Modelling with generalized stochastic Petri nets*, John Wiley Series in Parallel Computing - Chichester, 1995.
2. F. Aquilani, S. Balsamo, and P. Inverardi, *Performance analysis at the software architectural design level*, Performance Evaluation **45** (2001), 147–178.
3. S. Bernardi, *Building Stochastic Petri Net models for the verification of complex software systems*, Ph.D. thesis, Dipartimento di Informatica, Università di Torino, April 2003.
4. S. Bernardi, S. Donatelli, and J. Merseguer, *From UML sequence diagrams and statecharts to analysable Petri net models*, Proceedings of the ThirdInternational Workshop on Software and Performance (WOSP2002) (Rome, Italy), ACM, July 2002, pp. 35–45.
5. G. Berthelot, *Transformations and decompositions of nets*, Advances in Petri Nets 1986 - Part I (W. Brauer, W. Reisig, and G. Rozenberg, eds.), Lecture Notes in Computer Science, vol. 254, Springer-Verlag, Berlin, 1987, pp. 359–376.
6. F.P. Brooks, *Essence and accidents of software engineering*, IEEE Computer **20** (1987), no. 4, 10–19.
7. J. Campos, G. Chiola, and M. Silva, *Properties and performance bounds for closed free choice synchronized monoclass queueing networks*, IEEE Transactions on Automatic Control **36** (1991), no. 12, 1368–1382.
8. J. Campos, S. Donatelli, and M. Silva, *Structured solution of asynchronously communicating stochastic modules*, IEEE Transactions on Software Engineering **25** (1999), no. 2, 147–165.
9. G. Chiola, C. Anglano, J. Campos, J. M. Colom, and M. Silva, *Operational analysis of timed Petri nets and application to the computation of performance bounds*, Proceedings of the 5th International Workshop on Petri Nets and Performance Models (Toulouse, France), IEEE-Computer Society Press, October 1993, pp. 128–137.
10. J. M. Colom and M. Silva, *Improving the linearly based characterization of P/T nets*, Advances in Petri Nets 1990 (G. Rozenberg, ed.), Lecture Notes in Computer Science, vol. 483, Springer-Verlag, Berlin, 1991, pp. 113–145.
11. V. Cortellessa and R. Mirandola, *Deriving a queueing network based performance model from UML diagrams*, in Woodside et al. [46], pp. 58–70.
12. M. de Miguel, T. Lambolais, M. Hannouz, S. Betge, and S. Piekarec, *UML extensions for the specification of latency constraints in architectural models*, in Woodside et al. [46], pp. 83–88.
13. E. Domínguez, A.L. Rubio, and M.A. Zapata, *Dynamic semantics of UML state machines: A metamodelling perspective*, Journal of Database Management **13** (2002), 20–38.
14. R. Dumke, C. Rautenstrauch, A. Schmietendorf, and A. Scholz (eds.), *Performance engineering. state of the art and current trends*, Lecture Notes in Computer Science, vol. 2047, Springer-Verlag, Heidelberg, 2001.
15. H. Gomaa and D. Menascé, *Design and performance modeling of component interconnection patterns for distributed software architectures*, in Woodside et al. [46], pp. 117–126.
16. *The GreatSPN tool*, http://www.di.unito.it/~greatspn.
17. H. Hermanns, U. Herzog, and Katoen J.P., *Process algebra for performance evaluation*, Theoretical Computer Science **274** (2002), no. 1-2, 43 87.

18. P. Inverardi, S. Balsamo, and Selic B. (eds.), *Proceedings of the Third International Workshop on Software and Performance*, Rome, Italy, ACM, July 24-26 2002.

19. I. Jacobson, M. Christenson, P. Jhonsson, and G. Overgaard, *Object-oriented software engineering: A use case driven approach*, Addison-Wesley, 1992.

20. K. Kant, *Introduction to computer system performance evaluation*, Mc Graw-Hill, 1992.

21. P. King and R. Pooley, *Using UML to derive stochastic Petri nets models*, Proceedings of the Fifteenth Annual UK Performance Engineering Workshop (J. Bradley and N. Davies, eds.), Department of Computer Science, University of Bristol, July 1999, pp. 45–56.

22. J. P. López-Grao, J. Merseguer, and J. Campos, *From UML activity diagrams to stochastic PNs: Application to software performance engineering*, Proceedings of the Fourth International Workshop on Software and Performance (WOSP'04) (Redwood City, California, USA), ACM, January 2004, To appear.

23. J. Merseguer, *Software performance engineering based on UML and Petri nets*, Ph.D. thesis, University of Zaragoza, Spain, March 2003.

24. J. Merseguer, S. Bernardi, J. Campos, and S. Donatelli, *A compositional semantics for UML state machines aimed at performance evaluation*, Proceedings of the 6th International Workshop on Discrete Event Systems (Zaragoza, Spain) (A. Giua and M. Silva, eds.), IEEE Computer Society Press, October 2002, pp. 295–302.

25. J. Merseguer and J. Campos, *Exploring roles for the UML diagrams in software performance engineering*, Proceedings of the 2003 International Conference on Software Engineering Research and Practice (SERP'03) (Las Vegas, Nevada, USA), CSREA Press, June 2003, pp. 43–47.

26. J. Merseguer, J. Campos, and E. Mena, *A pattern-based approach to model software performance*, in Woodside et al. [46], pp. 137–142.

27. _____, *A performance engineering case study: Software retrieval system*, in Dumke et al. [14], pp. 317–332.

28. _____, *Analysing internet software retrieval systems: Modeling and performance comparison*, Wireless Networks: The Journal of Mobile Communication, Computation and Information **9** (2003), no. 3, 223–238.

29. M. K. Molloy, *Performance analysis using stochastic Petri nets*, IEEE Transactions on Computers **31** (1982), no. 9, 913–917.

30. Object Management Group, http:/www.omg.org.

31. Object Management Group, http:/www.omg.org, *UML Profile for Schedulability, Performance and Time Specification*, March 2002.

32. Object Management Group, http:/www.omg.org, *OMG Unified Modeling Language specification*, March 2003, version 1.5.

33. C. J. Pérez-Jiménez and J. Campos, *On state space decomposition for the numerical analysis of stochastic Petri nets*, Proceedings of the 8th International Workshop on Petri Nets and Performance Models (Zaragoza, Spain), IEEE Computer Society Press, September 1999, pp. 32–41.

34. D. Petriu and H. Shen, *Applying the UML performance profile: Graph grammar-based derivation of LQN models from UML specifications*, Computer Performance Evaluation, Modelling Techniques and Tools 12th International Conference, TOOLS 2002 (London, UK) (Tony Field, Peter G. Harrison, Jeremy Bradley, and Uli Harder, eds.), Lecture Notes in Computer Science, vol. 2324, Springer, April 14-17 2002, pp. 159–177.

35. R. Pooley and P. King, *The Unified Modeling Language and performance engineering*, IEE Proceedings Software, IEE, February 1999, pp. 2–10.

36. ArgoUML project, http://argouml.tigris.org.
37. J. Rumbaugh, M. Blaha, W. Premerlani, E. Frederick, and W. Lorensen, *Object oriented modeling and design*, Prentice-Hall, 1991.
38. A. Schmietendorf and A. Scholz, *Aspects of performance engineering - An overview*, in Dumke et al. [14], pp. IX–XII.
39. M. Silva, *Las redes de Petri en la automática y la informática*, Editorial AC, Madrid, 1985, In Spanish.
40. C. U. Smith, *Performance engineering of software systems*, The Sei Series in Software Engineering, Addison–Wesley, 1990.
41. C.U. Smith, *Increasing information systems productivity by software performance engineering*, Proceedings of the Seventh International Computer Measurement Group Conference (New Orleans, LA, USA) (D.R. Deese, R.J. Bishop, J.M. Mohr, and H.P. Artis, eds.), Computer Measurement Group, December 1-4 1981, pp. 5–14.
42. _____, *Origins of software performance engineering: Highlights and outstanding problems*, in Dumke et al. [14], pp. 96–118.
43. C.U. Smith and Ll.G. Williams, *Software performance antipatterns*, in Woodside et al. [46], pp. 127–136.
44. C.U. Smith, M. Woodside, and P. Clements (eds.), *Proceedings of the First International Workshop on Software and Performance*, Santa Fe, New Mexico, USA, ACM, October 12-16 1998.
45. P. Utton and B. Hill, *Performance prediction: an industry perspective*, Performance Tools 97 Conference (St Malo), June 1997, pp. 1–5.
46. M. Woodside, H. Gomaa, and D. Menascé (eds.), *Proceedings of the Second International Workshop on Software and Performance*, Ottawa, Canada, ACM, September 17-20 2000.
47. M. Woodside, C. Hrischuck, B. Selic, and S. Bayarov, *A wideband approach to integrating performance prediction into a software design environment*, in Smith et al. [44], pp. 31–41.
48. _____, *Automated performance modeling of software generated by a design environment*, Performance Evaluation **45** (2001), 107–123.

Enabling Simulation with Augmented Reality[*]

Erol Gelenbe[1], Khaled Hussain[2], and Varol Kaptan[1]

[1] Imperial College, London SW7 2BT, UK
e.gelenbe@imperial.ac.uk, v.kaptan@imperial.ac.uk
[2] University of Central Florida, Orlando, Florida 32816, USA
khaled@cs.ucf.edu

Abstract. In many critical applications such as airport operations, military simulations, and medical simulations, it is very useful to conduct simulations in accurate and realistic settings that are represented by real video imaging sequences. Furthermore, it is important that the simulated entities conduct autonomous actions which are realistic and which follow plans of action or intelligent behavior in reaction to current situations. We describe an approach to incorporate synthetic objects in a visually realistic manner in video sequences representing a real scene. We also discuss how the synthetic objects can be designed to conduct intelligent behavior within an augmented reality setting.

1 Introduction

Simulation is widely used to model real systems under varying synthetic conditions. It can be used to predict the capabilities of a system which is being designed, or to predict the performance of a system which is being modified. Simulation often concentrates on the algorithmic description and control of synthetic entities which are being modeled, and modern discrete event simulators often use a graphical interface to simplify and enrich the user's interaction both before, during and after simulation runs. A significant leap forward in simulation technology is to evaluate synthetic simulated conditions in truly realistic settings, and to ask questions about "what would happen if ..." in the context of a real environment and actual events. Our work mixes simulation with reality in real time, in order to examine how novel simulated conditions can interact with a real system's operation. This interaction can go in both directions: the course of the real world can be modified by virtual entities, and the virtual objects are constrained to operate in the real world.

Many simulation systems of interest to Defence Technologies are networked content delivery systems similar to the ones discussed elsewhere in this volume [11], as in distributed simulation for training purposes, and also in many cases where simulation is used for operational planning purposes where the simulation application will appear as a peer-to-peer application [14]. Thus security and reliability will be important considerations [19,26], and network quality-of-service will have a significant impact [11,15]. The

---

[*] This work has been supported by a contract for the **Defence Technology Centre on Information Fusion** from the British Ministry of Defence (MoD) via General Dynamics UK Ltd., to Imperial College.

wireless context [1,29] will obviously also be of great interest in such applications. Thus the networking perspective also raises many interesting research issues in the context of the problems that are addressed in this paper.

The rest of this paper is organized as follows. In Section 2, we present an overview of our approach. We discuss the architecture of the system we have designed and implemented in Section 2.2. In Section 2.3 we provide details about the new real-time moving-object injection algorithm which we have developed. Section 2.4 summarizes experimental results from the implementation of our algorithm. A simple model of a modular collaborating agent is also described in Section 3.

2   Visual Augmented Reality

Visual augmented reality creates a combination of a real and virtual scenes in which the user perceives a significant difference between the real and augmented world. Figure 1 shows an example of a view that the user might see in augmented reality system showing a live scene with a virtual tank. We refer to the real world as a "scene", even though we are dealing with a video sequence representing the world as it is being viewed in real-time. The scene has different visual representations depending on where it is being viewed from, and the viewing point is often referred to as the "aim-point".

Fig. 1. Live scene with virtual tank.

One of the difficult technical issues in augmented reality is the "registration problem", which refers to the need for determining the isomorphism between objects and features in a live scene with the corresponding features and the corresponding objects in an augmented version of that scene. Errors in registration will generate visual inconsistencies between real and virtual images with obvious consequences on the value of the augmented reality system for simulation purposes. Therefore several authors have addressed this issue, and in [4] it is indicated that registration using only information from a sensor based tracking system cannot achieve a perfect match. Thus several authors make use of an image processing based algorithm for improving registration. One approach is to detect features in the real image and uses them to enforce registration, while others have considered placing special marks (e.g. LEDs [4], circles [18], a calibration grid [16]) in the environment. Image-processing algorithms detect the locations of these marks and use them to enforce registration, assuming that one or more special marks are visible at all times. Another approach [18] uses a survey of the the live environment with real-time instrumentation, providing more information about objects and their distances in the live environment, but requires specific equipment and significant amounts of additional computation for the interpretation of the sensors' output.

Almost all augmented reality techniques assume that virtual objects and live objects have the same detailed shape. This assumption is only valid for rigid objects such as roads and buildings: many virtual object generators will use a simplified representation, and will even sometimes only make use of templates; e.g., a synthetic pine tree may be some idealized template of a pine tree, rather than the actual pine tree being represented at some location in a scene.

Figure 3 shows a simple example of a virtual vehicle behind a virtual tree. If we use current techniques (register the real tree with the virtual tree; then copy the visible part of the virtual vehicle into the live image), visual inconsistencies will happen as shown in Figure 3 (b). Also current techniques cannot tolerate any small errors in registration. Figure 4 (b, d) shows an example of visual inconsistencies due to small errors in registration.

In our work we use objects to represent live and virtual scenes. This representation provides the capability of our approach to deal with non-rigid objects and to tolerate small errors in registration. Since our system is expected to work in a wide variety of natural settings, we do not make use of special marks in the environment. Thus even if we obtain accurate registration, we may still have visual inconsistencies due to nonrigid objects (e.g., trees). To address this problem, our approach segments the live image into objects and then register them with the corresponding virtual objects. In general image segmentation techniques suffer from one or two of the following disadvantages:

1. They can't be used in real time applications.
2. They have an inaccurate segmentation output.

Figure 3 shows the same example of a virtual vehicle behind a virtual tree. Since in our approach we represent live and virtual word as objects, in the simple example of a vehicle behind a tree we have three objects: (1) a virtual tree object, (2) a virtual vehicle object (note that we mean by the virtual vehicle object the whole vehicle, not only the visible part of the vehicle), and (3) a real tree object. When we insert the virtual vehicle object into the live image, visual inconsistencies will not happen as shown in Figure 3 (c). Also our approach can tolerate any small errors in registration as shown in Figure 4(e).

2.1   Overview of Our Approach

Our approach is postulated on the premise that the real world is far more dense in significant objects, than in the number and type of synthetic objects we insert into it. This is of course totally different than an interactive computer game in which no real objects exist (i.e. everything is artificial), and in which there may be a large number of moving synthetic objects (e.g. aircrafts, robots, etc.). Our system design is based on the following assumptions:

- The 3-D real world of stationary objects, i.e., non-moving such as landmarks, terrain elevation, houses, trees, etc. is represented in a terrain database (TB) which has some desired level of precision.
- There are a limited number of moving real objects and virtual objects, which together cover only a small fraction of the scene being viewed at any time or being represented in the TB.
- Moving virtual objects are also represented in the TB; they are synthetically moved either manually (by the operator or learner), or using a simulation system.
- From the TB, and knowing a specific viewing location and direction (which we call the aim-point), it is possible to synthetically generate a graphics image representing the scene as it is viewed from the TB. This graphics based scene (GS) will include the synthetic objects.
- Inaccuracies in the TB are expected to exist and will often have to be compensated for in real-time.
- The real scene (RS) corresponding to a given aim-point is viewed through an appropriate video camera or sensor.

In have developed a new real-time moving-object injection method based on these assumptions with the purpose of inserting synthetic moving objects into live video in real-time, and involving techniques for image segmentation and registration. Our approach tolerates a certain level of inaccuracy in the TB, and avoids the expense of specific location and range finding instrumentation. The practical outcome of this work is a Target Overlay System (TOS) that will support the Inter-Vehicle Embedded Simulation Technology (INVEST) at the U.S. Army Simulation and Training Command.

2.2   System Architecture

A schematic representation of the system we have developed is shown in Figure 2. The first challenge challenge we faced in our design is to determine a representation of the live and virtual environments that could be compared to each other. The simplest method of capture of the **Live Environment** is through a video camera providing color RS-170 video. This provides a two-dimensional representation of the scene. Its main shortcoming is that it provides little information about the precise location and representation of the objects in the scene that could occlude the added virtual objects. Camera pointing angles and camera zoom information are available to a PC on its 1553 data bus. The position and speed of the platform supporting the camera are also provided by an instrumentation system to the 1553 bus. This positional information is used to determine the aim-point so that we may also generate a synthetic image that is equivalent to the live image.

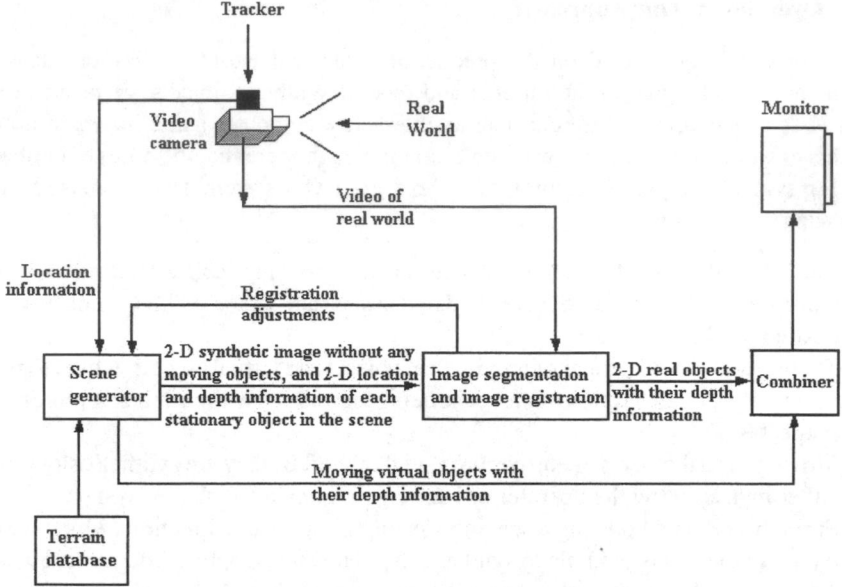

Fig. 2. System architecture.

The TB format of terrain we have used is the SAF (Semi-Automated Forces) [10,9] environment provided via topological terrain formats. Specifically, the Modular Semi-Automated Forces (ModSAF) simulation system uses the Compact Terrain Database (CTDB) format with terrain height and feature representations. It also provides detailed information about any virtual objects which are placed in the virtual TB. Another approach to representing the virtual environment is by visual databases which are used in stealth environment applications. These are often more detailed than topological databases because they were designed to be viewed as realistic renderings by a human user. Visual databases also provide a two-dimensional representation of the virtual environment, and are much closer to the available live and desired visual augmented representations. For the virtual representation, we have chosen OpenGL which renders three-dimensional virtual scenes in two dimensions. The rendering is often done by hardware, which is much faster than software rendering. As part of the rendering process in OpenGL, the depth (Z-buffer) of each object in the scene is determined. This depth is then used to calculate which objects are visible from the current aim-point.

The block diagram in Figure 2 outlines the flow of data in the system. A camera provides the live terrain image. A tracker is attached to the camera to provide the location and the direction of the camera. The scene generator uses this location information to generate the 2-D synthetic image, and 2-D location and depth information of each stationary object in the field of view. Our image segmentation algorithm uses this information to segment the live image into 2-D real stationary objects. Since we cannot obtain depth information from the live image in real time, we assume that the depth of each 2-D real stationary object is identical to the depth of the corresponding virtual stationary object.

The image registration algorithm uses the locations and the sizes of the virtual and real stationary objects to adjust the viewpoint of the virtual scene. The scene generator also generates the moving objects with their depth information. The combiner then inserts these moving objects based on their depth information between real stationary objects.

2.3 Moving-Object Injection in Real-Time

At the core of the system we have developed is an algorithm to insert synthetic moving objects into live images in real time. The essence of this approach is that it is purely computationally based, and that it does not make use of any direct video manipulation.

A primary concern is the proper placement of the virtual objects in front of, or behind, live objects. Thus the realistic representation of the inserted objects is tied to both the appropriate occlusions and the shapes and sizes of inserted objects. A good solution to the occlusion problem requires detailed knowledge of the objects and of their location in the live scene. Since the two-dimensional live images provide no prior information about the objects in the scene, we use an image segmentation technique to segment the live image into objects. We then use a registration technique to register objects in the live image with those in the virtual scene. Depth information from the virtual scene is used to associate relative depth information to each object in the live image, so that there is no need for additional instrumentation to calculate the depths of the live world surrounding the observer.

Given objects $(vo_1, vo_2, ..., vo_n)$ of the virtual scene (see Figure 6) with their position and color information, and given a live image, we segment this live image into the same number of objects $(lo_1, lo_2, , lo_n)$. Each object lo_i is equivalent to the corresponding virtual object vo_i, but they do not necessarily have the same exact shape, location, and color. Since we cannot obtain depth information from the live image, we assume that the depth of a live object is equal to the depth of the corresponding synthetic object. Representing the virtual live scene by objects provides approximate location and color information for each one of them. This in turn gives our image segmentation algorithm the capability to find accurately the corresponding objects in the live scene in real time.

Figure 3 shows a simple example of a virtual vehicle behind a virtual tree. Since in our approach, we represent live and virtual word as objects. Thus in the simple example of a vehicle behind a tree, we have three objects: (1) a virtual tree object, (2) a virtual vehicle object (note that we mean by the virtual vehicle object the whole vehicle, not only the visible part of the vehicle), and (3) a real tree object. When we insert the virtual vehicle object into the live image, visual inconsistencies will not happen as shown in Figure 3 (c). Also our approach can tolerate any small errors in registration as shown in Figure 4(e).

2.4 Experimental Implementation

To test our system, we considered several methods to achieve correlated live and virtual images. The first approach was to model terrain so that the virtual terrain model could be adapted to accurately model the real world. The main obstacle with this solution was the generation of both an OpenFlight database and a CTDB database for use with moving synthetic objects such as vehicles. Several OpenFlight databases exist for areas of the

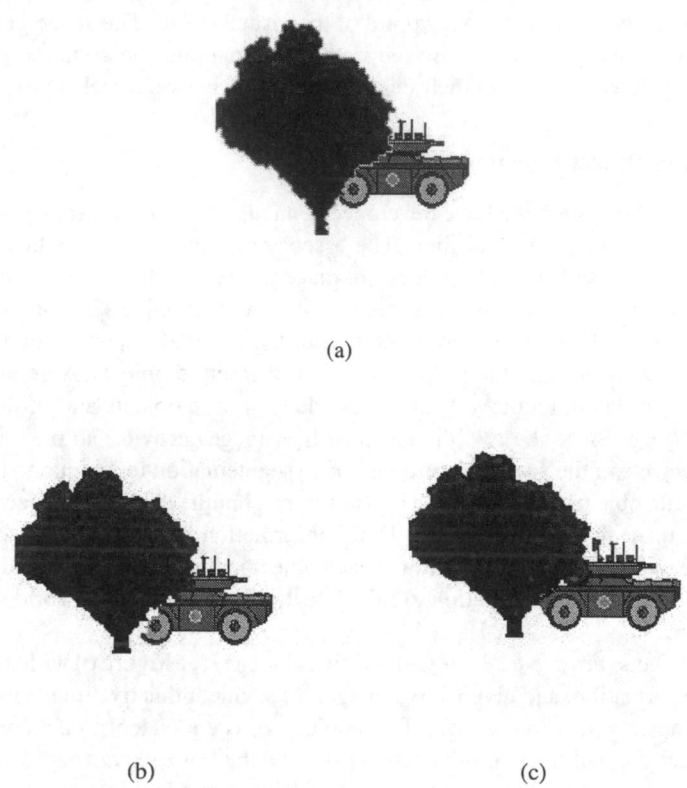

(a)

(b)   (c)

Fig. 3. A simple example of a virtual vehicle behind a tree for illustrating the limitations of current techniques. (a) A virtual vehicle behind a virtual tree. (b) Inserting the virtual vehicle into the real scene using current techniques causes visual inconsistencies. (c) Inserting the virtual vehicle into the real scene using our technique does not cause visual inconsistencies.

University of Central Florida (UCF) campus. However, there were no corresponding CTDB databases. Tools are available to convert from OpenFlight to CTDB format, but they require significant manual tweaking of input data. Another obstacle was the ability to isolate aspects of the live terrain. Variable weather would limit the times we could use the live image. The coming and going of cars, bikes, and pedestrians would also change the environment. Even larger structures change with ongoing new building construction at UCF.

We therefore developed a table-top model of the real world to be used for the live camera scene. The table-top represents a geographic area for which we already have correlated OpenFlight and CTDB databases. This approach allows the table-top to be correlated with the virtual terrain. We can also adjust the OpenFlight and CTDB databases to fix any inconsistencies between the real and synthetic views. The table-top also allow

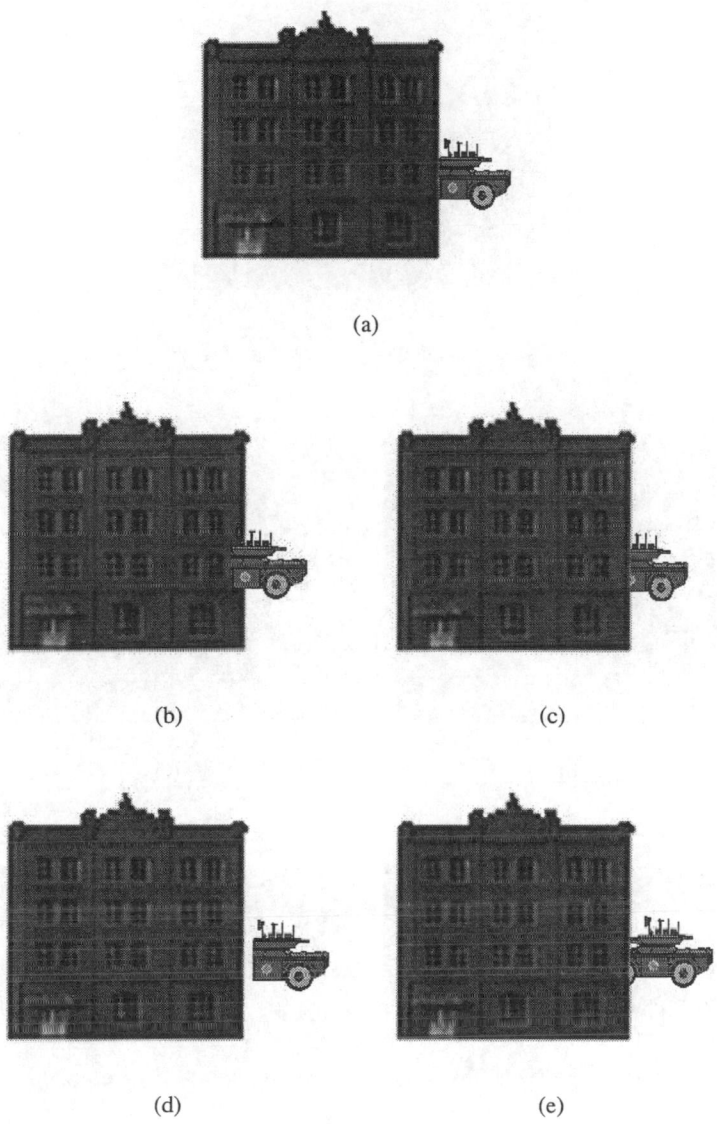

(a)

(b)   (c)

(d)   (e)

Fig. 4. A simple example of a virtual vehicle behind a building for illustrating the limitations of current techniques. (a) A virtual vehicle behind a virtual building. (b,d) small errors in registration cause visual inconsistencies because current techniques cannot tolerate any small errors in registration. (c,e) small errors in registration do not cause visual inconsistencies because our technique can tolerate small errors in registration.

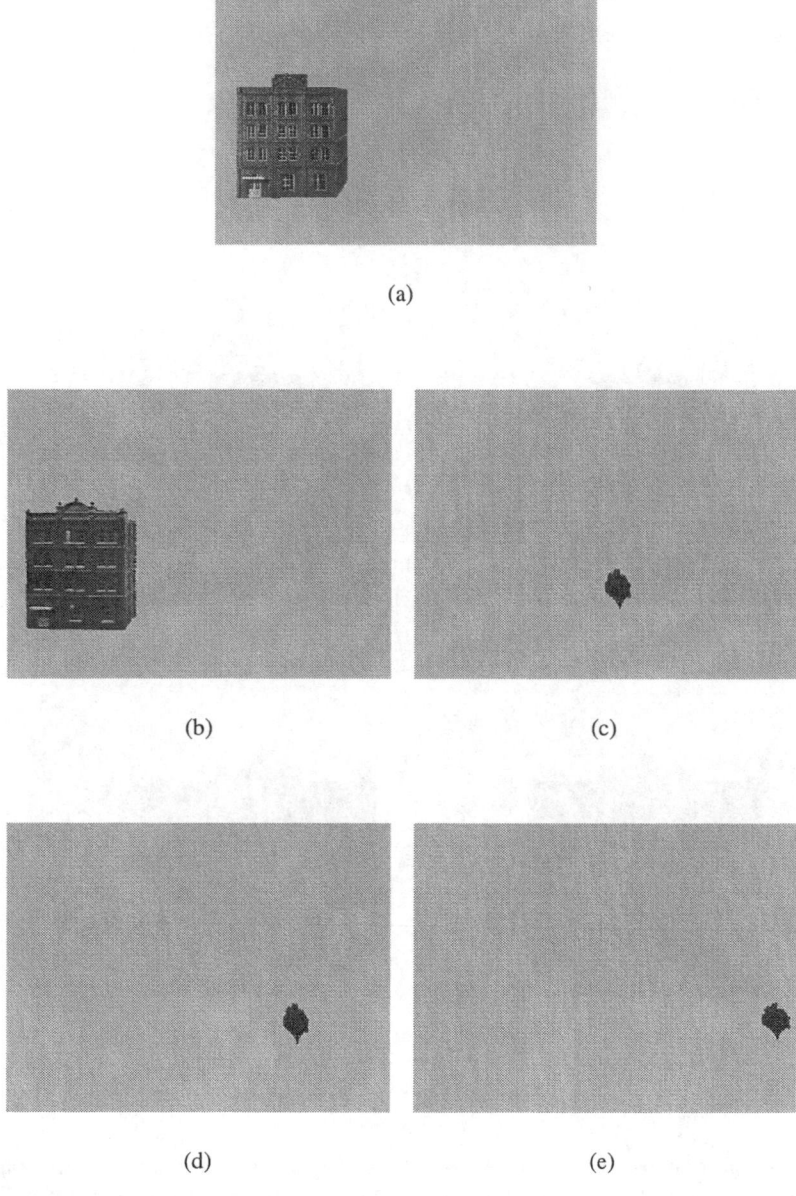

Fig. 5. The object representation of the virtual scene. (a) The synthetic image. (b) The synthetic building. (c) The left synthetic tree. (d) The middle synthetic tree. (e) The right synthetic tree.

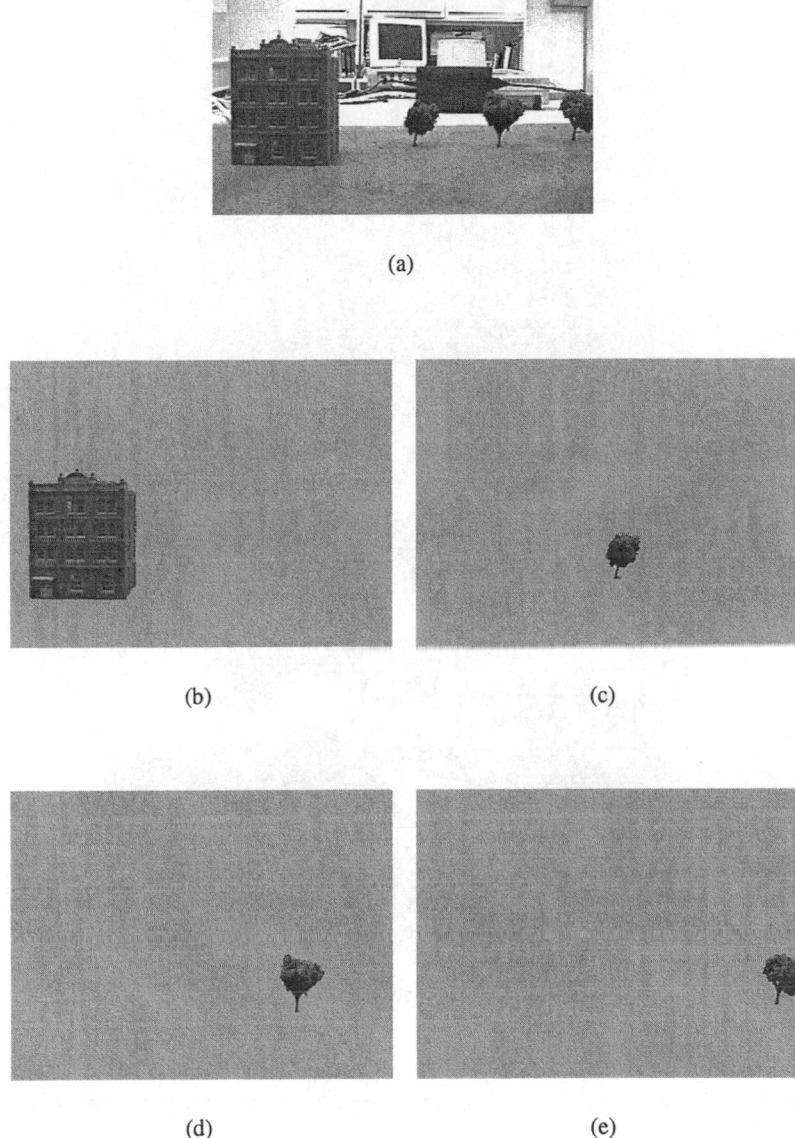

Fig. 6. The decomposition of the live image into objects. (a) Table-top camera scene. (b) The building. (c) The left tree. (d) The middle tree. (e) The right tree.

(a)

(b)

Fig. 7. (a) A synthetic moving-object, and the result (b) of inserting it into the live scene.

us to carry out the development work from "uncontrolled objects" such as unrelated vehicles, pedestrians, animals and adverse weather.

An example of the virtual scene that illustrates our approach is shown in Figure 5, and the corresponding live image and its decomposition into objects are shown in Figure 6. Once the object representation of the live scene is determined, the insertion of a synthetic moving-object will be based on the distance at which this object should be placed.

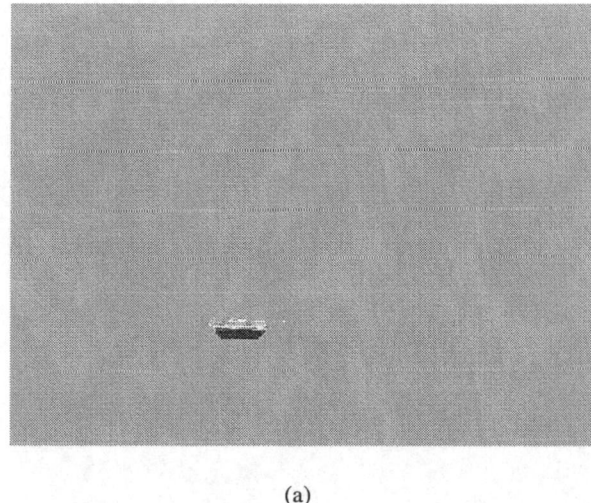

(a)

(b)

Fig. 8. (a) A synthetic moving-object, and the result (b) of inserting it into the live scene.

Figure 7, 8, 9, and 10 illustrate the insertion of synthetic-target objects into the live scene. Figure 11 shows some frames from an augmented video sequence that shows the capability of our approach to insert synthetic moving-tanks while the camera moves. Furthermore, our system can be used with real image data. Figure 12 shows a test of our system using a real world scene.

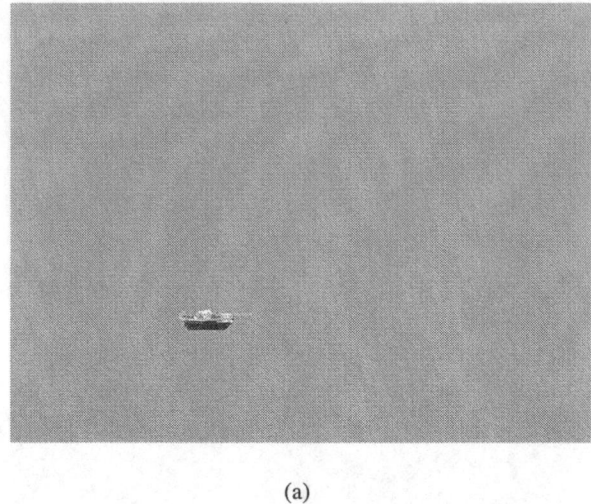

(a)

(b)

Fig. 9. (a) A synthetic moving-object, and the result (b) of inserting it into the live scene.

3   Autonomous Behavior of Injected Objects

The behavior of injected artificial entities can be as important as their appearance in a Visual Simulation. This is especially important in the context of simulations designed for training personnel or evaluating a "what if ..." situation. In such simulations, the behavior of agents will have an important effect on the final outcome in the form of "acquired

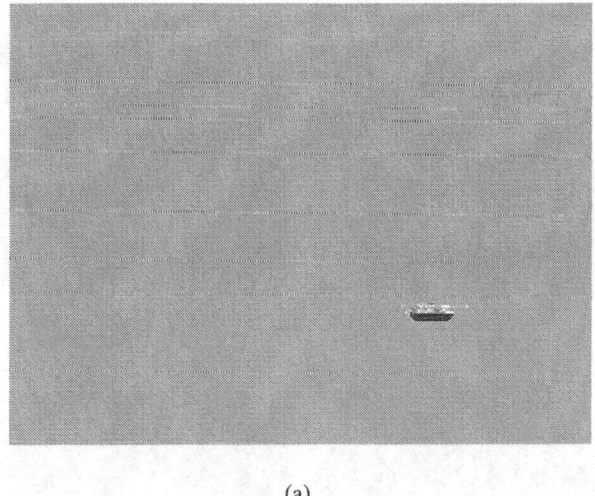

(a)

(b)

Fig. 10. (a) A synthetic moving-object, and the result (b) of inserting it into the live scene.

training experience". Unrealistic agent behavior, e.g., in the form of very limited or even extremely advanced intelligence may lead to poor performance of the trainees in a real-life situation.

Agent behavior in a sophisticated simulated environment can be very complex and may involve many entities. Intelligence can be employed at very different levels. A very simple example will be an agent that has to go from one position to another trying to

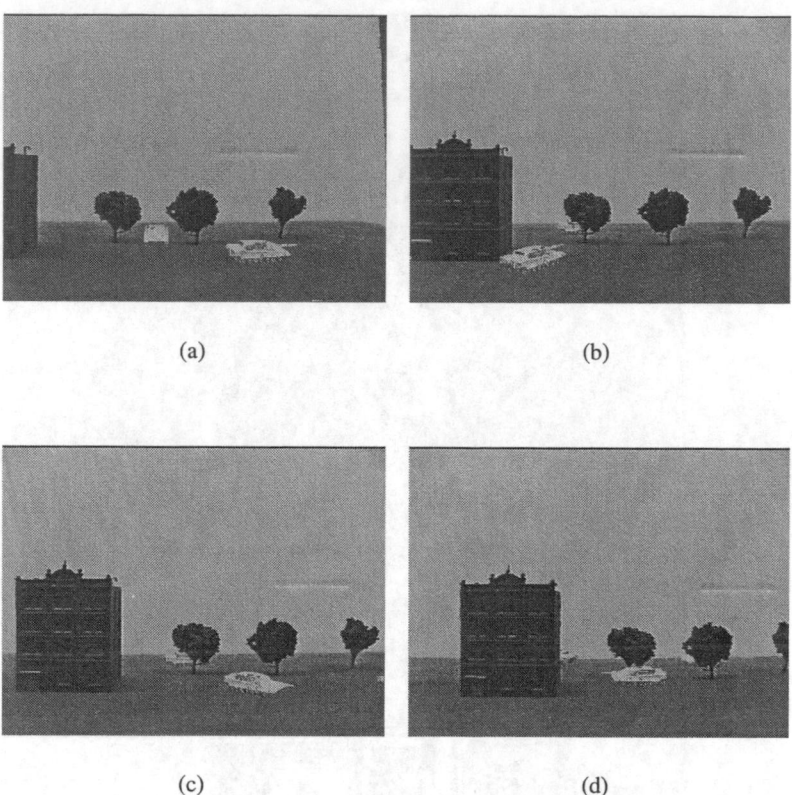

(a)   (b)

(c)   (d)

Fig. 11. Some frames from an augmented video sequence that shows the capability of our approach to insert synthetic moving-tanks while the camera moves. (a) Frame No. 620. (b) Frame No. 860. (c) Frame No. 880. (d) Frame No. 936.

minimize travel time. A very complex example of intelligent behavior can include the decision to cancel the mission of a group of entities and relocating them as a backup for another group. While the first problem can be easily solved by a single autonomous entity, the second will involve some authority that can make a higher-level decision based on information feedback from the lower-level autonomous agents.

3.1   Related Work

Multi-Agent systems are a natural way of dealing with problems of distributed nature. Such problems exist in a diversity of areas like military training, games and entertainment industry, management, transportation, information retrieval and many others. The classical approach to AI, until now, has been unable to provide a feasible approach for solving problems of this nature. The need for such tools has lead to the "alternative"

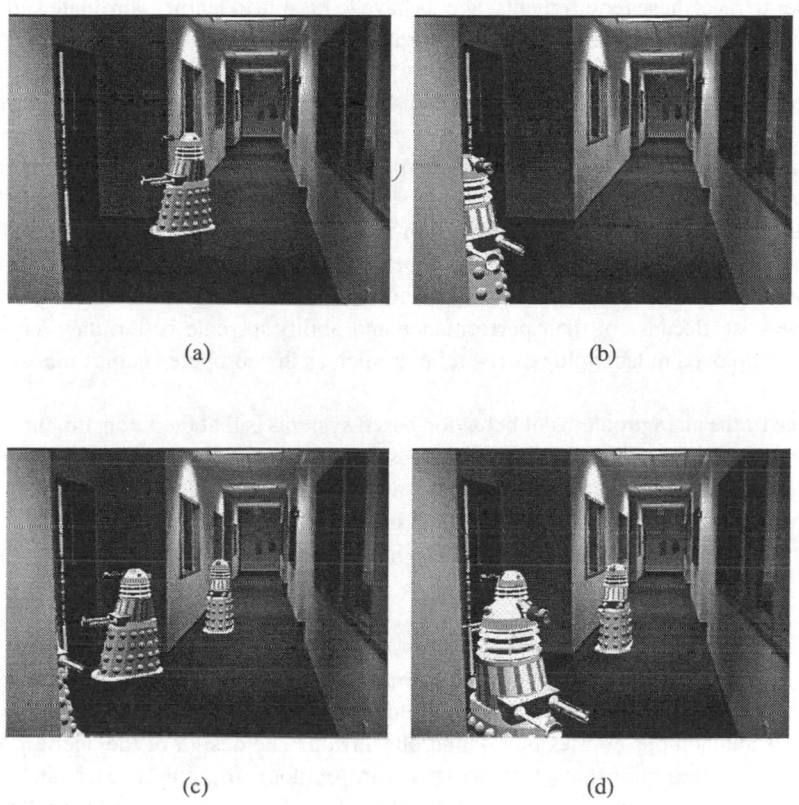

(a)   (b)

(c)   (d)

Fig. 12. Some frames from an augmented video sequence that shows the capability of our approach to deal with actual video from a real world scene. (a) Frame No. 595. (b) Frame No. 620. (c) Frame No. 830. (d) Frame No. 835.

approach of behavior-based systems, popularized by the works of Brooks [7] and Arkin [2]. This approach takes simple behavior patterns as basic building blocks and tries to implement and understand intelligent behavior through the construction of artificial life systems. Its inspiration comes from the way intelligent behavior emerges in natural systems studied by Biology and Sociology. Good discussions on the development of behavior-based AI can be found in [27,8] and an extensive treatment of the subject is given in [3].

Multi-agent systems interacting with the real world face some fundamental restrictions. Some of these are:

1. They have to deal with an unknown and dynamic environment
2. Their environment is inherently very complex
3. They have to act within the time frame of the real world
4. The level of their performance should be "acceptable"

In order to meet these requirements, agents have to be able to learn, coordinate and colla-borate with each other. Reinforcement Learning emerges as one of the "natural ways" of dealing with the dynamism and uncertainty of the environment. The complexity of the environment and the strict timing constraints however make the learning task extremely difficult. Even simple multi-agent systems consisting of only a few agents within a tri-vial environment can have prohibitively expensive computational requirements related to learning [28,20,21].

The behavior-based systems have been more successful at dealing with such pro-blems. Biology-inspired models of group behavior such as Reynold's "boids" [24] and approaches based on potential fields [23] are able to address group behavior at a rea-sonable cost. Because of their performance and ability to scale better, they have been widely employed in technology-driven fields such as the computer-games industry [25, 22].

One of the main problems of behavior-based systems is that their constituents can be very easily caught in local-minima. The question of how to combine different (possibly conflicting) behaviors in order to achieve an emergent intelligence is also very diffi-cult. Multi-Agent Reinforcement Learning in the behavior domain [5,17] is an actively explored approach to solve these problems in a robust way.

3.2 The Agent Model

Our agent model is designed with the primary application area being military visual simulations. The particular problem explored in this paper is goal-based navigation of a group of autonomous entities in a dangerous terrain. The design of the agent model is based on the assumption that agents will perform "outdoor" missions in a terrain which is relatively sparse with respect to obstacles and enemies. It is not very suitable for "indoor" missions like moving inside a building or a labyrinth, where a more specialized approach will be more successful. A "mission" in our model is defined as the problem of going from some position A to some other position B avoiding being hit by an enemy or crashing into the natural and artificial obstacles present in the terrain. The success of the mission is measured by the amount of time necessary for completing the goal and the survival rate of the agents.

Our approach is based on a hierarchical modular representation of agent behavior. This method allows for de-coupling the task of group navigation into simpler self-contained sub-problems which are easier to implement in a system having computational constraints due to interaction with real-life entities.

Different decision mechanisms are used to model different aspects of the agent behavior and a higher level coordination module is combining their output. Such an architecture allows "versatile agent personalities" both in terms of heterogeneity (agent specialization) within a group and dynamic (i.e. mission-context sensitive) agent beha-vior.

The hierarchical modularity of the system also facilitates the assessment of the per-formance of separate components and related behavior patterns on the overall success of the mission.

In our current model, we have three basic modules that we call the *navigation* module, *grouping* module and *imitation* module.

- The Navigation Module is responsible for leading a single agent from a source location to a destination location, avoiding danger and obstacles.
- The Grouping Module is responsible for keeping a group of agents together in particular formations throughout the mission.
- The Imitation Module is modeling the case when an inexperienced agent will try to mimic the behavior of the most successful agents in the group and thus increase its chances of success.

The decisions of these modules are combined at a higher-level module called the *Coordinator Module*. This particular agent model allows modeling of different parts of agent behavior using different approaches. Some of these approaches may incorporate memory (navigation) while others others can be purely reactive (grouping) and some may depend on the performance of other members in an agent group (imitation and grouping).

3.3   Navigation Module

For the purpose of simplicity and efficiency, the Navigational Module generates moves based on a quantized representation of the simulated environment in the form of a grid. Terrain properties are assumed to be uniform within each grid cell for the purpose of learning and storing information about the terrain. Each cell in the grid represents a position and an "agent action" is defined as the decision to move from a grid cell to one of the eight neighboring cells. A succession of such actions will result of a completion of a mission. The agents can also access terrain-specific information about features and obstacles of natural (trees, etc.), and artificial origin (buildings, roads, etc.) and also presence of other (possibly hostile) agents. The interaction between an enemy (a hostile agent) and an agent is modeled by an associated risk. This risk is expressed as a probability of being shot (for an agent) at a position, if the position is in the firing range of an enemy. The goal of the agent is to minimize a function G (which in this case is the estimated time of a safe transit to the destination). We use G to define the Reinforcement Learning Reward function as $R \propto 1/G$.

Successive measured values of R are denoted by $R_l, l = 1, 2, \ldots$. These values are used to keep track of a smoothed reward

$$T_l = bT_{l-1} + (1 - b)R_l, \quad 0 < b < 1$$

where b is close to 1. A Navigational Module of an agent has a so-called "cognitive map" which is a collection of latest and smoothed rewards for each decision taken at each visited grid cell.

The decision-making element of a Navigation Module is a fully-connected RNN network consisting of 8 neurons (each representing a possible decision). The training is performed by reinforcing the weights of each neuron, depending on the difference between the latest and smoothed rewards; positive difference indicates improvement and negative difference indicates deterioration. A detailed description of the network and the learning process are presented in [13].

By using previously acquired information and current sensory input, an agent can start with near-optimal estimates of the rewards and skip an otherwise prohibitively-long learning session and focus on adapting to the dynamic changes in the environment.

3.4   Grouping Module

Grouping behavior module is based on the idea of social potential fields [23] which is a simple distributed-control approach inspired by the attractive and repulsive forces between charged particle in physics. Although this method has been used in a broader domain (including path-planning), we restrict its usage only to model grouping behavior for which it is particularly well suited. Using potential fields methods for other purposes like generalized navigation and obstacle avoidance requires dealing with local minima problems and difficult to design force-configurations that can easily nullify the simplicity gained by using the method in the first place.

These behaviors are very similar and can be used to get the effect of the *collision avoidance* and *flock centering* rules as described by Reynolds [24]. By setting up a two-way mesh of forces between a number of agents, for example, a spatially localized group can be created that will try to stay together. Another simple example is a one-way mesh of forces from the leader of the group to the other members, suggesting that they should stay close and follow if necessary the leader, without having any effect on his decision making.

3.5   Imitation Module

The imitation module proposes a decision which is a weighted sum of the navigational decisions of some of the members of the agent group:

$$V_{imt_i} = \sum_{j \in S} w_j * V_{nav_j}$$

The weight distribution can be dynamic, in order to reflect the group members which are currently observable or known to be experienced, for example. The purpose of imitation is to efficiently take advantage of experience without going through the trouble of actually acquiring it - that is, it has a much lower computational cost, compared to the other methods.

The *velocity matching* flock behavior described in the work of Reynolds [24] which he defines as "attempt to match velocity with nearby flocks" is a very similar idea.

4   Conclusion

Modern discrete event simulators often require the representation of complex autonomous behaviors within a visually realistic setting. They often use a graphical interface both as an input and as an output medium to simplify and enrich the user's interaction with the simulation both before, during and after the simulation runs. Many simulation tools also provide an animated graphical interface which offers a real-time visual description of a simulation in real time.

A useful and very significant leap forward in simulation technology is to be able to evaluate synthetic simulated conditions in realistic settings. The idea here is to ask questions about "what would happen if ..." in the context of a real environment and actual events. This challenge is the focus of the work addressed in this paper where we mix simulation with reality in real time, in order to examine how novel simulated conditions can actually interact with a real system's operation. This interaction can go in both directions: the course of the real world can be modified by virtual entities, and the virtual objects are constrained to operate in the real world.

In this paper we have discussed the conceptual issues which arise in this key area of visual simulation, and we have presented some design principles and a practical implementation. Key issues we covered in this paper include a new method for injecting moving synthetic objects in real-time into real world video based on terrain databases, graphics rendering and image segmentation, and a novel approach to automatically control the motion of synthetic agents within the realistic live scene.

References

1. Al-Begain, K. 2004. "Performance Models for 2.5/3G Mobile Systems and Networks". In M. Calzarossa and E. Gelenbe, editors, *Performance Tools and Applications to Networked Systems*, volume 2965 of *Lecture Notes in Computer Science*. Springer.
2. Arkin, R. C. 1989. "Motor Schema-Based Mobile Robot Navigation", *International Journal of Robotics Research*, Vol. 8, No. 4, pp. 92-112.
3. Arkin, R.C. 1998. *Behavior-Based Robotics*, The MIT Press, Cambridge, Massachusetts.
4. Bajura, M. and U. Neumann. 1995. "Dynamic registration correction in video-based reality systems," *IEEE Computer Graphics and Applications*, Vol. 15, No. 5 (September), pp. 52-60.
5. Balch, T., 1998 "Behavioral Diversity in Learning Robot Teams", PhD. Thesis, Georgia Institute of Technology.
6. Bartolini, N.; E. Casalicchio; and S. Tucci. 2004 "A Walk Through Content Delivery Networks". In M. Calzarossa and E. Gelenbe, editors, *Performance Tools and Applications to Networked Systems*, volume 2965 of *Lecture Notes in Computer Science*. Springer.
7. Brooks, R. 1986. "A robust layered control system for a mobile robot", *IEEE Journal of Robotics and Automation*, Vol. RA-2, No. 1, pp. 14-23.
8. Brooks, R. 1999. *Cambrian Intelligence: The Early History of The New AI*, The MIT Press, Cambridge, MA.
9. Foss, B.; E. Gelenbe; K. Hussain; N. Lobo; and H. Bahr. 2000. "Simulation driven virtual objects in real scenes", Proc. ITSEC 2000, Orlando, FL (November).
10. Gelenbe, E. 1999. "Modeling CGF with learning stochastic finite-state machines", *Proc. 8 th Conference on Computer Generated Forces*, pp. 113-116, Orlando (May 11-13)
11. Gelenbe, E; R. Lent; M. Gellman; P. Liu; and P. Su. 2004. "CPN and QoS Driven Smart Routing in Wired and Wireless Networks". In M. Calzarossa and E. Gelenbe, editors, *Performance Tools and Applications to Networked Systems*, volume 2965 of *Lecture Notes in Computer Science*. Springer.
12. Gelenbe, E.; E. Şeref; and Z. Xu. 2000. "Discrete event simulation using goal oriented learning agents", *AI, Simulation & Planning in High Autonomy Systems*, SCS, Tucson, Arizona (March 6-8).
13. Gelenbe, E.; E. Şeref; and Z. Xu. 2001. "Simulation with learning agents", *Proceedings of the IEEE*, Vol. 89 (2), pp. 148-157 (Feb.).
14. Klemm, A.; C. Lindemann; and O.P. Waldhorst, O.P. 2004. "Peer–to–Peer Computing in Mobile Ad Hoc Networks". In M. Calzarossa and E. Gelenbe, editors, *Performance Tools and Applications to Networked Systems*, volume 2965 of *Lecture Notes in Computer Science*. Springer.

15. Kotsis, G. 2004. "Performance Management in Dynamic Computing Environments". In M. Calzarossa and E. Gelenbe, editors, *Performance Tools and Applications to Networked Systems*, volume 2965 of *Lecture Notes in Computer Science*. Springer.

16. Lawson, S. W. 1998. "Augmented reality for underground pipe inspection and maintenance," *SPIE Conference on Telemanipulator and Telepresence Technologies*, Boston, Massachusetts, Vol. 3524, pp. 98-104.

17. Mataric, A., 1997 "Reinforcement learning in the multi-robot domain", *Autonomous Robots*, Vol. 4, No. 1, pp. 73-83.

18. Mellor, J.P. 1995. "Enhanced reality visualization in a surgical environment," M. S. Thesis, Department of Electrical Engineering, MIT (January).

19. Naser H.; and H.T. Mouftah. 2004. "Modeling and Simulation of Mesh Networks with Path Protection and Restoration". In M. Calzarossa and E. Gelenbe, editors, *Performance Tools and Applications to Networked Systems*, volume 2965 of *Lecture Notes in Computer Science*. Springer.

20. Ono, N., and Fukumoto, K. 1996, "Multi-agent reinforcement learning: A modular approach", *Proc. of the 2nd Int. Conf. on Multi-Agent Systems*, AAAI Press, pp. 252-258.

21. Ono, N., and Fukumoto, K. 1997, "A modular approach to multi-agent reinforcement learning", In Gerhard Weiss, editor, *Distributed Artificial Intelligence Meets Machine Learning*, Springer-Verlag, p. 167.

22. Pottinger, D.C. 1999, "Implementing Coordinated Movement", *Game Developer Magazine*

23. Reif, J.H. and Wang, H. 1998. "Social Potential Fields: A Distributed Behavioral Control for Autonomous Robots" *International Workshop on Algorithmic Foundations of Robotics (WAFR)*, pages 431– 459. A. K. Peters, Wellesley, MA.

24. Reynolds, C.W. 1987, "Flocks, Herds, and Schools: A Distributed Behavioral Model", *Computer Graphics*, Vol. 21, No. 4, pp. 25-34.

25. Reynolds, C.W. 1999, "Steering Behaviors for Autonomous Characters", *Game Developers Conference*.

26. Serazzi , G.; and S. Zanero. 2004. "Computer Virus Propagation Models". In M. Calzarossa and E. Gelenbe, editors, *Performance Tools and Applications to Networked Systems*, volume 2965 of *Lecture Notes in Computer Science*. Springer.

27. Steels, L. 1993. "The Artificial Life Roots of Artificial Intelligence", *Artificial Life*, Vol. 1, pp. 75-110

28. Tan, M. 1993. "Multi-Agent Reinforcement Learning: Independent versus Cooperative Agents", *International Conference on Machine Learning*, pp. 330-337.

29. Williamson, C. 2004. "Wireless Internet: Protocols and Performance". In M. Calzarossa and E. Gelenbe, editors, *Performance Tools and Applications to Networked Systems*, volume 2965 of *Lecture Notes in Computer Science*. Springer.

PEPA Nets

Stephen Gilmore, Jane Hillston, and Leïla Kloul*

Laboratory for Foundations of Computer Science, The University of Edinburgh, Edinburgh EH9 3JZ, Scotland. {stg,jeh,leila}@inf.ed.ac.uk

Abstract. In this chapter we describe a formalism which uses the stochastic process algebra PEPA as the inscription language for labelled stochastic Petri nets. Viewed in another way, the net is used to provide a structure for linking related PEPA systems. The combined modelling language naturally represents such applications as mobile code systems where the PEPA terms are used to model the program code which moves between network hosts (the places in the net). We demonstrate the modelling capabilities of the formalism on a number of examples, including a mobile server running MobileIP.

1 Introduction

Over the last decade mobility has had a major impact on the way we design, implement and manage many computer systems. Mobility may be manifest in the form of devices which change location and spontaneously connect/disconnect, or in the form of executable code which is moved around the network for a variety of reasons. In either case the effect is that the context in which computation is taking place is dynamically changing, and these changes will have consequences for the performance of the system. In this chapter we introduce the modelling formalism PEPA nets which have been designed to capture information about mobility and so allow performance models of such systems be readily and naturally developed.

The rest of the chapter is organised as follows. Section 2 introduces the notation and terminology of PEPA nets, after a brief introduction to PEPA. (Readers are assumed to be familiar with the basic ideas of Stochastic Petri Nets, and are referred to [1] for basic definitions.) In Section 3 we present two small examples, a simple mobile agent system, and a secure web service, all modelled as PEPA nets. Section 4 is a more detailed case study of a mobile host accessible via MobileIP. In Section 5 we discuss tool support for PEPA nets. Related work is discussed in Section 6. Concluding remarks and further work are presented in Section 7.

2 PEPA Nets

In this section we present the concepts and definitions used in PEPA nets. First we give an intuitive overview of the formalism, followed by a more detailed account of its definition and its formal properties.

---

* L. Kloul is on leave from PRISM, Université de Versailles, 45, Av. des Etats-Unis 78000 Versailles, France.

M.C. Calzarossa and E. Gelenbe (Eds.): MASCOTS 2003, LNCS 2965, pp. 311–335, 2004.
© Springer-Verlag Berlin Heidelberg 2004

A PEPA net is a stochastic Petri net with coloured tokens. The tokens represent *mobile* objects with state and behaviour, where we use the term mobile loosely to characterise objects which may find themselves in different contexts during execution. The tokens are described using a stochastic process algebra, Hillston's Performance Evaluation Process Algebra (PEPA).

The use of stochastic Petri nets for performance models is well-established [1] and coloured variants, e.g. Stochastic Well-Formed Nets (SWN) [4], have also been developed. However the use of colours in PEPA nets offers something quite distinct — the possibility of differentiating between two types of change of state within a system. Unlike SWN where tokens remain indistinguishable within their colour classes, tokens within PEPA nets are autonomous components. Firings of the net will typically be used to model macro-step (or *global*) changes of state, whereas transitions within the PEPA tokens are typically used to model micro-step (or *local*) changes of state as components undertake activities. Thus modelling with PEPA nets uses both Petri nets and process algebras together as a single, structured performance modelling formalism. There is some reason to believe that these two formalisms complement each other [7]. In particular, we have previously demonstrated [12] that PEPA nets offer some expressivity which is not directly offered by either PEPA or Petri nets.

2.1 Summary of the PEPA Language

In the following paragraphs we give a brief overview of PEPA. Readers are referred to [14] for a more detailed introduction.

The PEPA language provides a small set of combinators. These allow language terms to be constructed defining the behaviour of components, via the activities they undertake and the interactions between them. In particular we develop a model as a *model component* constructed using *static combinators*, from a number of *sequential components* constructed using *dynamic combinators*.

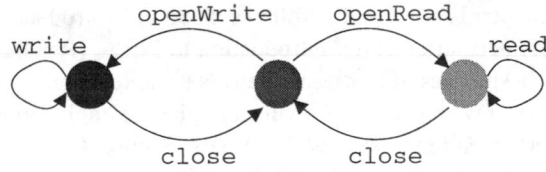

Fig. 1. Simple file protocol

Consider a File class with methods openRead(), openWrite(), read(), write() and close(). The order in which the methods can be applied defines a *protocol* for a File object. This could be represented as shown in Figure 1. We can express this as a PEPA component as below:

$$File \stackrel{def}{=} (openRead, r_o).InStream + (openWrite, r_o).OutStream$$

$$InStream \stackrel{def}{=} (read, r_r).File + (close, r_c).File$$

$$OutStream \stackrel{def}{=} (write, r_w).File + (close, r_c).File$$

Here $\stackrel{def}{=}$ denotes definitional equality and a *constant* or *identifier* (e.g. *File*) is used to assign a name to a pattern of behaviour associated with a component. Each activity such as $(openRead, r_o)$ has an *action type* $(openRead)$ and an exponentially distributed duration, denoted by its parameter (r_o). The basic mechanism for describing the behaviour of a system is to give a component a designated first action using the *prefix* combinator, denoted ".". However, the life cycle of a component may be more complex than any behaviour which can be expressed using the prefix combinator alone. The choice combinator, denoted by $+$, captures the possibility of competition between different possible activities: only one will succeed and the other alternative will be discarded (for the moment). The competition is resolved via the race policy.

File is a sequential component and prefix and choice are the dynamic combinators.

Now consider a `File` being accessed by a `FileReader` object. The two components must interact via the file protocol. In PEPA this is would be denoted as

$$File \bowtie_{L} FileReader$$

where L denotes the set of action types $\{openRead, read, close\}$. The combinator $\bowtie_{L}$ denotes *cooperation* between the components. This is the basis of compositionality within PEPA. The set which is used as the subscript of the cooperation symbol, the *cooperation set L*, determines those activities on which the two components are forced to synchronise. For action types not in L, the components proceed independently and concurrently with their enabled activities. However, if a component enables an activity whose action type is in the cooperation set it will not be able to proceed with that activity until the other component also enables an activity of that type. The two components then proceed together to complete the *shared activity*. The rate of the shared activity may be altered to reflect the work carried out by both components to complete the activity (for details see [14]). We write $P \parallel Q$ as an abbreviation for $P \bowtie_{L} Q$ when L is empty.

In some cases, when an activity is known to be carried out in cooperation with another component, a component may be *passive* with respect to that activity. This means that the rate of the activity is left unspecified (denoted $\top$) and is determined upon cooperation, by the rate of the activity in the other component. All passive actions must be synchronised in the final model.

In the case of the `File` object cooperating with a `FileReader` we may wish to impose that the `File` object may only be read by this `FileReader`. This is denoted using the *hiding* combinator, $/$:

$$(File \bowtie_{L} FileReader)/\{read\}$$

Hiding provides the possibility to abstract away some aspects of a component's behaviour. In a component P/L, the set L of visible action types will be considered internal or private to the component and will appear as the unknown type τ. Hiding and cooperation are the dynamic combinators.

The syntax of PEPA may be formally introduced by means of the grammar shown in the lower part of Figure 2. In that grammar S denotes a *sequential component* and P denotes a *model component* which executes in parallel. I stands for a constant which denotes either a sequential or a model component, as defined by a defining equation.

Model components capture the structure of the system in terms of its *static* components. The dynamic behaviour of the system is represented by the evolution of these components, either individually or in cooperation. The form of this evolution is governed by a set of formal rules which give an operational semantics of PEPA terms. The semantic rules, in the structured operational style, are presented in Figure 8 in the Appendix.

The semantics of each term in PEPA is given via a labelled transition system. In the transition system a state corresponds to each syntactic term of the language, or *derivative*, and an arc represents the activity which causes one derivative to evolve into another. The complete set of reachable states is termed the *derivative set* of a model and these form the nodes of the *derivation graph* which is formed by applying the semantic rules exhaustively.

The timing aspects of components' behaviour are represented on each arc as the parameter of the negative exponential distribution governing the duration of the corresponding activity. The interpretation is as follows: when enabled an activity $a = (\alpha, r)$ will delay for a period sampled from the negative exponential distribution which has parameter r. If several activities are enabled concurrently, either in competition or independently, we assume that a *race condition* exists between them. The evolution of the model will determine whether the other activities have been *aborted* or simply *interrupted* by the resulting state change. In either case the memoryless property of the distribution eliminates the need to record the previous execution time.

When two components carry out an activity in cooperation the rate of the shared activity will reflect the working capacity of the slower component. We assume that each component has a capacity for performing an activity type α, which cannot be enhanced by working in cooperation, unless the component is passive with respect to that activity type. For a component P and an action type α, this capacity is termed the *apparent rate* [14] of α in P. It is the sum of the rates of the α type activities enabled in P. The apparent rate of α in a cooperation between P and Q over α will be the minimum of the apparent rate of α in P and the apparent rate of α in Q.

The derivation graph is the basis of the underlying Continuous Time Markov Chain (CTMC) which is used to derive performance measures from a PEPA model. The graph is systematically reduced to a form where it can be treated as the state transition diagram of the underlying CTMC. Each derivative is then a state in the CTMC. The *transition rate* between two derivatives P and Q in the derivation graph is the rate at which the system changes from behaving as component P to behaving as Q. It is denoted by $q(P, Q)$ and is the sum of the activity rates labelling arcs connecting node P to node Q. In order for the CTMC to be *ergodic* its derivation graph must be strongly connected. Some necessary conditions for ergodicity, at the syntactic level of a PEPA model, have been defined [14]. These syntactic conditions are imposed by the grammar in Figure 2.

2.2   Introduction to PEPA Nets

In outlining a framework of design paradigms for mobile code systems in [8], Fuggetta *et al.* emphasise three architectural concepts: *components*, *interactions* and *sites*. The importance of including an explicit notion of *location* has also been recently highlighted by Köhler *et al.* [17], who additionally choose *entities* and *movement* as primitives. In PEPA nets the places of the PEPA net represent sites or locations, the PEPA components represent components or entities. Interactions are represented by PEPA cooperations, which may only take place between co-located components, and movement is represented by firings of the PEPA net which move a token from one place to another. Thus in PEPA nets we have two distinct types of change of state: *firings* and *transitions*.

A firing in a PEPA net causes the transfer of one token from one place to another. The token which is moved is a PEPA component, which causes a change in the subsequent evaluation both in the source (where existing cooperations with other components now can no longer take place) and in the target (where previously disabled cooperations are now enabled by the arrival of an incoming component which can participate in these interactions). Firings have global effect because they involve components at more than one place in the net.

A transition in a PEPA net takes place whenever a transition of a PEPA component can occur (either individually, or in cooperation with another component). Components can only cooperate if they are resident in the same place in the net. The PEPA net formalism does not allow components at different places in the net to cooperate on a shared activity. An analogy is with message-passing distributed systems without shared-memory where software components on the same host can exchange information without incurring a communication overhead but software components on different hosts cannot. Additionally we do not allow a firing to coincide with a transition which is shared, i.e. it is not possible for two components in one place to cooperate *and* transfer to another place as an atomic action. Thus transitions in a PEPA net have local effect because they involve only components at one place in the net.

There are distinct alphabets for transitions and firings, meaning that the same action type cannot be used for both. Thus there can be no ambiguity between them.

A PEPA net is made up of PEPA *contexts*, one at each place in the net. A context consists of a number of *static* components (possibly zero) and a number of *cells* (at least one). Like a memory location in an imperative program, a cell is a storage area to be filled by a datum of a particular type. In particular in a PEPA net, a cell is a storage area dedicated to storing a PEPA component. For example, if we consider the *FileReader* introduced earlier, it might reside in a fixed location of the system and interact with any files copied into its scope. This would be denoted

$$File[\_] \underset{L}{\bowtie} FileReader$$

The components which fill cells can circulate as the tokens of the net. In contrast, the static components cannot move. Most variants of Petri nets do not include static tokens, the closest concept being "self loops" where a token is deleted from a place and then immediately replaced. Here static components provide the infrastructure of the place and act as cooperation partners in synchronisation activities with tokens. Contexts have previously been used in both classical process algebras [20], and in the stochastic process algebra PEPA [5].

We use the notation $Q[\_]$ to denote a context which could be filled by the PEPA component Q or one with the same alphabet. If Q has derivatives Q' and Q'' only and no other component has the same alphabet as Q then there are four possible values for such a context: $Q[\_], Q[Q], Q[Q']$ and $Q[Q'']$. $Q[\_]$ enables no transitions. $Q[Q]$ enables the same transitions as Q. $Q[Q']$ enables the same transitions as Q', etc. As usual with

PEPA components we require that the component has an ergodic definition so that it is always possible to return to a state which one has previously reached.

We use capitalised names to denote PEPA components (such as P and Q) and lower-case for PEPA transitions (such as a and b). We use bold capitalised names for places (such as $\mathbf{P_1}$ and $\mathbf{P_2}$) and bold lowercase for firings (such as $\mathbf{a}$ and $\mathbf{b}$).

Note that the expression of the structural information contained in a PEPA net could be represented by any transition-based modelling formalism. Indeed it would be possible to use a PEPA component to control the possible "firings" (macro-steps) of the model. However, we feel that there are some advantages to using a Petri net in this role. Firstly, using a different formalism gives a clearer separation of concerns within our model making it both easier to construct and to understand. Furthermore, this macro-level is often of a size that can benefit from graphical representation, to give an intuitive understanding of the coarse structure of the model. Finally, the *movement* of components—to a new host or context—has resonance with the systems we study.

Markings in a PEPA net. The *marking* of a classical Petri net records the number of tokens which are resident at each place in the net. Since the tokens of a classical Petri net are indistinguishable it is sufficient to record their number and one could present the marking of a Petri net with places P_1, P_2 and P_3 as $(P_1 : 2, P_2 : 1, P_3 : 0)$. If an ordering is imposed on the places of the net a more compact representation of the marking can be used. Place names are omitted and the marking can be written using vector notation thus, $(2, 1, 0)$.

Consider now a PEPA net with places $\mathbf{P_1}$, $\mathbf{P_2}$ and $\mathbf{P_3}$ as shown below.

$$\mathbf{P_1}[Q] \stackrel{def}{=} Q[Q] \bowtie_{L} R$$

$$\mathbf{P_2}[Q] \stackrel{def}{=} Q[Q] \bowtie_{K} S$$

$$\mathbf{P_3}[Q] \stackrel{def}{=} Q[Q] \bowtie_{K \cup L} (R \parallel S)$$

From its use in the contexts at each place we see that Q is a component which can move as a token around the net whereas R and S are static components which cannot move. There is a copy of R at place $\mathbf{P_1}$ and another at $\mathbf{P_3}$. There is a copy of S at place $\mathbf{P_2}$ and another at $\mathbf{P_3}$. Clearly the marking of a PEPA net needs to record the current location of the tokens circulating in the net, and their current state. However, this is not sufficient to establish the state of the system — the local state captured by each place's marking will also depend on the current state of the static components in the place. Thus to identify these states we allow place definitions to specify a particular state of each of the static components. Thus, in the example above, if S can evolve to S' we can define $\mathbf{P'_2}[Q] \stackrel{def}{=} Q[Q] \bowtie_{K} S'$.

Net-level transitions in a PEPA net. A labelling function ℓ is used to associate an activity, consisting of an action type and a rate, (α, r), with each net level transition t. Note that it is possible that $\ell(t_i) = \ell(t_j)$ but $t_i \neq t_j$. The first element of a pair (α, r) specifies an *activity* which must be performed in order for a component to move from the input place of the transition to the output place. The activity type records formally the activity which must be performed if the transition is to fire. The second element is an exponentially-distributed random variable which quantifies the rate at which the activity can progress in conjunction with the component which is performing it. For a firing to be enabled there must be a token in the input place of the net-level transition, that token must enable the corresponding activity, and there must be an empty cell at the output place of the transition of the correct token type.

As an example, suppose that Q is a component which is currently at place $\mathbf{P_1}$ and that it can perform an activity α with rate r_1 to produce the derivative Q'. Further, say that the net has a transition between $\mathbf{P_1}$ and $\mathbf{P_2}$ labelled by (α, r_2). If Q performs activity α in this setting it will be removed from $\mathbf{P_1}$ (leaving behind an empty cell) and Q' will be deposited into $\mathbf{P_2}$ (filling an empty cell there).

A priority function π maps action types to the natural numbers, and can be used to eliminate some firings from the labelled multi-transition system: only enabled firings with the highest priority value are considered eligible to fire. For example, suppose that Q is a component which is currently at place $\mathbf{P_1}$ and that it can perform activities of types α, β and γ where $\pi(\alpha) = \pi(\beta) = 2$ whereas $\pi(\gamma) = 1$. Further, suppose that there are net transitions between $\mathbf{P_1}$ and each of $\mathbf{P_2}$, $\mathbf{P_3}$ and $\mathbf{P_4}$ labelled by α, β and γ respectively. Assuming that there are appropriate empty cells in all places, Q may perform activity α and be deposited in place $\mathbf{P_2}$, or activity β and be deposited in place $\mathbf{P_3}$ but it cannot perform activity γ and be deposited in place $\mathbf{P_4}$. Only if there are no empty cells in places $\mathbf{P_2}$ and $\mathbf{P_3}$ will activity γ become enabled.

Net structure of a PEPA net. The class of nets that we currently use for modelling the net structure of a PEPA net is restricted to *structural state machines*, i.e. nets whose transitions can have only one input place and one output place[1] . This means that we can represent conflicts at the net level, while synchronisations are not allowed. This is consistent with the fact that PEPA components cannot cooperate on a shared activity when they are resident in different places.

It is usual with coloured Petri nets to associate functions with arcs, offering a generalisation of the usual, basic "functions" offered by arc multiplicities. In PEPA nets the arc functions are implicit. The modification of a token which takes place when it is fired is wholly specified by the action type of the firing, the definition of the token and the semantics. Furthermore, although we allow multiple tokens within net places, only one token can move at each firing. Thus arc multiplicities greater than one are not allowed.

Formal definition. The introduction of contexts requires an extension to the syntax of PEPA. This extension is presented in Figure 2.

We assume that there is a set $\mathcal{A}$ of PEPA action types which can be partitioned into disjoint subsets $\mathcal{A}_f$ and $\mathcal{A}_t$ corresponding to firings and local transitions respectively.

Definition 1. *A PEPA net $\mathcal{N}$ is a tuple $\mathcal{N} = (\mathcal{P}, \mathcal{T}, I, O, \ell, \pi, \mathcal{C}, D, M_0)$ such that*

- *$\mathcal{P}$ is a finite set of places;*
- *$\mathcal{T}$ is a finite set of net transitions;*

---

[1] This restriction has recently been relaxed by allowing more general net structures [10].

$$N ::= D^+ M \qquad \text{(net)}$$

(definitions and marking)

$M ::= (M_{\mathbf{P}}, \ldots)$ (marking) $D ::= I \overset{def}{=} S$ (component defn)

$M_{\mathbf{P}} ::= \mathbf{P}[C, \ldots]$ (place marking) $| \quad \mathbf{P}[C] \overset{def}{=} P[C]$ (place defn)

                                           $| \quad \mathbf{P}[C, \ldots] \overset{def}{=} P[C] \underset{L}{\bowtie} P$ (place defn)

(marking vectors)   (identifier declarations)

$S ::= (\alpha, r).S$ (prefix) $P ::= P \underset{L}{\bowtie} P$ (cooperation) $C ::= \text{`\_'}$ (empty)

$| \quad S + S$ (choice) $| \quad P/L$ (hiding) $| \quad S$ (full)

$| \quad I$ (identifier) $| \quad P[C]$ (cell)

                             $| \quad I$ (identifier)

(sequential components)   (concurrent components)   (cell term expressions)

Fig. 2. The syntax of PEPA extended with contexts

- $I : \mathcal{T} \to \mathcal{P}$ *is the input function;*
- $O : \mathcal{T} \to \mathcal{P}$ *is the output function;*
- $\ell : \mathcal{T} \to (\mathcal{A}_f, \mathbb{R}^+ \cup \{\top\})$ *is the labelling function, which assigns a PEPA activity ((type, rate) pair) to each transition. The rate determines the negative exponential distribution governing the delay associated with the transition;*
- $\pi : \mathcal{A}_f \to \mathbb{N}$ *is the priority function which assigns priorities (represented by natural numbers) to firing action types;*
- $\mathcal{C} : \mathcal{P} \to P$ *is the place definition function which assigns a PEPA context, containing at least one cell, to each place;*
- D *is the set of token component definitions;*
- M_0 *is the initial marking of the net.*

PEPA nets are governed by the semantic rules for PEPA and few additional rules — both sets are provided in the Appendix. Informally the new rules can be interpreted as follows:

- The Cell rule defines that a cell which is filled by a component Q is able to make the same transitions as Q itself. There are no rules to infer transitions for an empty cell because an empty cell enables no transitions.
- The Transition rule states that the net has local transitions which change only a single component in the marking vector and that these transitions are exactly the transitions generated by the PEPA semantics (including the extension for contexts).
- The Firing rule takes one marking of the net to another marking by performing a PEPA activity and moving a PEPA component from the input place to the output

place. This has the effect that two entries in the marking vector change simultaneously. In order for a firing to take place it must be the case that the type of the enabled firing has the highest priority level in the set of enabled firings. In other words, for a firing to occur there must not be any other firing satisfying the Enabling rule (empty destination cell) which has a higher priority.

- The Enabling rule determines when a transition is considered to be enabled. In addition to ensuring that there is a valid token to fire we must also check that there is an empty cell in the destination place into which the token can be transferred. The Enabling rule ensures that this is the case, and defines a transition relation, decorated with the priority level of the corresponding activity type. The rate at which the activity is enabled is calculated as in the PEPA semantics of cooperation.

From the operational semantics a derivation graph and underlying CTMC can be extracted from any PEPA net, and this is the basis on which performance analysis is carried out.

The net bisimulation relation. PEPA nets are equipped with an equivalence relation called *net bisimulation*. This relation is important both in theory and in practice. In the evolution of the state space of a model by our tool we only store states up to net bisimulation, i.e. we carry out automatic aggregation over equivalent states. This provides a dramatic reduction in the state space of the model under certain conditions.

Our relation is defined in the style of Larsen and Skou [19], based on a conditional transition rate between markings. The *conditional transition rate* from marking M to marking M' via action type α, denoted $q(M, M', \alpha)$, is the sum of the activity rates labelling arcs connecting the corresponding nodes in the derivation graph which are labelled by the action type α. The *total conditional transition rate* from a marking M to a set of markings E is defined as

$$q\lfloor M, E, \alpha \rfloor = \sum_{M' \in E} q(M, M', \alpha)$$

Definition 2. *An equivalence relation over markings, $\mathcal{R} \subseteq M \times M$, is a net bisimulation if whenever $(M, M') \in \mathcal{R}$ then for all $\alpha \in A$ and for all equivalence classes $E \in M/\mathcal{R}$,*

$$q[M, E, \alpha] = q[M', E, \alpha]$$

3   Examples

3.1   A Mobile Agent System

We present a small example to reinforce the reader's understanding of PEPA nets. In this example a roving agent visits three sites. It interacts with static software components at these sites and has two kinds of interactions. When visiting a site where a network probe is present it interrogates the probe for the data gathered on recent patterns of network traffic. When it returns to the central co-ordinating site it dumps the data which it has harvested to the master probe. The master probe performs a computationally expensive statistical analysis of the data. The structure of the system allows this computation to be overlapped with the agent's communication and data gathering. The marshalling and unmarshalling costs for mobile code applications are a significant expense so overlapping this with data processing allows some of this expense to be offset.

The structure of the application is as represented by the PEPA net in Figure 3. This marking of the net shows the mobile agent resident at the central co-ordinating site. In this example the activities which can cause a firing of the net are **go** and **return**.

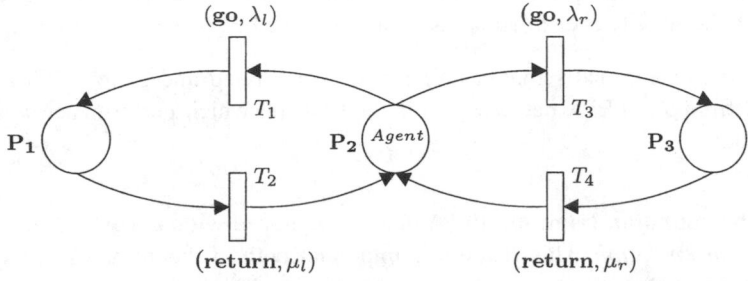

Fig. 3. A simple mobile agent system

Formally, we define the places of the net as shown in the PEPA context definitions below. We denote the local state of the context $\mathbf{P_2}$ by $\mathbf{P_2'}$. This local state is arrived at when the static component *Master* has evolved to *Master'*.

$$\mathbf{P_1}[Agent] \stackrel{def}{=} Agent[Agent] \underset{\{interrogate\}}{\bowtie} Probe$$

$$\mathbf{P_2}[Agent] \stackrel{def}{=} Agent[Agent] \underset{\{dump\}}{\bowtie} Master$$

$$\mathbf{P_2'}[Agent] \stackrel{def}{=} Agent[Agent] \underset{\{dump\}}{\bowtie} Master'$$

$$\mathbf{P_3}[Agent] \stackrel{def}{=} Agent[Agent] \underset{\{interrogate\}}{\bowtie} Probe$$

The initial marking of the net is $(\mathbf{P_1}[\_], \mathbf{P_2}[Agent], \mathbf{P_3}[\_])$ The behaviour of the components is given by the following PEPA definitions.

$$Agent \stackrel{def}{=} (\mathbf{go}, \lambda).Agent' \qquad\qquad Master \stackrel{def}{=} (dump, \top).Master'$$
$$Agent' \stackrel{def}{=} (interrogate, r_i).Agent'' \qquad Master' \stackrel{def}{=} (analyse, r_a).Master$$
$$Agent'' \stackrel{def}{=} (\mathbf{return}, \mu).Agent''' \qquad\quad Probe \stackrel{def}{=} (monitor, r_m).Probe +$$
$$Agent''' \stackrel{def}{=} (dump, r_d).Agent \qquad\qquad\qquad (interrogate, \top).Probe$$

The derivation of the transition system underlying the model (Figure 4) gives us its underlying CTMC. This CTMC is solved for its stationary distribution and performance measures are calculated from that.

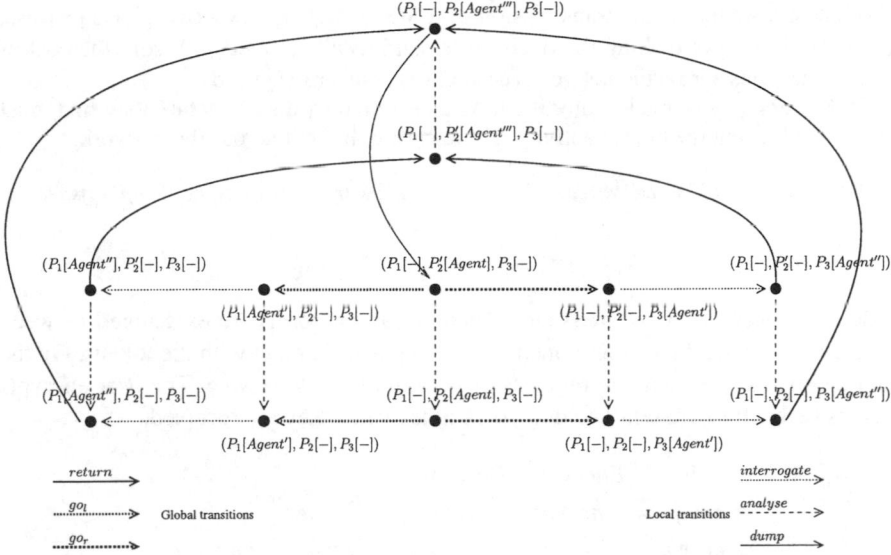

Fig. 4. The transition system of the mobile agent example

3.2 Secure Web Service

Our second example is a model of a mobile object system where a client sends SOAP message objects to a remote Web service. Our scenario is that a financial tycoon is sending requests for stocks and share price information to a remote Web service which provides this information. These requests for information are encrypted. An eavesdropper could make use of the information in the messages if they were sent as clear text. The Web service itself is protected by a firewall.

The token type. The tokens exchanged in the system are SOAP messages in various formats. These may either be sent across the network as clear text or encrypted to preserve their contents. A SOAP message may be parsed to build an in-memory data structure which can be read and modified as needed. This data structure is a DOM tree (a Document Object Model tree).

$$SoapMessage \stackrel{def}{=} (send_{clr}, r_{sc}).SentClearMessage + (encrypt, r_e).EncryptedMsg$$
$$+ (parse, r_p).DOMtree$$

$$SentClearMessage \stackrel{def}{=} (\textbf{copyClear}, \top).SoapMessage$$

Encrypted messages can be decrypted to recover their initial contents or sent across the network in encrypted form.

$$EncryptedMsg \stackrel{def}{=} (decrypt, r_d).SoapMessage + (send_{enc}, r_{se}).SentEncMessage$$

$$SentEncMessage \stackrel{def}{=} (\textbf{copyEncrypted}, \top).EncryptedMsg$$

In both cases, we model the transmission of a SOAP message as a two-phase process, separating the cost of making the decision to send ($send_{clr}$, $send_{enc}$) from the cost of copying the bytes across the network (**copyClear**, **copyEncrypted**).

DOM trees may be read or modified. As an in-memory data structure they first must be serialised (using the *export* activity) if they are to be sent across the network.

$$DOMtree \stackrel{def}{=} (read, r_r).DOMtree + (modify, r_m).DOMtree + (export, r_x).SoapMessage$$

Static components. SOAP message tokens of various forms are exchanged between the places of the net. Static components at these places interact with the tokens. On the client side is the user, making requests of the remote Web Service. The user encrypts requests before they are sent and decrypts replies when they are received.

$$User \stackrel{def}{=} Encrypt + Decrypt$$
$$Encrypt \stackrel{def}{=} (encrypt, \top).(send_{enc}, \top).User$$
$$Decrypt \stackrel{def}{=} (decrypt, \top).(parse, \top).(read, \top).Request$$
$$Request \stackrel{def}{=} (modify, \top).(export, \top).User$$

Running on the firewall is a gatekeeper process which performs three distinct functions: decrypting user requests, bouncing flawed requests and encrypting replies from the server. The gatekeeper receives requests from the user and decrypts them. The decrypted message might be a well-formed request, in which case it is forwarded to the server. Alternatively it might be in an invalid format, request a non-existent service, or have suspicious attachments. In this case it is bounced back to the user with a diagnostic error message attached. This decomposition of responsibilities means that the load on the server is reduced.

All of the communication with the user is sent in encrypted format. Behind the firewall the messages are exchanged in the clear. Thus $send_{enc}$ always sends to the user and $send_{clr}$ always sends to the server.

$$GateKeeper \stackrel{def}{=} FilterIn + Bounce + FilterOut$$
$$FilterIn \stackrel{def}{=} (decrypt, \top).(send_{clr}, \top).GateKeeper$$
$$Bounce \stackrel{def}{=} (decrypt, \top).FilterOut$$
$$FilterOut \stackrel{def}{=} (encrypt, \top).(send_{enc}, \top).GateKeeper$$

Behind the firewall the share price Web service runs on the server. Its life cycle is parsing, reading, modifying, serialising and returning requests.

$$WebService \stackrel{def}{=} (parse, \top).(read, \top).(modify, \top).(export, \top).(send_{clr}, \top).WebService$$

The PEPA net. The PEPA net of the system sites the above static components at places of the net and specifies the communication between different places of the net by naming the transitions which must be fired for tokens to move from place to place[2]. The *User* is

---

[2] Note that for convenience we represent places as large rectangular boxes.

operating the client machine, the *GateKeeper* process runs on the firewall, the *WebService* on the server behind the firewall.

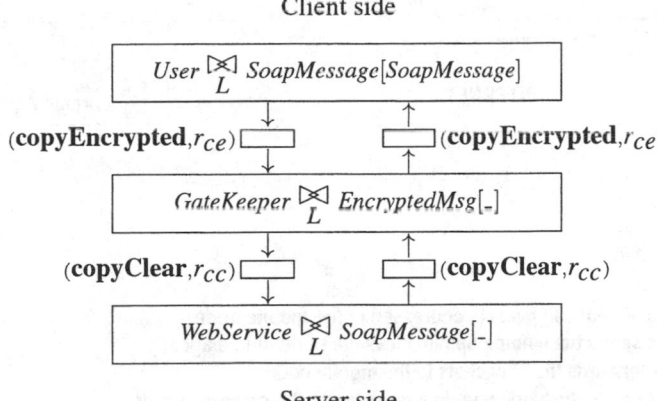

The synchronisation set used at each place, L, is { *decrypt*, *send$_{clr}$, parse, read, modify, export, encrypt, send$_{enc}$* }.

4   Case Study: Mobile IP

In this section we present a larger case study of modelling with PEPA nets. Based on the Internet protocol, Mobile IP is a standard protocol that makes user mobility transparent to applications and higher level protocols like TCP. It allows a *mobile node* to freely roam between network links and to remain always accessible. To achieve this the mobile node uses two IP addresses: a *home address* which is statically assigned on its *home network* and a *care-of address* which changes at each new point of attachment. On the home network, a proxy known as a *home agent* is responsible for forwarding all packets which are addressed to the mobile node on to its current care-of address.

Whenever the mobile node moves, it sends to its home agent a binding update message containing its home address, its current care-of address and the lifetime for which the binding should be honoured [16]. The home agent may refresh a binding cache entry by regularly requesting the transmission of the latest care-of address.

When the mobile node receives, via its home agent, a packet from a correspondent, it sends a binding update message to this correspondent. The correspondent may maintain a binding cache allowing its transmit function to redirect the packets to the mobile node's current care-of address [16]. The mobile node maintains a list of all its current correspondents and has to send them a binding update message each time it changes its point of attachment. Figure 5 summarises the main steps of the Mobile IP protocol.

In this study we assume that the system is composed of N domain or network hosts, besides the home network and the correspondent network. We assume that the home agent sends requests to the mobile node to update the care-of address. Once the correspondent has the care-of address, its transmit function redirects the packet to this address, saving one network hop relative to the route through the home agent [16].

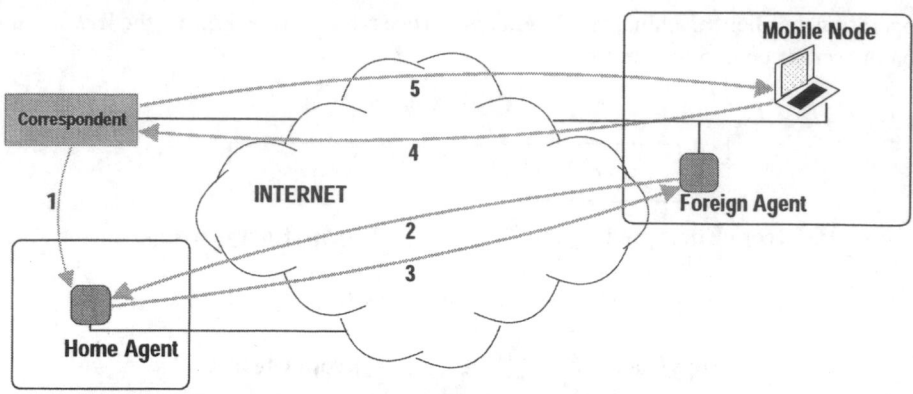

1 Correspondent sends IP packets addressed to the mobile node
2 Mobile Node sends the binding update message to the home agent
3 Home agent forwards the IP packets to the mobile node
4 Mobile Node sends the binding update message to the correspondent
5 Correspondent sends the following IP packets directly to the mobile node

Fig. 5. The Mobile IP Protocol

4.1   The PEPA Net Model

The system is modelled using the PEPA net model depicted in Figure 6 where the home network of the mobile node is represented using a place called **HOME_M**. The networks to which the mobile node may move are modelled by places **DOMAIN**$_i$ where $i = 1, \ldots, N$. Place **HOME_C** models the home network of the mobile node's correspondent.

Note that for the sake of readability in Figure 6 the rates of the activities labelling the firings are omitted. Moreover only the arcs between **DOMAIN**$_1$ and **HOME_M** on one hand, and **HOME_C** on the other, are depicted. The arcs between the other domains and **HOME_M** and **HOME_C** are analogous.

To model the part of the protocol which manages the interaction between the home agent and the mobile node during its stay in place **DOMAIN**$_i$, we use two components *ProtoMA* and *CommuMA*. Similarly, components *ProtoMC* and *CommuMC* are used to model the protocol interactions between the mobile node and the correspondent, and *ProtoAC* and *CommuAC* for the exchanges between the home agent and the correspondent. Components *Mobile*, *Agent* and *Corresp* model the behaviour of the mobile node, the home agent and the current correspondent of the mobile node respectively. In contrast to *Mobile*, the last two components are static. All these components and places are explained in detail in the following.

Component *ProtoMA*$_i$ models the protocol part which consists of updating the care-of address at the home agent level. When the mobile node changes its point of attachment, it generates a binding update message, action *generate*$_{bua}$. Here, the rate associated with

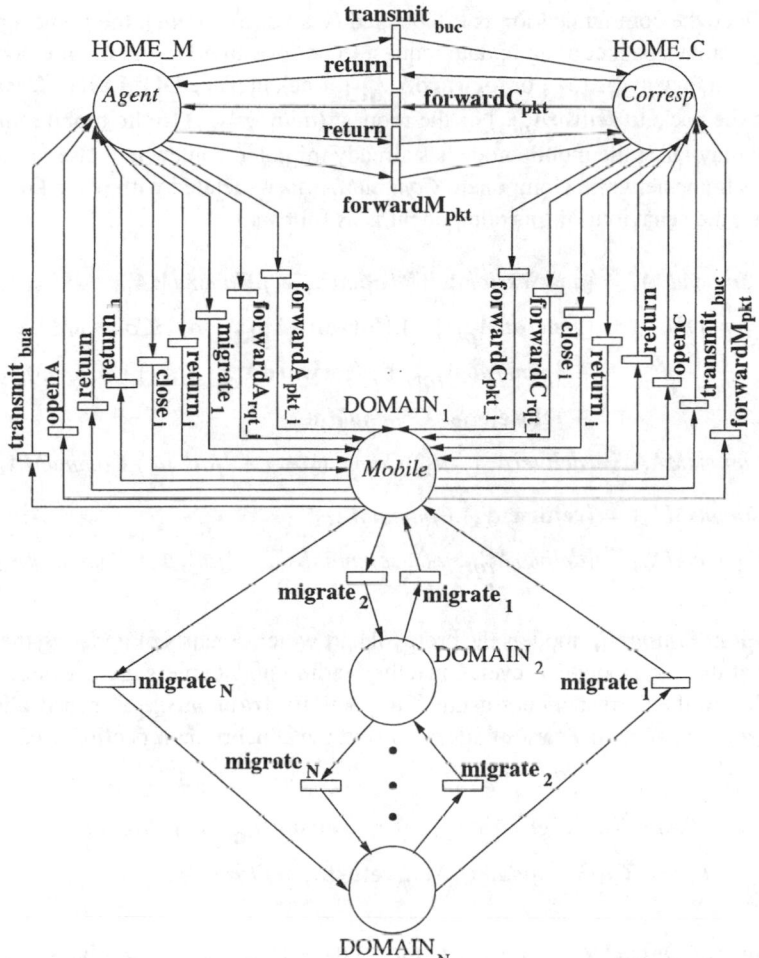

Fig. 6. The PEPA net model

this action is unspecified since *ProtoMA* does not generate this message, but has just to transmit it. This is modelled using the firing action **transmit$_{bua}$** with rate r_1. Once component *ProtoMA* is in **HOME_M**, it allows the home agent to update the care-of address using the synchronizing action *updateA* with rate λ. A *ProtoMA$_i$* component is associated with each place **DOMAIN$_i$**, $i = 1, \ldots, N$.

$$ProtoMA_i \stackrel{def}{=} (generate_{bua}, \top).(\textbf{transmit}_{\textbf{bua}}, r_1).ProtoMA_{i1}$$
$$ProtoMA_{i1} \stackrel{def}{=} (updateA, \lambda).(\textbf{return}_i, r_2).ProtoMA_i$$

Component *CommuMA$_i$* allows us to model the exchanges between the home agent and the mobile node once the binding update message has been received by the home agent. Once the communication is established (via *newSessionA*), the home agent will either generate a packet or an update request to be forwarded to the mobile node using firing action **forwardA$_{\mathbf{pkt\_i}}$** or **forwardA$_{\mathbf{rqt\_i}}$**. Back in place **DOMAIN**$_i$, *CommuMA$_i$* delivers the packet (*deliverA$_{pkt}$*) or the request (*deliverA$_{rqt}$*) to the mobile node. The delivery may fail if the mobile node has already migrated somewhere else or just returned to its home network. Component *CommuMA$_i$* is associated with place **DOMAIN**$_i$. Formally, the behaviour of this component is as follows.

$$CommuMA_i \stackrel{def}{=} (newSessionA, \top).(\mathbf{openA}, \alpha_1).CommuMA_{i0}$$

$$CommuMA_{i0} \stackrel{def}{=} (generateA_{pkt}, \top).(\mathbf{forwardA_{pkt\_i}}, \alpha_3).CommuMA_{i2}$$
$$+ (generateA_{rqt}, \top).(\mathbf{forwardA_{rqt\_i}}, \alpha_5).CommuMA_{i4}$$
$$+ (\mathbf{close}_i, \alpha_2).CommuMA_i$$

$$CommuMA_{i2} \stackrel{def}{=} (deliverA_{pkt}, \mu_1).CommuMA_{i3} + (fail, \mu_2).CommuMA_{i3}$$

$$CommuMA_{i3} \stackrel{def}{=} (\mathbf{return}, \alpha_4).CommuMA_{i0}$$

$$CommuMA_{i4} \stackrel{def}{=} (deliverA_{rqt}, \mu_3).CommuMA_{i3} + (fail, \mu_4).CommuMA_{i3}$$

Component *ProtoMC$_i$* models the protocol part which consists of updating the care-of address at the correspondent level. Once the binding update message is generated, it is forwarded to the correspondent using firing activity **transmit$_{\mathbf{buc}}$**. Then it allows the home agent to update the care-of address using the synchronizing action *updateC* with rate λ_1.

$$ProtoMC_i \stackrel{def}{=} (generate_{buc}, \top).(\mathbf{transmit_{buc}}, r_3).ProtoMC_{i1}$$
$$ProtoMC_{i1} \stackrel{def}{=} (updateC, \lambda_1).(\mathbf{return}_i, r_4).ProtoMC_i$$

Component *CommuMC$_i$* allows us to model the effective exchanges between the mobile node and the correspondent. Once the communication is established (*newSessionC*) by the mobile node, the correspondent may either generate a packet or an update request that will be forwarded using firing transition **forwardC$_{\mathbf{pkt\_i}}$** or **forwardC$_{\mathbf{rqt\_i}}$** respectively. The delivery may succeed or fail if the mobile has already moved. In the first case, the mobile node may either stay silent (*silent*) or generate a packet (*generateM$_{pkt}$*) which is forwarded, using firing transition **forwardM$_{\mathbf{pkt}}$**, to the correspondent. The communication is considered finished when transition **close$_i$** is fired at the correspondent level. A component *CommuMC$_i$* is associated with each place **DOMAIN**$_i$, $i = 1, \ldots, N$.

$$CommuMC_i \stackrel{def}{=} (newSessionC, \top).(\mathbf{openC}, \beta_1).CommuMC_{i0}$$

$$CommuMC_{i0} \stackrel{def}{=} (generateCM_{pkt}, \top).(\mathbf{forwardC_{pkt\_i}}_{,\beta_3}).CommuMC_{i1}$$
$$+(generateCM_{rqt}, \top).(\mathbf{forwardC_{rqt\_i}}, \beta_5).CommuMC_{i4}$$
$$+(\mathbf{close}_i, \beta_2).CommuMC_i$$

$$CommuMC_{i1} \stackrel{def}{=} (deliverC_{pkt}, \nu_1).CommuMC_{i0} + (fail, \nu_2).CommuMC_{i5}$$

$$CommuMC_{i2} \stackrel{def}{=} (generateM_{pkt}, \top).(\mathbf{forwardM_{pkt}}, \beta_4).CommuMC_{i3}$$
$$+(silent, v_3).CommuMC_{i5}$$
$$CommuMC_{i3} \stackrel{def}{=} (deliverM_{pkt}, v_4).CommuMC_{i0}$$
$$CommuMC_{i4} \stackrel{def}{=} (deliverC_{rqt}, v_5).CommuMC_{i5} + (fail, v_6).CommuMC_{i5}$$
$$CommuMC_{i5} \stackrel{def}{=} (\mathbf{return}, \beta_6).CommuMC_{i0}$$

Component *ProtoAC* models the case where the mobile node returns to its home network and has to send a binding update message to its current correspondent. This component is associated with place **HOME_M** and returns to it once the care-of address has been updated at the correspondent level using activity *updateCh* with rate λ_2.

$$ProtoAC \stackrel{def}{=} (generateH_{buc}, \top).(\mathbf{transmit_{buc}}, r_5).ProtoAC_1$$
$$ProtoAC_1 \stackrel{def}{=} (updateCh, \lambda_2).(\mathbf{return}, r_6).ProtoAC$$

Component *CommuAC* models the communication between the correspondent and the home agent if the mobile node has left its home network or with the mobile node itself if not. It is associated with place **HOME_C** and forwards the packets generated by the correspondent to the mobile node in its home network. This is modelled using the firing action **forwardC_{pkt}** with rate γ_1. If the mobile node is present, the packets are delivered to it with action *deliverC_{pkt}*. Component *CommuAC* then forwards the packets generated by the mobile node to the correspondent. This is done with the firing action **forwardM_{pkt}** with rate γ_2. If the mobile node is not in its home network, the correspondent's packets are saved (*saveC_{pkt}*) by the home agent and component *CommuAC* goes back to **HOME_C** using firing action **return** at rate γ_3.

$$CommuAC \stackrel{def}{=} (generateCA_{pkt}, \top).(\mathbf{forwardC_{pkt}}, \gamma_1).CommuAC_1$$
$$CommuAC_1 \stackrel{def}{=} (deliverC_{pkt}, w_1).CommuAC_2 + (saveC_{pkt}, w_2).CommuAC_4$$
$$CommuAC_2 \stackrel{def}{=} (generateM_{pkt}, w_3).(\mathbf{forwardM_{pkt}}, \gamma_2).CommuAC_3$$
$$+(silent, w_4).CommuAC_4$$
$$CommuAC_3 \stackrel{def}{=} (deliverMh_{pkt}, w_5).CommuAC$$
$$CommuAC_4 \stackrel{def}{=} (\mathbf{return}, \gamma_3).CommuAC$$

Component *Mobile* models the behaviour of the mobile node whatever its current point of attachment. It may generate packets, receive packets or simply choose to move to another network, modelled using actions *generateM_{pkt}*, *deliverC_{pkt}* and **migrate_i** respectively. When the mobile node changes its point of attachment, it first generates a binding update message for its home agent with action *generate_{bua}* at rate τ_2 and then opens a new communication session (*newSessionA*). It may then receive from its home agent either an update request (*deliver A_{rqt}*) or a correspondent's packet (*deliver A_{pkt}*).

In this last case, it generates a binding update message for the correspondent with action $generate_{buc}$ at rate τ_3 and establishes a new communication session ($newSessionC$). In network i, the mobile node may stay in the current network or, with probability p_j, move to another network j with firing action **migrate**$_j$, $j \neq i$. It may also return to its home network with the firing action **return$_h$**.

$$Mobile \stackrel{def}{=} (generateM_{pkt}, \tau_1).Mobile + (deliverC_{pkt}, \top).Mobile$$

$$+ \sum_{i=1}^{N}(\textbf{migrate}_i, p_i \times \delta_1).Mobile_1$$

$$Mobile_1 \stackrel{def}{=} (generate_{bua}, \tau_2).(newSessionA, s_1).Mobile_2$$

$$Mobile_2 \stackrel{def}{=} (deliverA_{pkt}, \top).Mobile_3 + (deliverA_{rqt}, \top).Mobile_1 + (\textbf{return}_h, \delta_2).Mobile$$

$$Mobile_3 \stackrel{def}{=} (generate_{buc}, \tau_3).(newSessionC, s_2).Mobile_4$$

$$Mobile_4 \stackrel{def}{=} (deliverA_{rqt}, \top).Mobile_7 + (deliverA_{pkt}, \top).Mobile_4$$

$$+(deliverC_{rqt}, \top).Mobile_8 + (deliverC_{pkt}, \top).Mobile_4$$

$$+(generateM_{pkt}, \tau_4).Mobile_4 + (\textbf{return}_h, \delta_2).Mobile_5$$

$$+ \sum_{j=1/j \neq i}^{N}(\textbf{migrate}_j, p_j \times \delta_5).Mobile_6$$

$$Mobile_5 \stackrel{def}{=} (generateh_{buc}, \tau_5).Mobile$$

$$Mobile_6 \stackrel{def}{=} (generate_{bua}, \tau_6).(newSessionA, s_1).Mobile_3$$

$$Mobile_7 \stackrel{def}{=} (generate_{bua}, \tau_7).Mobile_4$$

$$Mobile_8 \stackrel{def}{=} (generate_{buc}, \tau_3).Mobile_4$$

Component *Agent* models the home agent's behaviour. It is given by the following PEPA equations:

$$Agent \stackrel{def}{=} (updateA, \top).Agent_1 + (saveC_{pkt}, \top).Agent_3$$

$$Agent_1 \stackrel{def}{=} (saveC_{pkt}, \top).Agent_2 + (generateA_{rqt}, \nu_1).Agent_1 + (updateA, \top).Agent_1$$

$$Agent_2 \stackrel{def}{=} (generateA_{pkt}, \nu_2).Agent_1 + (updateA, \top).Agent_2 + (saveC_{pkt}, \top).Agent_2$$

$$Agent_3 \stackrel{def}{=} (saveC_{pkt}, \top).Agent_3 + (updateA, \top).Agent_2$$

Component *Corresp* models the behaviour of a correspondent of the mobile node. It may generate packets to send to the mobile node in its home network ($generateCA_{pkt}$) and receive packets from the mobile node when it is still there ($deliverMh_{pkt}$). It may also receive a binding update message from the mobile node ($updateC$). In this case, the correspondent may generate packets ($generateCM_{pkt}$) or update requests ($generateCA_{rqt}$)

to send to the mobile node in its current attachment point. Action type *updateCh* models the case where the correspondent receives a binding update message from the mobile node back in its home network.

$$Corresp \stackrel{def}{=} (generateCA_{pkt}, c_1).Corresp + (deliverMh_{pkt}, \top).Corresp$$
$$+(updateC, \top).Corresp_1$$
$$Corresp_1 \stackrel{def}{=} (generateCM_{pkt}, c_2).Corresp_1 + (generateCM_{rqt}, c_3).Corresp_1$$
$$+(deliverM_{pkt}, \top).Corresp_1 + (updateC, \top).Corresp_1$$
$$+(updateCh, \top).Corresp$$

The Places. The places of the PEPA net are defined as follows:

$$\textbf{HOME\_M} \stackrel{def}{=} \left(ProtoAC[ProtoAC] \underset{L_1}{\bowtie} \left(Mobile[Mobile] \underset{L_2}{\bowtie} \left(CommuAC[\_] \underset{L_3}{\bowtie} \right. \right. \right.$$
$$\left. \left. \left. ((CommuMA_1[\_] \| \dots \| CommuMA_N[\_]) \underset{L_4}{\bowtie} \left(Agent \underset{L_5}{\bowtie} \right. \right. \right. \right.$$
$$\left. \left. \left. \left. (ProtoMA_1[\_] \| \dots \| ProtoMA_N[\_]) \right) \right) \right) \right)$$

$$\textbf{HOME\_C} \stackrel{def}{=} \left(\left(\left(\left(ProtoAC[\_] \underset{L_6}{\bowtie} Corresp \right) \underset{L_7}{\bowtie} (CommuMC_1[\_] \| \dots \| CommuMC_N[\_]) \right) \right. \right.$$
$$\left. \left. \underset{L_8}{\bowtie} (ProtoMC_1[\_] \| \dots \| ProtoMC_N[\_]) \right) \underset{L_9}{\bowtie} CommuAC[CommuAC] \right)$$

$$\textbf{DOMAIN}_i \stackrel{def}{=} \left(\left(\left(\left(ProtoMA_i[ProtoMA_i] \underset{L_{10}}{\bowtie} Mobile[\_] \right) \underset{L_{11}}{\bowtie} CommuMA_i[CommuMA_i] \right) \right. \right.$$
$$\left. \left. \underset{L_{12}}{\bowtie} ProtoMC_i[ProtoMC_i] \right) \underset{L_{13}}{\bowtie} CommuMC_i[CommuMC_i] \right)$$

where $i = 1 \dots N$ and the synchronizing sets are defined as follows

$$L_1 = \{generateH_{buc}\} \qquad\qquad L_7 = \{deliverM_{pkt}, generateCM_{pkt}, generateCM_{rqt}\}$$
$$L_2 = \{deliverC_{pkt}, generateM_{pkt}\} \quad L_8 = \{updateC\}$$
$$L_3 = \{saveC_{pkt}\} \qquad\qquad\qquad L_9 = \{generateCA_{pkt}, deliverMh_{pkt}\}$$
$$L_4 = \{generateA_{pkt}, generateA_{rqt}\} \quad L_{10} = \{generate_{bua}\}$$
$$L_5 = \{updateA\} \qquad\qquad\qquad L_{11} = \{newSessionA, deliverA_{pkt}, deliverA_{rqt}\}$$
$$L_6 = \{updateCh\} \qquad\qquad\qquad L_{12} = \{generate_{buc}\}$$
$$L_{13} = \{newSessionC, generateM_{pkt}, deliverC_{pkt}, deliverC_{rqt}\}$$

5   Tool Support

The PEPA stochastic process algebra is supported by a range of tools including the PEPA Workbench [9] and the Möbius Modelling Framework [6]. We have implemented the PEPA nets formalism as an extension of the PEPA Workbench. The PEPA modelling tools, together with user documentation, papers and examples are available from the PEPA Web page which is located at http://www.dcs.ed.ac.uk/pepa.

We have provided tool support for PEPA nets in two forms. The PEPA Workbench for PEPA nets is a dedicated tool which can be used to generate the Markov process underlying a PEPA net in a format suitable for solution by a number of solvers. In contrast, the PEPA net compiler allows the existing tool support for PEPA to be exploited by translating a PEPA net model into an equivalent PEPA model [10]. For example, this gives us the ability to use the PRISM probabilistic symbolic model checker [18] which has been extended to support PEPA.

```
PEPA Workbench for PEPA Nets Version 0.83 "Hanover Street"
Compiling the model
Generating the derivation graph
The model has 982740 states
The model has 2059174 transitions
The model has 3379038 firings
Writing the hash table file to modelIPv6NA1.hash
Exiting PEPA Workbench.
```

Fig. 7. The PEPA Workbench for PEPA nets processing the MobileIP example

The input language of the tool is an extension of the concrete syntax used for storing PEPA language models. The topology of the net is specified by providing a textual description of the places and the arcs connecting them. The use of the PEPA Workbench for PEPA nets is illustrated in Figure 7.

6   Related Work

Stochastic Petri nets and stochastic process algebras have complementary strengths. A comparison of the two formalisms [7] concludes that "there is scope for future work incorporating the attractive characteristics of the formalisms ... from one paradigm into the other". Some work has been done in this area in beginning to develop a structural theory for process algebras [11] on the one hand and in importing composition operations from stochastic process algebras into net formalisms on the other [23,15,13]. In contrast the work on PEPA nets aims to use both Petri nets and process algebras together as a single, structured performance modelling formalism.

Complementary to own work is Valk's work on *Elementary Object Systems* [24]. In this work an extension of Petri nets is presented in which the tokens circulating in the net structure (called the *System net*) are themselves Petri nets (termed *Object nets*). Object nets move like ordinary tokens and they can change their markings but not their structure. Three different types of transitions are defined. Transitions occurring in the Object net (i.e. in the marking) are called *system autonomous* and represent the object internal behaviour. An *interaction* takes places when both the Object and the System net enable transitions with the same attached label. A third type of transition causes a change in the System net only and it is called *transport*. In PEPA nets we do not allow such transitions, since a firing cannot occur without modifying the state of a component.

Despite the superficial similarities there are some quite strong differences between the work on PEPA nets and that on Elementary Object Systems (EOS). Fundamentally, EOS are without any timing considerations, other than the relative timing imposed by the Petri net causality relation. In PEPA nets, in addition to this implicit timing information we have explicit time delays integrated into behaviour at both the net level and the token level. Moreover the origins of the works are distinct. Valk's work is motivated by a desire to provide a fundamental model of object-oriented programming, and the development of EOS has been strongly influenced by this goal. Our motivation has been to develop a convenient high-level modelling language for Markov processes, for systems exhibiting mobility. Recent work by Köhler *et al.* has examined the possibility of using Object Systems for modelling mobility and mobile agents [17].

Several process calculi have been developed specifically for modelling mobile computation, primarily for the purpose of functional verification, the most notable being the π-calculus [21] and the calculus of mobile ambients [3]. The π-calculus, and Priami's subsequent extension, the stochastic π-calculus [22], have a very different style of representing systems [2], which does not satisfy our criterion of clearly separating state changes into distinct types related, in the case of mobile computation, to concepts of location and mobility. In this respect our formalism is closer to the work on mobile ambients.

The calculus of mobile ambients is intended to capture notion of *locations*, *mobility* and *authority for movement*. This is achieved by introducing the concept of *ambient*, i.e. a bounded place where computation happens. Ambients can be nested into other ambients and can be moved as a whole. Mobility primitives are provided by considering *capabilities*: it is possible to *enter* into another ambient, to *exit* from an ambient, to *open* an ambient. Processes are executed within ambients and a simple asynchronous communication mechanism that works within a single ambient is chosen. Communication across ambients is modelled as the movement of 'messenger' agents that must cross ambient boundaries. (This is similar in style to our own representation of the Mobile IP protocol.) The most pronounced differences between PEPA nets and the ambient calculus are the lack of timing information in the ambient calculus and the ability to nest ambients which gives a hierarchical structure to locations which cannot be matched by the places in PEPA nets.

7   Conclusions

In this tutorial we have introduced the PEPA nets modelling language and focussed on the use of PEPA nets as a performance modelling formalism tailored for systems with inherent mobility. We have applied the PEPA nets modelling language to modelling examples of mobile agent systems where the agents in the system are mobile objects under a discipline of dynamic binding of names. We have also used PEPA nets to build larger case studies of mobile system protocols such as MobileIP.

We have implemented a tool set to explore PEPA net models and to generate the corresponding continuous-time Markov chain representation of the model. These can be solved by standard numerical procedures for solving continuous-time Markov chains. Future work remains to investigate efficient state-space generation procedures for PEPA nets and efficient solution techniques for the generated models. Both of these programmes of work will take advantage of the hierarchical structure of a PEPA net as a network of PEPA models.

Acknowledgements. Stephen Gilmore, Jane Hillston and Leïla Kloul are supported by the DEGAS (Design Environments for Global ApplicationS) IST-2001-32072 project funded by the FET Proactive Initiative on Global Computing.

References

1. M. Ajmone Marsan, A. Bobbio, and S. Donatelli. Petri nets in performance analysis: An introduction. In Reisig, W. and Rozenberg, G., editors, *Lectures on Petri Nets I: Basic Models*, volume 1491 of *LNCS*, pages 211–256. Springer-Verlag, 1998.
2. L. Brodo, S. Gilmore, J. Hillston, and C. Priami. A stochastic π-calculus semantics for PEPA nets. In *Proc. of the Workshop on Process Algebras and Stochastically Timed Activities*, pages 1–17. LFCS, University of Edinburgh, June 2002.
3. L. Cardelli and A.D. Gordon. Mobile ambients. In M. Nivat, editor, *Foundations of Software Science and Computational Structures*, volume 1378 of *LNCS*, pages 140–155. Springer Verlag, 1998.
4. G. Chiola, C. Dutheillet, G. Franceschinis, and S. Haddad. Stochastic well-formed coloured nets and symmetric modeling applications. *IEEE Transactions on Computers*, 42:1343–1360, nov 1993.
5. G. Clark. *Techniques for the Construction and Analysis of Algebraic Performance Models*. PhD thesis, The University of Edinburgh, 2000.
6. G. Clark, T. Courtney, D. Daly, D. Deavours, S. Derisavi, J. M. Doyle, W. H. Sanders, and P. Webster. The Möbius modeling tool. In *Proc. of 9th Int. Workshop on Petri Nets and Performance Models*, pages 241–250, Aachen, Germany, September 2001.
7. S. Donatelli, J. Hillston, and M. Ribaudo. A comparison of Performance Evaluation Process Algebra and Generalized Stochastic Petri Nets. In *Proc. 6th International Workshop on Petri Nets and Performance Models*, Durham, North Carolina, 1995.
8. A. Fuggetta, G.P. Picco, and G. Vigna. Understanding code mobility. *IEEE Transactions on Software Engineering*, 24(5):342–361, may 1998.
9. S. Gilmore and J. Hillston. The PEPA Workbench: A Tool to Support a Process Algebra-based Approach to Performance Modelling. In *Proc. of 7th Int. Conf. on Modelling Techniques and Tools for Computer Performance Evaluation*, number 794 in LNCS, pages 353–368, Vienna, May 1994. Springer-Verlag.
10. S. Gilmore, J. Hillston, L. Kloul, and M. Ribaudo. Software performance modelling using PEPA nets. to appear in proc. of Int. Workshop on Software Performance, 2004.
11. S. Gilmore, J. Hillston, and L. Recalde. Elementary structural analysis for PEPA. Technical Report ECS-LFCS-97-377, Laboratory for Foundations of Computer Science, Department of Computer Science, The University of Edinburgh, 1997.
12. S. Gilmore, J. Hillston, M. Ribaudo, and L. Kloul. PEPA nets: A structured performance modelling formalism. *Performance Evaluation*, 54(2):79–104, October 2003.
13. H. Hermanns, U. Herzog, V. Mertsiotakis, and M. Rettelbach. Exploiting stochastic process algebra achievements for generalized stochastic Petri nets. In *Proc. of 7th International Workshop on Petri Nets and Performance Models*, Saint Malo, June 1997. IEEE CS Press.
14. J. Hillston. *A Compositional Approach to Performance Modelling*. Cambridge University Press, 1996.

15. R. Hopkins and P. King. A visual formalism for the composition of stochastic Petri nets. In T. Field *et al.*, editor, *Proc. of 12th Int. Conf. on Modelling Techniques and Tools for Computer Performance Evaluation*, volume 2324 of *LNCS*, pages 239–258. Springer, 2002.
16. D. Johnson, C. Perkins, and J. Arkko. Mobility support in IPv6. IETF Mobile IP Working Group Internet-Draft draft-ietf-mobileip-ipv6-20, January 2003.
17. M. Köhler, D. Moldt, and H. Rölke. Modelling mobility and mobile agents using nets within nets. In W.M.P. van der Aalst and E. Best, editors, *Proc. of Int. Conf. on Applications and Theory of Petri Nets*, volume 2679 of *LNCS*, pages 121–139. Springer-Verlag, june 2003.
18. M. Kwiatkowska, G. Norman, and D. Parker. PRISM: Probabilistic symbolic model checker. In T. Field *et al.*, editor, *Proc. of 12th Int. Conf. on Modelling Techniques and Tools for Computer Performance Evaluation*, volume 2324 of *LNCS*, London, 2002. Springer.
19. K. Larsen and A. Skou. Bisimulation through Probabilistic Testing. *Information and Computation*, 94(1):1–28, September 1991.
20. K.G. Larsen. Compositional theories based on an operational semantics of contexts. In *REX Workshop on Stepwise Refinement of Parallel Systems*, volume 430 of *LNCS*, pages 487–518. Springer-Verlag, May 1989.
21. R. Milner. *Communicating and Mobile Systems: the π-calculus*. Cambridge University Press, 1999.
22. C. Priami. Stochastic π-calculus. In S. Gilmore and J. Hillston, editors, *Proceedings of the Third International Workshop on Process Algebras and Performance Modelling*, pages 578–589. Special Issue of *The Computer Journal*, 38(7), December 1995.
23. I. C. Rojas M. *Compositional Construction and Analysis of Petri net Systems*. PhD thesis, The University of Edinburgh, 1997.
24. R. Valk. Petri nets as token objects—an introduction to Elementary Object Nets. In J. Desel and M. Silva, editors, *Proc. of 19th Int. Conf. on Application and Theory of Petri Nets*, volume 1420 of *LNCS*, pages 1–25, Lisbon, Portugal, 1998. Springer-Verlag.

A   Semantics of PEPA

The semantic rules, in the structured operational style, are presented in Figure 8; the interested reader is referred to [14] for more details. The rules are read as follows: if the transition(s) above the inference line can be inferred, then we can infer the transition below the line. The notation $r_\alpha(E)$ which is used in the third cooperation rule denotes the apparent rate of α in E.

A.1   Additional Semantic Rules for PEPA Nets

For any token component its action type set can be partitioned in distinct subsets corresponding to transitions and firings respectively. Thus for a component Q, $\mathcal{A}_t(Q)$ is the set of local transitions currently enabled in Q and $\mathcal{A}_f(Q)$ is the set of firings currently enabled for Q.

We use the notation

$$\mathbf{P_1} \xrightarrow{\ (\alpha, r)\ }\!\!\!| \longmapsto \mathbf{P_2}$$

to capture the information that there is a transition connecting place $\mathbf{P_1}$ to place $\mathbf{P_2}$ labelled by (α, r). This relation captures static information about the structure of the net, not dynamic information about its behaviour. The semantic rules for PEPA nets are those for PEPA plus the additional rules presented in Figure 9;

Prefix

$$(\alpha, r).E \xrightarrow{(\alpha,r)} E$$

Cooperation

$$\frac{E \xrightarrow{(\alpha,r)} E'}{E \bowtie_L F \xrightarrow{(\alpha,r)} E' \bowtie_L F} \ (\alpha \notin L) \qquad \frac{F \xrightarrow{(\alpha,r)} F'}{E \bowtie_L F \xrightarrow{(\alpha,r)} E \bowtie_L F'} \ (\alpha \notin L)$$

$$\frac{E \xrightarrow{(\alpha,r_1)} E' \quad F \xrightarrow{(\alpha,r_2)} F'}{E \bowtie_L F \xrightarrow{(\alpha,R)} E' \bowtie_L F'} \ (\alpha \in L) \quad \text{where } R = \frac{r_1}{r_\alpha(E)} \frac{r_2}{r_\alpha(F)} \min(r_\alpha(E), r_\alpha(F))$$

Choice

$$\frac{E \xrightarrow{(\alpha,r)} E'}{E + F \xrightarrow{(\alpha,r)} E'} \qquad \frac{F \xrightarrow{(\alpha,r)} F'}{E + F \xrightarrow{(\alpha,r)} F'}$$

Hiding

$$\frac{E \xrightarrow{(\alpha,r)} E'}{E/L \xrightarrow{(\alpha,r)} E'/L} \ (\alpha \notin L) \qquad \frac{E \xrightarrow{(\alpha,r)} E'}{E/L \xrightarrow{(\tau,r)} E'/L} \ (\alpha \in L)$$

Constant

$$\frac{E \xrightarrow{(\alpha,r)} E'}{A \xrightarrow{(\alpha,r)} E'} \ (A \stackrel{def}{=} E)$$

Fig. 8. The operational semantics of PEPA

The Cell rule conservatively extends the PEPA semantics to define that a cell which is filled by a component Q has the same transitions as Q itself. A healthiness condition on the rule (a *typing judgement*) requires a context $Q[\_]$ to be filled with a component which has the same alphabet as Q. We write $Q =_a Q'$ to state that Q and Q' have the same alphabet. There are no rules to infer transitions for an empty cell because an empty cell enables no transitions.

The Transition rule states that the net has local transitions which change only a single component in the marking vector. This rule also states that these transitions agree with the transitions which are generated by the PEPA semantics (including the extension for contexts). Recall that the transition and firing alphabets are distinct. We do not give priority to one alphabet of actions over the other; the highest-priority firings and the transitions compete based on a race policy.

The Firing rule takes one marking of the net to another by performing a PEPA activity and moving a PEPA component from the input place to the output place. This has the effect that two entries in the marking vector change simultaneously. In order to take account of the priorities we define a number of supplementary transition relations, one for each priority level. A net level transition's eligibility for firing depends on two conditions. Firstly there must be an empty cell in the destination place into which the token can be transferred. The Enabling rule ensures that this is the case, and defines a transition relation, decorated with the priority level of the corresponding activity type. The rate at which the activity is enabled is calculated as in the PEPA semantics of cooperation. In order for a firing to take place it must also be the case that the type of the enabled firing has the highest priority level in the set of the enabled firings. This is imposed by the Firing rule in which we discard those enabled firings which do not have the highest priority. In other words for a firing to occur there must not be any other firing satisfying the Enabling rule (empty destination cell) which has a higher priority.

Cell:

$$\frac{Q' \xrightarrow{(\alpha, r)} Q''}{Q[Q'] \xrightarrow{(\alpha, r)} Q[Q'']} \quad (Q =_a Q')$$

Transition:

$$\frac{M_{\mathbf{P}} \xrightarrow{(\alpha, r)} M_{\mathbf{P}}'}{(\ldots, M_{\mathbf{P}}, \ldots) \xrightarrow{(\alpha, r)} (\ldots, M_{\mathbf{P}}', \ldots)} (\alpha \in \mathcal{A}_t)$$

Enabling:

$$\frac{Q \xrightarrow{(\alpha, r_1)} Q' \qquad \mathbf{P_i} \xrightarrow{(\alpha, r_2)} \mathbf{P_j}}{(.., \mathbf{P_i}[.., Q, ..], .., \mathbf{P_j}[.., -, ..], ..) \xrightarrow{(\alpha, R)}_{\pi(\alpha)} (.., \mathbf{P_i}[.., -, ..], .., \mathbf{P_j}[.., Q', ..], ..)} (\alpha \in \mathcal{A}_f)$$

Firing:

$$\frac{M \xrightarrow{(\alpha, r)}_n M' \qquad M \xrightarrow{(\beta, s)}_m M''}{M \xrightarrow{(\alpha, r)} M'} (n \geq m)$$

Fig. 9. Additional semantic rules for PEPA nets

Scheduling in Distributed Systems

Helen Karatza

Department of Informatics, Aristotle University of Thessaloniki,
54124 Thessaloniki, Greece
karatza@csd.auth.gr

Abstract. This paper considers the problem of scheduling in distributed systems. Several scheduling policies employed in distributed server environments are presented, each of which addresses different aspects of the scheduling problem. The objective of this paper is to summarize research on distributed systems scheduling that is being performed worldwide. ...

1 Introduction

Scheduling is fundamental to the success of distributed systems. Even the most powerful high performance computing environments require proper scheduling in order to efficiently serve the users. For this reason, many researchers have been seeking scheduling methods to optimize distributed system performance. Furthermore, recent developments in computing environments, such as grid environments and heterogeneous clusters, present new challenges to schedulers. The heterogeneity of these systems and changes in execution environment often require dedicated scheduling techniques for each group of resources in the distributed computing environment.

Distributed server systems are common because they offer considerable computational power, are cost-effective and are easily scalable. A distributed system shares resources among a number (community) of users. Unfortunately, it is not always possible to efficiently execute jobs. Good scheduling improves system performance and also preserves individual application's performance so that some jobs do not suffer unbounded delays.

Distributed system scheduling has been studied extensively while applying different performance goals to many distributed system types and parallel workload types. It has been clearly established that scheduling has a substantial impact on performance. The references section lists a few examples taken from the literature.

Many types of distributed systems are currently available, some consisting of large numbers of processors. The availability of different distributed systems as well as a diversity of hardware and software makes the management of distributed resources among users very difficult. Scheduling often is a composite problem of deciding where and when to execute an application (i.e., on which processors, and in which order). Some scheduling policies offer over-all solutions to the composite problem, while others merely focus on one of the sub-problems.

M.C. Calzarossa and E. Gelenbe (Eds.): MASCOTS 2003, LNCS 2965, pp. 336–356, 2004.
© Springer-Verlag Berlin Heidelberg 2004

Scheduling is a fundamental issue regarding performance of distributed applications and distributed systems. However, many related problems remain unsolved. Hence, research continues. Techniques at the application level and at the system level are of interest.

Traditional performance methodologies, i.e., analytic modeling, experimental measurements, and simulation modeling are useful when evaluating distributed systems. Analytical techniques used to evaluate distributed system performance are comprised of queuing networks and stochastic Petri Nets. Other modeling techniques combine approximate solutions and analytical methods. However, the high level of distributed system complexity limits the applicability of these techniques.

Experimental measurement is another method used to evaluate distributed system performance. However, measurements can only be gathered on existing systems. Many believe that the most straightforward way to evaluate scheduling algorithms, without building a full-scale implementation is through modeling and simulation. Simulation models can help identify problems such as performance bottlenecks inherent in the architecture, and provide a basis for refining the system configuration. Some papers related to simulation of parallel and distributed systems scheduling are listed in [1].

The objective of this paper is to present further research into the area of scheduling in distributed systems. The referenced research papers represent significant effort into distributed scheduling that is being expended worldwide. These papers provide insight into current and future trends in distributed systems performance.

2   Scheduling Issues

The range of research topics covered next is quite broad, reflecting on the importance of scheduling in distributed systems.

2.1   Task Assignment in Distributed Systems

The performance analysis of task assignment policies is a very interesting research topic. However, analyzing even the simplest policy is difficult, leaving many open questions.

The analysis of task assignment is addressed in [2]. This paper considers a particular distributed server system model in which hosts are homogeneous and the execution of jobs is non-preemptive. The authors consider both the case where jobs are immediately dispatched upon arrival to one of the host machines for processing, and the case where jobs are held in a central queue until requested by a host machine. In this research short jobs are separated from long jobs, but short jobs may be run in the long job partition if it is idle (cycle stealing). It is considered that short and long jobs have generally distributed service requirements, and that arrivals are Poisson.

The authors present the first analysis of cycle stealing under the central-queue model, while cycle stealing under the immediate dispatch model is analyzed by the same authors in their past work. Their analysis uses a technique which is referred to as *busy period transitions*. It involves creating a Markov chain that includes transitions that correspond to various types of busy periods. The authors denote that their analysis is an approximation, since it depends on approximating these busy periods by a finite number of moments, but that this approximation can be made as precise as desired by using more moments. Actually, they refer that even with just three moments their analysis agrees well with simulation.

The results of this research reveal that cycle stealing can significantly reduce mean response time for short jobs, while long jobs are not significantly penalized. It is also shown that the central queue approach is superior than immediate dispatch, regarding both the benefit to short jobs and the penalty to long jobs.

2.2 Scheduling in Distributed Systems with Variable Workload

Most research in the area of parallel job scheduling assumes that the number of tasks per job is defined by a specific distribution (for example uniform or normal) and also that job inter-arrival times and task service demands are defined by specific distributions (for example by an exponential or a hyper-exponential distribution). However, in real systems the variability of job parallelism and also the variability of job inter-arrival time and task service demands can vary depending on which jobs run in the different time intervals.

Karatza in [3] studies scheduling in a distributed system with a time varying workload. This paper considers time varying distributions for job inter-arrival times, the parallelism of jobs and for task service demand. These distributions represent real parallel system workloads. The impact of different workload parameters on performance metrics is examined to identify conditions that produce good overall system performance, and maintain fairness of individual job execution times. Simulation generates results used to compare different scheduling configurations. Four known scheduling policies are analyzed and compared in this paper. The *First-Come-First-Served* (FCFS), the *Shortest-Task-First* (STF), the *Limited STF* (LSTF-*l*, where the STF method is applied *l* times and then the oldest task in the queue is scheduled), and the *Job with the Smallest Number of Incomplete Tasks First* (JSNITF).

The technique used to evaluate the performance of the scheduling disciplines in this paper is experimentation using synthetic workload simulation. An open queuing network model of a distributed server system is considered. There are P homogeneous and independent processors each serving their own queue and interconnected by a high-speed network with negligible communication delays. The model configuration is shown in Fig. 1.

Job tasks are considered to be independent so that they can execute at any time, in any order, and on any processor. The number of tasks that a job consists of is referred to as the job's *degree of parallelism*. Each task is dispatched randomly to processors with equal probability. Tasks in processor queues are executed according to the scheduling (temporal) method that is currently employed. No migration or pre-emption is permitted.

On completion of execution, a task waits at the join point for sibling tasks of the same job to complete execution. Therefore, synchronization among tasks is required. The price paid for increased parallelism is a synchronization delay that occurs when tasks wait for siblings to finish execution.

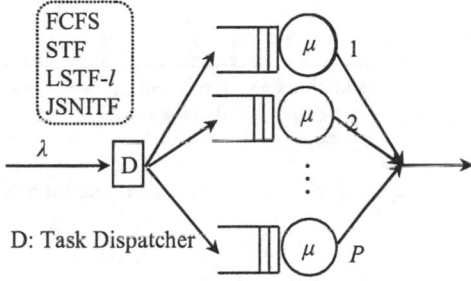

Fig. 1. The queuing network model

The workload considered in this paper is characterized by the following three parameters: Distribution of job inter-arrival times, distribution of the number of tasks per job, and distribution of task service demand. There is no correlation between the different parameters. For example, a job with a small number of tasks may have a long execution time.

Distribution of Job Inter-arrival Times. The distribution of job inter-arrival time changes in exponentially distributed time intervals b_1, b_2, ..., b_k from exponential to Branching Erlang and vice versa (Fig. 2). The mean time interval for distribution change is b. In both the exponential and Branching Erlang cases, the mean job inter-arrival time is $1/\lambda$ where λ is the mean job arrival rate.

It is obvious that with this variable distribution, over some time intervals the arrival process is Poisson, over other time intervals there are no job arrivals, while over still other time intervals inter-arrival times are so small that there are almost bulk arrivals. The parameter that represents the variability of job inter-arrival times is the coefficient of variation of job inter-arrival time (CV). In the exponential distribution case, $CV = 1$ while in the Branching Erlang distribution case, $CV > 1$.

Distribution of the Number of Tasks per Job. A time varying distribution for job parallelism is assumed. It changes from uniform to normal and vice versa at the end of exponentially distributed time intervals d_1, d_2, , d_n (Fig. 2). The mean time interval for distribution change is d. In the uniform distribution case, the number of job tasks is uniformly distributed in the range of $[1..P]$. The mean number of tasks per job is $n = (1+P)/2$. In the normal distribution case we assume a "bounded" normal distribution for the number of tasks per job in the range of $[1..P]$ with mean $n = (1+P)/2$ and standard deviation $\sigma = n/4$.

Jobs that all arrive at processors within the same time interval d_i have the same distribution for their tasks. However, during the same time interval, if some jobs arrived at the processors during previous time intervals, they may have different distributions for their tasks. These jobs may wait at the processor queues or they may be served.

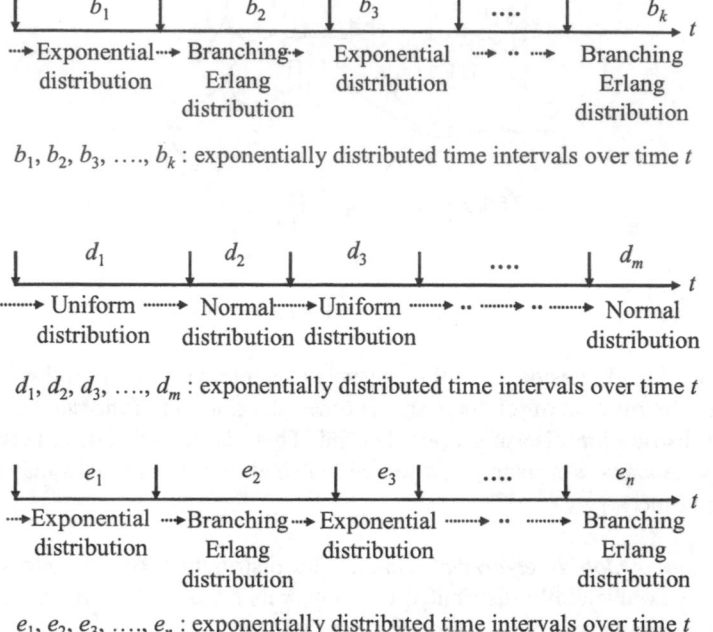

Fig. 2. Time varying distributions for job inter-arrival time, job parallelism and task service demand

It is obvious that jobs in the uniform distribution case present larger variability in their degree of parallelism than jobs in the normal distribution case. In the second case, most of the jobs have a moderate degree of parallelism (close to the mean n). Since the distribution of job parallelism changes with time, for some time intervals, jobs have a highly variable degree of parallelism, while over other time intervals, the majority of the arriving applications have moderate parallelism in comparison to the number of processors.

Distribution of Task Service Demand. The distribution of task service demand changes in exponentially distributed time intervals e_1, e_2, ..., e_m from exponential to Branching Erlang and vice versa (Fig. 2). The mean time interval for distribution change is e. In both the exponential and Branching Erlang cases, mean task service demand is $1/\mu$ where μ is the mean service rate of each processor.

Jobs that arrive at the processors within the same time interval e_i have the same distribution for their task service demand. However, during the same time interval if some jobs arrived during previous time intervals, they may have a different task service demand distribution. These jobs are either waiting at the processor queues or are being served.

Tasks of the Branching Erlang distribution case have larger variability in their service demand than exponential tasks. A high variability in task service demand implies that there are a proportionately high number of service demands that are very small as compared with the mean service demand, and that there are a comparatively low number of service demands that are very large. When a task with a large service demand begins execution, it occupies the processor for a long time interval and, depending on the scheduling policy, it may introduce inordinate queuing delays for other tasks waiting for service. The parameter that represents the variability of task execution time is the coefficient of variation of execution time. The same notation CV is considered for the coefficient of variation of task execution time and job inter-arrival time because in all the experiments conducted in this paper the two parameters have the same values.

The three groups of time intervals $\{b_1, b_2, ..., b_k\}$, $\{d_1, d_2, ..., d_m\}$, and $\{e_1, e_2, ..., e_n\}$ are independent of one another. Therefore, the task service demand distribution change is independent of the job parallelism distribution change as well as the job arrival distribution change (Fig. 2).

From the simulation results that are presented in the paper it is shown that the choice of a scheduling policy depends on whether the performance goal is to achieve only good overall performance or to also provide some guarantee for fairness in terms of individual job service. Simulation results also indicate that the relative performance of the different algorithms depends on the workload, and that all methods have merit. It is also shown that STF-l for a proper l value can combine good performance and fairness.

Time varying workload is also examined in [4] where a closed queuing network model of a distributed system is considered. The model includes P homogeneous and independent processors and the I/O subsystem. Since a system with a balanced program flow is considered, the I/O subsystem has the same service capacity as the processors. The I/O subsystem may consist of an array of disks (multi-server disk center) but it is modeled as a single I/O node with a given mean service time.

In the simulation experiments of [4], it is assumed that a fixed number of jobs N are repeatedly executed in the closed circle of parallel processors and an I/O unit shown in Fig. 3. N is called the degree of multiprogramming of a simulation experiment. x and z represent the mean processor and the mean I/O service time respectively.

Two cases for the assignment of tasks to processors are examined. One case takes communication overhead into account that is incurred when the scheduler collects global system information about processor queue assignment decisions. In the other case it is assumed negligible overhead when assigning job tasks to processor queues. In Fig. 3, the scheduler is modeled as a single server queuing system, with mean service time Co, that represents the effects of communication overhead. $Co = 0$ in the cases where overhead is ignored. In the latter case, the scheduler dispatches job tasks immediately to processor queues upon job arrival.

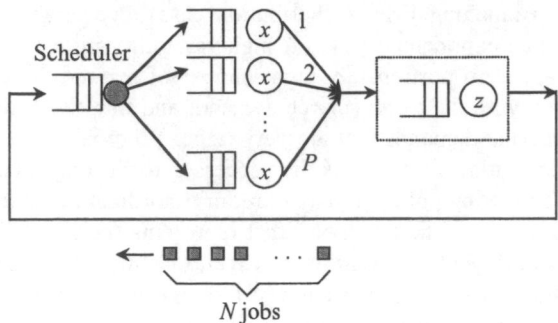

Fig. 3. The queuing network model

As in the [3] model case, jobs are partitioned into independent tasks that can run in parallel. The workload considered here is characterized by the following four parameters: The distribution of the number of tasks per job, the distribution of task service demand, the distribution of I/O service time, and the degree of multiprogramming.

This paper considers that job parallelism and task service demand are not defined by a specific distribution but that the distribution changes with time in the same way with that depicted in [3]. The notations $\{d_1, d_2, ..., d_m\}$, and $\{e_1, e_2, ..., e_n\}$ in Fig. 3 also hold for this case. Each task is assigned to one of the queues accordingly to the routing policy that is applied. Tasks in processor queues are executed according to the temporal scheduling method that is currently employed. No migration or pre-emption is permitted.

After a job leaves the processors, it requests service on the I/O unit. The I/O service times are exponentially distributed with mean z. Each time a job returns from I/O service for scheduling on distributed processors, it is partitioned into a different number of tasks even if it arrives during the same time interval d_i in which case it executed last.

The following scheduling policies are examined: *Probabilistic routing – FCFS, Shortest Queue routing – FCFS (SQFCFS), Shortest Queue routing – FCFS – Co (SQFCFS-Co), Probabilistic routing – STF, Probabilistic routing – Task of the job with the Smallest Number of Incomplete Tasks First, and Probabilistic routing – Limited STF* .

In the SQFCFS case, it is assumed that $Co = 0$. However, the shortest queue method invokes the scheduler every time a job demands processing service. Invoking the scheduler requires considerable communication overhead to collect load information from all processors. For this reason, the SQFCFS–Co policy is used where the overhead of collecting global system information is not negligible. The time required by the scheduler to make an assignment decision is considered to be uniformly distributed over an interval $[a, b]$ with a mean of $Co > 0$. For the I/O subsystem, the FCFS policy is employed.

The impact of different workload parameters on performance metrics is examined in this paper. Similarly to the objective of [3], the aim of this paper is to identify conditions that produce good overall system performance while maintaining fair individual job execution times. Also, simulation is used to generate results for different configurations.

Simulation results of this research indicate that the SQFCFS policy performs much better than the other methods examined and also is the most fair. It is more superior at lower degrees of multiprogramming and when task service demand involves time intervals with large differences in the service demand variability. Furthermore, results show that, up to a certain value of the mean communication overhead, performance of the SQFCFS method is only marginally affected. Generally, it can be concluded that even though overhead can degrade SQFCFS, in many cases its performance remains high in comparison to other methods.

2.3   Epoch Load Sharing and Epoch Scheduling

Load sharing is an important topic that has been studied by many authors. Load sharing and also job scheduling in a network of workstations are studied in [5]. Along with traditional methods of load sharing and temporal job scheduling, this paper also examines two methods referred to as *epoch load sharing* and *epoch scheduling* respectively. Epoch load sharing evenly distributes the load among workstations with job migration that occurs only at the end of predefined intervals. The time interval between successive load sharing events is called an *epoch*. In the epoch scheduling case, job priorities in processor queues are recalculated only at the end of epochs when queue rearrangements take place.

The paper investigates probabilistic, deterministic and adaptive load sharing policies. In the probabilistic case (Pr), scheduling policies are described by state independent branching probabilities. Jobs are dispatched randomly to workstations with equal probability. In the deterministic case, jobs join the Shortest workstation Queue (SQ method) since routing decisions are based on system state. The adaptive case uses a variation of the probabilistic policy. When workstations become idle, jobs can migrate from heavily loaded workstation queues to idle workstations (PrM). This is a receiver initiated adaptive load sharing method. It balances the job load and can improve overall system performance. Epoch load sharing which is also an adaptive load sharing method is described next.

Epoch Load Sharing (ELS). This policy uses migration at the end of epochs to distribute the load evenly among workstations (Fig. 4). At the end of epochs, the scheduler collects information about the status of all workstation queues, evaluates the mean of all queue lengths, and places processor queue lengths into increasing order in a table. Then, it moves jobs from the most heavily loaded to lightly loaded processors until either all processors have queue lengths equal to the mean or differ at most by one job.

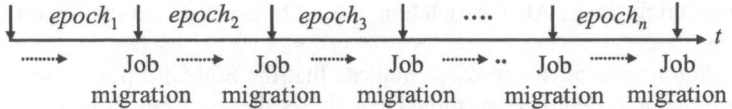

Fig. 4. Epoch load sharing over time t

The transfer of jobs to remote workstations incurs communication costs. In this model, only queued jobs are transferred. The communication channel is modeled as a single server queuing system, whose service time is an exponentially distributed random variable with mean Co, in order to deal with the effects of communication overhead.

Most distributed system load sharing research focuses on improving some performance metrics such as mean response time where scheduling overhead is assumed to be negligible. However, scheduling overhead can seriously degrade performance. Therefore, the number of times the scheduler is called to make load sharing decisions can be a factor that degrades performance.

Epoch load sharing is compared with the other load sharing methods when the FCFS scheduling strategy is employed. The objective is to reduce the number of times that global system information is needed to make allocation decisions, while at same time achieving good performance.

In the case where global system state information is not available, the performance of three temporal scheduling methods is examined when probabilistic routing is employed. The FCFS method is the simplest scheduling strategy. It is fair to individual jobs but often yields sub-optimal performance. This method incurs no overhead. The Shortest-Job-First (SJF) policy usually performs best but it has the following two disadvantages: a) It involves a considerable amount of overhead because processor queues are rearranged each time new jobs are added. b) It is possible to starve a job if its service time is large in comparison to the mean service time. The third method is epoch scheduling described next.

Epoch Scheduling (ES). Processor queues are only rearranged at the end of epochs with this policy (Fig. 5). Then, the scheduler recalculates the priorities of all jobs in system queues using the SJF criterion. The goal of ES is to decrease as much as possible the number of queue rearrangements in comparison with the SJF method, while being fair and producing good system performance.

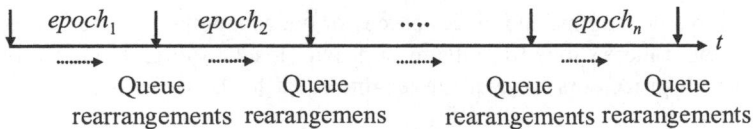

Fig. 5. Epoch queue rearrangements over time t

An open queuing network models the network of workstations. Simulation is used to generate results used to compare different configurations. The performance of load sharing methods and job scheduling strategies are compared for various workloads and different epoch sizes (epoch sizes which are multiples of the mean processor service time are considered). Fairness among competing jobs is required. Simulation results of this paper reveal the following:

PrM, SQ and ELS for small epoch sizes perform better than SJF and ES with probabilistic routing in terms of overall performance and fairness.

From the methods that are examined, SQ is the best policy only if overhead is not taken into account. ELS with small epoch size is preferred as it performs close to SQ and produces much less overhead than PrM and SQ.

However, when global system state information is not available, probabilistic routing combined with ES with a small epoch size is a good policy to use, as it performs close to SJF, and involves less overhead than SJF does.

2.4   Scheduling in Clusters

The increasing performance and the decreasing cost of commodity components for networking makes clusters attractive for high-performance computing. Clusters are frequently the platform of choice for scientific, commercial and industrial applications. Numerous large clusters have been deployed to support computation-intensive applications, so there is considerable interest in improving cluster scheduling and resource management. Indeed much research already exists, such as by Dandamudi [6]. He discusses the performance of a hierarchical scheduling policy for cluster systems that includes a load-balancing component to cope with the local load fluctuations. Other examples related to cluster scheduling are summarized below.

Flexible CoScheduling in a Cluster of Workstations. There is a diversity of different types of cluster computing environments, and workstation clusters are one of them. Workstation clusters provide attractive scalability in terms of computation power and memory size and they have become viable platforms for a wide range of applications. However, usually workstation cluster utilization remains low and this results in suboptimal performance. Therefore, new policies are needed to efficient allocate workstation resources to competing applications.

In [7] it is shown that it is possible to significantly increase the resource utilization in a cluster of workstations and to effectively perform system-level load balancing. A special scheduling mechanism, the Flexible CoScheduling (FCS) is designed. The aim is to alleviate gang scheduling related problems, such as fragmentation, load imbalance and hardware heterogeneity. The authors propose a classification of the processes according to their characteristics (needs and behavior), and dynamic scheduling based on this classification. Gang scheduling is used only for those jobs that need it, while processes from other jobs, which can be scheduled without any constraints are used to reduce fragmentation.

The authors have implemented FCS on top of the STORM resource management framework. They have tested FCS on a 32-node/64-processor cluster using both synthetic and real applications. The results of their research demonstrate that FCS is competitive with gang scheduling and implicit coscheduling.

Exploiting Heterogeneity for Cluster Based Parallel Multithreading. PC clusters connected by a high-speed network have been extensively used for the execution of computation intensive parallel multithreaded applications. However, the efficient scheduling in PC clusters requires two difficult research issues to be explored. The first issue is about finding a way to reduce the communication overhead of executing a multithreaded application on the cluster. The second issue is about finding a manner to exploit the heterogeneity in PC clusters, for the application. Heterogeneity characterizes almost all PC networks.

In [8], Kwok proposes to use a duplication based approach in scheduling tasks/threads to a heterogeneous cluster of PCs. According to this approach, critical tasks are redundantly scheduled to more than one machine, in order to reduce the number of inter-task communication operations. The start times of the succeeding tasks are also reduced. The task duplication process is based on the system heterogeneity where the critical tasks are scheduled or replicated in faster machines. The authors implemented their algorithm in their experimental application parallelization system for generating multithreaded parallel code executable on a cluster of Pentium PCs. The experiments use several real applications. The results presented in this paper reveal that heterogeneity of PC cluster is useful for optimizing the execution of parallel multithreaded programs.

Scheduling Parallel Jobs with CPU and I/O Resource Requirements. Most research papers found in the literature about job scheduling consider scheduling in processors only. They do not explicitly model the I/O processing, even though it can significantly influence the overall system performance. However, since I/O is also a critical resource for many jobs, the allocation of processor and I/O resources must be coordinated to offer better service to jobs and to improve system performance.

An on-line coordinated allocation of processor and I/O resources in large-scale shared heterogeneous cluster computing systems is studied in [9]. The authors present a combined CPU and I/O resource scheduling policy and study its performance under various system and workload parameters. They also compare the performance of their proposed policy with static space time sharing policy. Their objectives are to minimizing the average job completion times and maximizing the system utilization.

The proposed scheduling policy shares the system resources both spatially and temporally. The proposed strategy is based on a federation of schedulers, where each scheduler is responsible for a set of nodes in the system. It is demand driven and it combines self-scheduling and affinity-scheduling. In this way data locality is taken into account when mapping jobs to resources for balancing the load among the clusters in the system. Furthermore, another important characteristic of this approach is that the decision making process is also distributed among several schedulers which, as a consequence improves fault-tolerance and scalability of the system.

The authors used discrete event simulation to compare their scheduling policy with the static space time sharing policy. The simulation results show that the proposed policy performs significantly better than the static space time sharing policy under all system and workload parameters.

2.5 Scheduling in Grid Systems

The need for high computation power is continuously increasing. One solution to accommodate a huge demand for resources is provided by Grid computing. Grid differs from conventional distributed computing because it focuses on the large-scale sharing of Internet-connected resources.

Grid computing could evolve into the world's next major computing infrastructure. Computational Grids are comprised of a large number of various resources with different owners. They make it possible for different users to run computational intensive applications at remote sites. The sites are made up of heterogeneous resources where heterogeneity depends on the characteristics of resources.

Grids could someday coordinate millions of heterogeneous resources distributed over the world. The management and scheduling of resources in computational grids is complex, demanding specialized tools and a significant amount of effort.

Simulation Tools for Grid Systems. The efficiency of scheduling policies is crucial to Grid performance. Simulation currently is the only feasible way many large-scale distributed systems of heterogeneous resources can be evaluated. Simulation is effective when working with very large problems that involve a large number of resources and users. Simulation makes it possible to explore different types of systems operating under varying workloads.

In [10] the SimGrid framework is presented which enables the simulation of distributed applications in distributed computing environments for the specific purpose of developing and evaluating scheduling algorithms. In this paper the authors focus on SimGrid v2, which improves on the first version of the software with more realistic network models and topologies. Another feature of the second version of SimGrid is it enables the simulation of distributed scheduling agents, which is an important research topic for scheduling in large distributed environments.

GridSim [11, 12] is a toolkit that allows for the modeling and simulation of entities in parallel and distributed computing systems - users, applications, resources, and resource brokers (schedulers) for design and evaluation of scheduling algorithms. It provides a comprehensive facility for creating different classes of heterogeneous resources that can be aggregated using resource brokers to solve computation and data intensive applications. A resource can be a single processor or multi-processor with shared or distributed memory and managed by time or space-shared schedulers. The processing nodes within a resource can be heterogeneous in terms of processing capability, configuration, and availability. The resource brokers use scheduling algorithms or policies for mapping jobs to resources to optimize system or user objectives depending on their goals.

A Decoupled Scheduling Approach for Grid Environments. While many users could benefit from the extensive resources offered by Grids, application development remains a difficult problem. Each application has unique resource requirements that must be considered for deploying applications on Grid resources. In [13] an adaptive scheduling approach is proposed which is designed to improve the performance of parallel applications in Grid environments. The approach is decoupled, that is it explicitly separates general-purpose scheduling components from application-specific information and components. As part of the scheduler, the authors also developed an application-generic resource selection procedure that identifies desirable resources.

Two applications from the class of iterative, mesh-based applications have been selected for test purposes. The authors used a prototype of their approach with these applications to perform validation experiments in production Grid environments. They expect the largest performance advantage of their approach under the following conditions:

. Resource information is available.

. An application performance model and mapping strategies are already available or can be easily created.

. The shared Grid environments considered do not include dedicated machines or batch schedulers.

. This scheduling approach is most effective for applications that have moderate computation to communication ratios.

The authors note though that it is not expected that general purpose software will achieve the performance of a highly tuned application-specific scheduler. Instead, the purpose is to provide consistently improved performance relative to conventional scheduling algorithms. The results of experiments that they have conducted reveal that this scheduler succeeds in this pursuit and it gracefully handles degraded levels of availability of application and Grid resource information. They also show that their methodology involves reasonable overheads.

Scheduling on Non-Dedicated Computational Farms. Grid users are provided with a dynamically configured, parallel computing environment, but, without guarantees on contiguous resource availability. The users of a grid are either hosts or guests. A guest can use a resource if no host is using it. The performance of a grid depends on the ability of system software to exploit idle intervals on shared grid resources.

The utilization of idle memory in non-dedicated computational farms with limited memory resources is studied in [14]. The authors present an adaptive scheduling strategy that prevents thrashing, enables adaptation to memory pressure at arbitrary points of execution in a program, and takes into account the relative priorities of jobs. The thrashing prevention scheduling module is integrated with a dynamic co-scheduling module. The two modules may work in co-operating way or they may work as competitors.

To investigate the performance of the scheduling extensions, the authors use workloads consisting of mixes of parallel and sequential jobs. They use three sequential workloads, an interactive workload, an I/O-intensive workload with short-lived processes and a CPU and memory-intensive workload with long-lived processes.

The paper presents experiments in which the parallel jobs run with either of two priorities. In the first case, parallel jobs are executed with guest priorities and sequential jobs are executed with host priorities. Guest priorities are always lower than host priorities. If there is at least one host job running in the system, no guest job can use CPU time. In the second case, parallel jobs and sequential jobs compete using dynamic priorities.

The goal of the experiments is the performance of the parallel jobs on a non-dedicated cluster of privately owned workstations running general-purpose workloads. With the experiments the authors show that thrashing prevention is more important than co-scheduling, in cases of limited memory resources. The paper reveals that co-scheduling provides a limited performance improvement at low memory utilization levels, and that thrashing prevention should be enforced in all other cases due to the cost of paging.

An Approach to Assemble Grids with Equitable Resource Sharing. Most grid computing users do not have access to an efficient number of shared grid resources because most times gaining access to resources requires negotiations with the resources owners. To address this problem, Andrade et al. ([15]) are developing the OurGrid resources sharing system, a peer-to-peer network of sites that share resources "equitably" in order to form a grid to which they all have access. The resources are shared accordingly to a "network of favors" model in which a site donates its idle resources as a favor, expecting to be prioritized when it asks for favors from the community. This has a result that peers that contribute more to the community are prioritized when they request resources. This design aims to provide this prioritization in a completely decentralized manner.

The type of applications for which OurGrid intends to provide resources are those parallel applications whose tasks are loosely coupled known as bag-of-tasks (BoT) applications. BoT are those parallel applications composed of a set of independent tasks that need no communication among them during execution.

For the environment in which the system will operate the authors assume that: (i) there are at least two peers in the system willing to share their resources in order to obtain access to more resources and (ii) the applications that will be executed using OurGrid need no quality of service guarantees. With these assumptions, they aim to build a resource sharing network that promotes equity in the resources sharing. By equity they mean that those participants in the network which have donated more resources are prioritized when they ask for resources.

Simulation results of this approach reveal that it is a promising design. The authors expect to evolve the present approach into a solution for the real grid user requirements.

Parallel Job Scheduling in a Heterogeneous Multi-Site Environment. Different sites in computational grids have generally different configurations of their processors and have generally different performance characteristics. Much of the research on job scheduling for heterogeneous systems considers the scheduling of independent sequential jobs or precedence constrained task graphs where each task is sequential. The scheduling of parallel jobs in a heterogeneous multi-site environment is studied in [16]. The authors consider a system where each site hosts a homogeneous cluster of processors, but processors at different sites have different speeds. They start this research first with a simple greedy scheduling strategy, and they propose and evaluate several enhancements using trace driven simulations. They consider the use of multiple simultaneous reservations at different sites, use of relative job efficacy as a queuing priority, and compare the use of conservative versus aggressive backfilling. Contrary to the single-site case, their simulation results reveal that for the heterogeneous multi-site environment conservative backfilling performs consistently better than aggressive backfilling.

2.6 Web Server Scheduling

There is a demand for high-performance web servers that can provide fast response and high throughput to client requests. This section investigates various methods and techniques designed to enhance distributed web server performance and provide quality of service to millions of web clients.

Size-based Scheduling for Web Performance Improvement. In [17] Harchol-Balter et al. investigate the possibility of reducing the expected response time of every request at a web server, by changing the order in which the requests are scheduled. The authors propose a method for performance improvement of web servers that service static HTTP requests. Their idea is to give preference to requests for small files or those requests with short remaining file size, in accordance with the well-known scheduling algorithm preemptive Shortest-Remaining-Processing-Time-first (SRPT). As far as it concerns the limitation that SRPT requires a-priori knowledge of the time to service the request, their experiments show that the time to service a request is well-approximated by the size of the file requested, which is well-known to the server. Also, a primary goal of the authors is to investigate whether SRPT "starves" requests for large files or does not in the case of web servers serving typical web workloads.

Their implementation is at the kernel level and involves controlling the order in which socket buffers are channeled into the network. They execute the experiments in a LAN and a WAN environment. They use the Linux operating system and the Apache and Flash web servers. They are concerned with response time which is defined to be the time from when the client sends out the SYN-packet requesting to open a connection until the client receives the last byte of the file requested.

The results presented in the paper indicate that SRPT-based scheduling of connections yields significant reductions in delay at the web server. These result in a significant reduction in mean response time and mean slowdown for both the LAN and WAN environments. Furthermore, the authors show that their method hardly penalizes or not at all penalizes requests for large files.

Quantifying the Properties of SRPT Scheduling. In [18] a probe-based sampling approach is used to study the behavioral properties of Web server scheduling strategies, such as Processor Sharing (PS) and Shortest Remaining Processing Time

(SRPT). The approach can be used to estimate the mean and variance of the job response time, for any arrival process, service time distribution, and Web scheduling policy.

The authors apply the approach to trace-driven simulation of Web server scheduling to compare and contrast the PS and SRPT scheduling policies. They identify two types of unfairness in a Web server scheduling system: endogenous unfairness that a job can suffer because of its own size, and exogenous unfairness that a job can suffer because of the state of the Web server at the time it arrives. The authors quantify each, mainly examining the mean and variance of slowdown, conditioned on job size, for different system loads.

The approach of this paper is used to show the asymptotic convergence of slowdown for the largest jobs, providing confirmation of previous theoretical results obtained by Harchol-Balter et al. [19]. Typical performance results are shown for a practical range of job sizes from an empirical Web server workload. The range of job sizes for which the "crossover effect" is evident is identified, where "crossover effect" is defined in this paper as the region where SRPT provides worse performance than PS.

The authors conclude that their results are encouraging for the SRPT deployment in Internet Web servers.

Load Balancing Using Mobile Agents. Load balancing is an important technique to improve the performance of distributed web servers. Arriving client requests should be evenly distributed among the web servers to achieve fast service. The jobs on an overloaded server should be transferred to an underloaded server to improve overall system performance.

In [20], a framework called Mobile Agent based LoaD balancing (MALD) is proposed that uses mobile agents technology to implement scalable load balancing on distributed web servers. The web servers can dispatch mobile agents to collect information about the load of the system and perform load redistribution on all servers.

The MALD framework provides a basis to implement different flexible load-balancing schemes for scalable distributed web server systems. Two load-balancing schemes are developed based on this framework, respectively, intended for the web servers in cluster and in WAN. The performance evaluations that have been conducted in this paper demonstrate that the two schemes can outperform the load-balancing approaches based on message-passing paradigm in the cases of large num- bers of servers and client requests.

2.7 Scheduling in Networks

Scheduling on Shared Networks. Shared networks, varying from workstation clusters to computational grids constitute a very important platform for high performance computing. In these distributed computing environments, applications performance strongly depends on the dynamically changing availability of resources. Therefore it is important to be able to predict the performance of an application under given resource conditions.

[21] studies automatic development of application performance models that can estimate application execution behavior under different network conditions. The authors develop a framework to model the performance of parallel applications executing in a shared network computing environment. The actual performance is predicted for the case of sharing of a single computation node or network link, while performance bounds are developed for the case of sharing of multiple nodes and links. The methodology that the authors employ is based on monitoring an application's execution behavior and resource usage under controlled dedicated execution.

The procedure can be applied to applications developed with any programming model since it is based on system level measurements and does not require access to the source code. This paper shows that detailed measurement of the resources that an application needs and uses can be used to build an accurate model for the performance prediction of the same application under different network conditions.

Agent Based Routing in Telecommunication Networks. Agent-based routing in telecommunication networks appears as an appealing research topic. The metaphor of the behavior of social insects and their self-organizing capabilities, called ant-based routing technique is used in order to create an efficient load balancing and co-scheduling for large volume of calls.

In nature ants can find the shortest path between two nodes (food-source, nest-destination) by exploiting pheromone information onto ground. Ants simply deposit pheromone on the route while walking, which influence the rest of them to follow the same route. In Fig. 6 there is an illustration of this principle. Two ants are leaving from their source node the same time by following different paths, which are laid, with pheromone trails. The shortest route will influence more ants afterwards because it contains a double quantity of pheromone (the quantity of pheromone while spread on the shortest route is higher in density). In such a way a distributed load balancing is achieved and the positive feedback mechanism that forms a continuous circle is enforced, so the shortest path is strongly marked. In Fig. 7 the positive feedback mechanism is shown for the reinforcement of path.

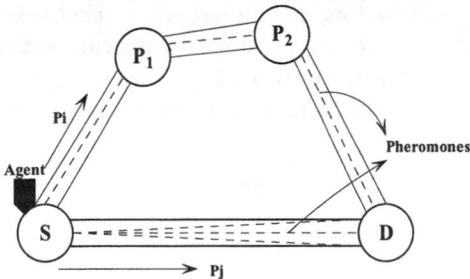

Fig. 6. Migration strategy using edge pheromones-trail laying principle. Pi and Pj are the corresponding edge routing probabilities

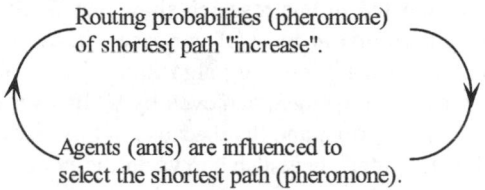

Fig. 7. Positive feedback mechanism for reinforcement of the agents

In a network using a typical decentralized-distributed routing model, each node maintains a routing table indicating where the call has to go in order to reach the final destination. Artificial ants adjust the table entries continually affecting the current network state. Thus, routing tables are represented with pheromone tables having the likelihood of each path to be followed by the artificial ant.

In general ant colonies are constructive metaheuristics algorithms that share specific behaviours of ants for solving congestion problems. There are many variants of ant algorithms but most of them are mainly based on the amount of pheromone is spread while ant is moving.

An agent-based routing control system and its behavior while network topology changes to an extended version, is being studied in [22]. In real time networks, for various reasons some of the nodes collapse or are taken off-line for maintenance. This has as a consequence a significant reduction in performance. Inspired from this, the authors examine the behavior of an agent-based routing which is applied to a realistic Synchronous Digital Hierarchy complex topology network that tends to increase-or decrease- the number of nodes randomly/under conditions or even activate certain nodes that were inactive. Algorithmically complex problems like routing in telecommunication networks, needs a dynamic adaptive approach. Agent-based routing method continually modifies the routing tables in response to congestion and overload in a network, to two different network topologies, measuring performance for the assumptions made. This decentralized routing scheme is shown not be substantially affected from any topological changes, when subjected to a certain frequency limit in variations. Through various newly introduced metrics and simulation of the proposed scenarios, the paper demonstrates that agent-based routing offers a decentralized routing control in the network and an efficient way to stabilize responsiveness and improve overall performance.

2.8 Other Scheduling Issues in Distributed Systems

Scheduling Divisible Workloads. [23] presents an algorithm named UMR (Uniform Multi-Round) for minimizing the makespan of parallel divisible workload applications on homogeneous and heterogeneous distributed computing platforms with a master/worker strategy. UMR dispatches work in multiple rounds, which makes it possible to overlap communication and computation. The authors use "uniform rounds": during each round a fixed amount of work is sent to each worker. This makes it possible to compute an optimal number of rounds.

The experiments conducted in this research show that UMR leads to better schedules than competing algorithms in most of the cases. In those cases where UMR is outperformed, it is close to the competing algorithms. The main contribution of this paper is this performance improvement achieved by UMR over previous work in spite of the "uniform" round restriction, and that because of this restriction an optimal number of rounds can be computed. It is also shown in the paper that UMR is robust to platform heterogeneity.

Energy Aware Scheduling for Real-Time Systems. Energy aware computing has become popular in mobile computing systems and in large systems consisting of multiple servers. In [24] static and dynamic power management schemes are proposed for a set of real-time tasks with precedence constraints which execute on a distributed system. First, static power management for parallelism allocates global static slack, which is defined as the difference between the length of the static schedule and the deadline, to different sections of the schedule based on their degree of parallelism. Second, when pre-emption is considered, the gap-filling technique enhances the greedy algorithm by allowing out-of-order execution. This is the case where if there is some slack and the next expected task is not ready, the processor will run the future ready tasks which are assigned to it. The execution of an out-of-order task will be preempted by the next expected task when it receives all its data and is ready.

By comparing their static technique with the simple static power management, which distributes global static slack proportionally to the length of the schedule, the authors find that their static scheme can save an average of 10% more energy. Also they find that their schemes significantly improve energy savings when are combined with dynamic schemes.

Integrated Scheduling. In [25] a paradigm for parallel job scheduling is introduced called integrated scheduling or iScheduling. In contrast to a general-purpose system scheduler the iScheduler is an application-aware job scheduler. It dynamically controls resource allocation among a set of competing applications and it can interact directly with an application during execution in order to optimize resource allocation. As shown in the paper, an iScheduler may add or remove resources from a running application in order to improve the performance of other applications. The paper also shows that both individual applications and the system as a whole can benefit from this type of resource management.

The authors propose a framework for building iSchedulers and evaluate the potential of the iScheduler paradigm on several workload traces obtained both from supercomputer centers and from a set of real parallel jobs. Their results indicate that integrated scheduling can significantly reduce both waiting time and the end-to-end finishing time for these workload classes, outperforming other standard scheduling strategies.

3 Summary

Distributed systems pose challenging problems and require efficient methods to evaluate the performance of scheduling algorithms. This paper summarizes several recent contributions to distributed system scheduling. The objective is to focus on important research topics that have improved various aspects of distributed systems scheduling. Hopefully, material in this paper will stimulate research that will further improve the performance of distributed systems.

References

1. Karatza, H.: Simulation of Parallel and Distributed Systems Scheduling, Concepts, Issues and Approaches. In Papadimitriou, G., Obaidat M. (eds.): Applied System Simulation: Methodologies and Applications, Kluwer Academic Publisher, New York (2003) 61-80

2. Harchol-Balter, M., Li, C., Osogami, T., Scheller-Wolf, A., Squillante, M.: Analysis of Task Assignment with Cycle Stealing under Central Queue, 23rd International Conference on Distributed Computing Systems, May 19 - 22, Providence, Rhode Island, IEEE Computer Society, Los Alamitos, CA (2003) 628-637

3. Karatza, H.: Simulation Study of Multitasking in Distributed Server Systems with Variable Workload. Simulation Modelling Practice and Theory Journal, Elsevier, Amsterdam (to appear).

4. Karatza, H.D., Hilzer, R.C.: Parallel Job Scheduling in Distributed Systems. Simulation: Transactions of the Society for Modeling and Simulation International, Sage Publications, Thousand Oaks, CA, 79 (2003) 287-298.

5. Karatza, H.D.: A Comparison of Load Sharing and Job Scheduling in a Network of Workstations. International Journal of Simulation: Systems, Science Technology, UK Simulation Society, Nottingham, UK, 4: 3&4 (2003) 4-11.

6. Dandamudi, S.: Hierarchical Scheduling in Parallel and Cluster Systems. 1rst edn. Kluwer Academic/Plenum Publishers, New York (2003)

7. Frachtenberg, E., Feitelson, D.G., Petrini F., Fernandez, J.: Flexible CoScheduling: Mitigating Load Imbalance and Improving Utilization of Heterogeneous Resources. Proceedings of the International Parallel and Distributed Processing Symposium (IPDPS'03), April 22 – 26, Nice, France, IEEE Computer Society, Los Alamitos, CA (2003) 85-94

8. Kwok, Y.-K.: On Exploiting Heterogeneity for Cluster Based Parallel Multithreading Using Task Duplication. The Journal of Supercomputing, Kluwer Academic Publishers, The Netherlands, Amsterdam, 25 (2003) 63–72

9. Abawajy, J.H., Dandamudi, S.P.: Scheduling Parallel Jobs with CPU and I/O Resource Requirements in Cluster Computing Systems. Proceedings of the 11th IEEE/ACM International Symposium on Modeling, Analysis and Simulation of Computer and Telecommunications Systems (Mascots'03), 12-14 October 12-14, Orlando, Florida, USA, IEEE Computer Society, Los Alamitos, CA (2003) 336-343

10. Legrand, A., Marchal, L. Casanova, H.: Scheduling Distributed Applications: the SimGrid Simulation Framework. Proceedings of the 3rd IEEE/ACM International Symposium on Cluster Computing and the Grid (CCGRID'03), May 12-15, Tokyo, Japan, IEEE Computer Society, Los Alamitos, CA (2003) 145-152

11. Buyya, R., Murshed, M.: GridSim: A Toolkit for the Modeling and Simulation of Distributed Resource Management and Scheduling for Grid Computing. The Journal of Concurrency and Computation: Practice and Experience (CCPE), Wiley Press, USA, 14 (2002) 1175-1220

12. Murshed, M., Buyya, R.: Using the GridSim Toolkit for Enabling Grid Computing Education. Proceedings of the International Conference on Communication Networks and Distributed Systems Modeling and Simulation (CNDS'02), January 27-31, San Antonio, Texas, SCS, San Diego, CA (2002) 18-24

13. Dail, H., Berman, F., Casanova, H.: A Decoupled Scheduling Approach for Grid Application Development Environments. J. Parallel Distrib. Comput., Academic Press, Amsterdam, The Netherlands, 63 (2003) 505–524

14. Nikolopoulos, D.S., Polychronopoulos, C.D.: Adaptive Scheduling Under Memory Constraints on Non-Dedicated Computational Farms. Future Generation Computer Systems, Elsevier, Amsterdam, 19 (2003) 505–519

15. Andrade, N., Cirne, W., Brasileiro, F., Roisenberg, P.: OurGrid: An Approach to Easily Assemble Grids with Equitable Resource Sharing. In: Feitelson, D., Rudolph, L., Schwiegelshohn, W. (eds.): Job Scheduling Strategies for Parallel Processing. Lecture Notes in Computer Science, Vol. 2862. Springer-Verlag, Berlin Heidelberg (2003) 61–86.

16. Sabin, G., Kettimuthu, R., Rajan, A., Sadayappan, P.: Scheduling of Parallel Jobs in a Heterogeneous Multi-Site Environment. In: Feitelson, D., Rudolph, L., Schwiegelshohn, W. (eds.): Job Scheduling Strategies for Parallel Processing. Lecture Notes in Computer Science, Vol. 2862. Springer-Verlag, Berlin Heidelberg (2003) 87-104

17. Harchol-Balter, M., Schroeder, B., Bansal, N., Agrawal, M.: Size-based Scheduling to Improve Web Performance. ACM Transactions on Computer Systems, Association for Computing Machinery, New York, N.Y., 21:2 (2003) 1-27

18. Gong, M., Williamson, C.: Quantifying the Properties of SRPT Scheduling. Proceedings of the 11th IEEE/ACM International Symposium on Modeling, Analysis and Simulation of Computer and Telecommunications Systems (Mascots'03), 12-14 October 12-14, Orlando, Florida, USA, IEEE Computer Society, Los Alamitos, CA (2003) 126-135

19. Harchol-Balter, M., Sigman, K., Wierman, A.: Asymptotic Convergence of Scheduling Policies with Respect to Slowdown. Proceedings of IFIP Performance 2002, September 22-27, Rome, Italy, Performance Evaluation, Elsevier B.V., Amsterdam, The Netherlands, 49 (2002) 241-256

20. Cao, J., Sun, Y., Wang, X., Das, S.: Scalable Load Balancing on Distributed Web Servers Using Mobile Agents. J. Parallel Distrib. Comput., Academic Press, Amsterdam, The Netherlands, 63 (2003) 996–1005

21. Subhlok, J., Venkataramaiah, S.: Performance Estimation for Scheduling on Shared Networks. In: Feitelson, D., Rudolph, L., Schwiegelshohn, W. (eds.): Job Scheduling Strategies for Parallel Processing. Lecture Notes in Computer Science, Vol. 2862. Springer-Verlag, Berlin Heidelberg (2003) 148–165.

22. Mavromoustakis, C.X., Karatza, H.D.: On the Extensibility Properties and Performance Measures of Circuit Switched Telecommunication Networks, using Agent-based Distributed Routing Algorithm. Proceedings of 2003 International Symposium on Performance Evaluation of Computer and Telecommunication Systems (SPECTS), July 20-24, Montreal, Canada, SCS, San Diego, CA (2003) 240-247

23. Yang, Y., Casanova, H.: UMR: A Multi-Round Algorithm for Scheduling Divisible Workloads. Proceedings of the International Parallel and Distributed Processing Symposium (IPDPS'03), April 22 – 26, Nice, France, IEEE Computer Society, Los Alamitos, CA (2003) 24-32

24. Mishra, R., Rastogi, N., Zhu, D., Mosse, D., Melhem, R.: Energy Aware Scheduling for Distributed Real-Time Systems. Proceedings of the International Parallel and Distributed Processing Symposium (IPDPS'03), April 22 – 26, Nice, France, IEEE Computer Society, Los Alamitos, CA (2003) 21-29

25. Weissman, J.B., Abburi, L.R., England, D.: Integrated Scheduling: the Best of Both Worlds, J. Parallel Distrib. Comput., Elsevier Science, New York, USA, 63 (2003) 649–668

Data Allocation and Scheduling in Disks and Disk Arrays

Alexander Thomasian*

Computer Science Department
New Jersey Institute of Technology
Newark, NJ 07102, USA
athomas@cs.njit.edu

1   Introduction

Magnetic disks, which together with disk arrays constitute a multibillion dollar industry, were developed in 1950s. Disks were an advance over magnetic drums, which had a dedicated read/write head per track, since much higher amounts of data could be accessed in a cost effective manner due to the sharability of the movable read/write heads. DRAM memories, which are *volatile*, were projected to replace disks a decade ago (see Section 2.4 in [33]). This did not materialize due to the inherent volatility of DRAM, i.e., a power source is required to ensure that DRAM contents are not lost, but also due to recent dramatic increases in areal recording density and hence disk capacity, which is estimated at 60% *compound annual growth rate - CAGR*. This has resulted in a rapid decrease in cost per megabyte of disk capacity, so that it is lower than DRAM by a factor of 1000 to one.

After a brief introduction to disk organization in Section 2, in Section 3 we review disk scheduling policies, since they have a major effect on disk performance. In Section 4 we discuss several techniques to improve disk performance, e.g., by reorganizing disk data.

What makes an array of disks more interesting than a *just a bunch of disks - JBOD* is that the array can provide a capability for fault tolerance and parallel access and load balancing by striping. *Redundant array of independent disks - RAID* paradigm defined in [86] provides these capabilities and defines five RAID levels. In Section 5 we discuss general concepts associated with RAID. In Sections 6, 7, and 8 we discuss the three most important RAID levels: RAID1 or mirrored disks, RAID5 or rotated parity disk arrays, and RAID6 and more generally *two disk failure tolerant - 2DFT* arrays, which include the EVENODD [12] and RM2 [87] schemes.

In Section 9 we review analytic and simulation studies evaluating the performance of disks and disk arrays. Conclusions, in Section 10, briefly discusses some emerging areas.

* We acknowledge the support of NSF through Grant 0105485 in Computer System Architecture

M.C. Calzarossa and E. Gelenbe (Eds.): MASCOTS 2003, LNCS 2965, pp. 357–384, 2004.
© Springer-Verlag Berlin Heidelberg 2004

2   Magnetic Disks

A disk drive consists of multiple disks or *disk platters*, which are mounted on a *spindle*. More platters (within limits of disk drive height) can be added to the spindle to increase the capacity of the disk drive, but this requires more power to rotate the disks. Disks are covered with magnetic material and data is recorded on each disk surface in concentric circles called *disk tracks*.

The data bits recorded on tracks can be read/written only when they are placed underneath a *read/write - R/W* head. Attached to a *disk arm* there are as many R/W heads as active disk surfaces. Only tracks constituting a hypothetical *disk cylinder* are accessible to the heads. Other cylinders can be accessed by rotating the arm to position the R/W heads. *Head switching* selects a single head for activation. Disks are rotated at a *constant angular velocity*, specified as *rotations per minute - RPM*, to make the data recorded on a track accessible to the R/W heads. Disk RPMs have been increasing over the years, e.g., 15,000 RPM for Maxtor's ATLAS 15K.

The time to select the R/W head by head switching is called *head selection time*. The time to move the R/W heads to the desired track, i.e., the *seek time* $t(d)$, is a function of the seek distance expressed in the number of disk tracks being bypassed. The *seek time characteristic* $t(d), 1 \leq d \leq C - 1$, where C is the number of disk cylinders, can be approximated as a line with a higher slope for short distance seeks and a line with a lower slope for long seeks.

The increase in disk recording density is a result of more *tracks per inch - TPI* and more *bits per inch - BPI*, i.e., increased linear recording density. The former allows more tracks per disk and a reduction in actual seek distances, but there is an associated increase in *head settling time* following seeks [83]. Bits on a track are grouped into 512 byte *sectors* and there are more sectors on outer tracks than inner tracks, e.g., 455 versus 650 sectors in Maxtor's ATLAS 15K. This is because *Zoned Bit Recording - ZBR* maintains about the same BPI at all tracks.

The increased BPI and RPMs have resulted in an increased *data transfer rate*, which is higher for outer tracks than inner tracks in disks with zoning. The time to transfer a file varies based on its placement and it is advantageous to place frequently accessed files on outer disk cylinders. Disk zoning complicates the assignment of files for guaranteed throughput [76].

To ensure uninterrupted reading of files extending across tracks, *track skew* and *cylinder skew* are introduced to mask the head switching delay and single cylinder seeks, respectively. *Zero latency* capability allows a block to be read starting at any sector boundary, rather than just the first.

Disk drives are equipped with an *onboard disk cache* or *track buffer* to deal with the mismatch in the bandwidth of the disk, which varies from zone to zone, and the I/O bus. The onboard cache is also used in prefetching, e.g., the remainder of a data track is prefetched after the reading of a block.

The reader is referred to [96], [83] for a more detailed discussion of modern disk drives. The validation of a detailed disk simulator is also described in [96].

3   Disk Arm Scheduling

FCFS scheduling of disk requests provides a rather poor performance in the case of accesses to small randomly placed data blocks, since the ratio of transfer time to service time is orders of magnitude smaller than the effective disk transfer rate. It has been argued that disk queue-lengths are short, so that little is to be gained by more sophisticated disk scheduling policies, but evidence to the contrary is given in [53].

Disk transfer efficiency can be improved by minimizing positioning time. The high seek time in early disk drives served as the motivation for the *shortest seek time first - SSTF* and the *SCAN* policies, which are aimed at reducing the seek time [28]. The $V(R)$ disk scheduling algorithm ranges from V(0)=SSTF to V(1)=SCAN, so that it provides a continuum of algorithms [31]. To bias the arm to continue its movement in the direction of the current scan, $d_{bias} = R \times C$, (C is the number of disk cylinders), is subtracted from the seek distance in that direction. Simulation results show that for lower arrival rates SSTF is best, i.e., $R = 0$, while at higher arrival rates $R = 0.2$ provides the best performance.

The *shortest access* (resp. *positioning time first - SATF* (resp. *SPTF*) policy processes a request next which will minimize the access (resp. positioning) time, respectively [53]. When the sizes of transferred blocks are small or requests have the same size then SATF and SPTF policies yield the same scheduling decisions. We therefore use SATF and SPTF interchangeably. Several simulation experiments have shown that SPTF outperforms FCFS, SSTF, and SCAN in minimizing the mean response time [142], [121], [122]. In what follows we discuss some variations of SPTF.

The SATF policy is extended in [121], [122] to allow one class of requests (say reads) to be processed at a higher priority than another class (say writes). According to the *conditional priority - CP-SATF* policy, SATF winner read requests are processed unconditionally, while an SATF winner write request is processed only if the following relationship holds among service times $x_{write}^{best}/x_{read}^{best} < t$, $0 \leq t \leq 1$. Simulation results show that $t \approx 0.7$ improves the response time of high priority requests, at the cost of a small reduction in overall throughput.

SATF performance can be improved by applying lookahead [121], [122]. Consider the scheduling of the next request after a given request is completed. Let A be the SATF winner request, whose processing will be followed by request X, if there are no new arrivals which are even better. Let B be another pending request such that $T_B > T_A$, but which can be followed by requests Y such that $T_B + \alpha T_Y < T_A + \alpha T_X$. Since X or Y may not be processed at all, their service times are discounted with a factor $0 \leq \alpha \leq 1$. An SATF policy looking ahead two requests is denoted SATF-LA2 and its cost is $O(n^2)$ versus $O(n)$ for SATF.

Starvation in SATF can be dealt with by artificially reducing the positioning time of a request according to its waiting time w. For example, an adjusted positioning time $T_{pos}^{adjusted} = T_{pos}(1 - w/W_M)$, where W_M is the "maximum" waiting time is utilized in [102]. The difficulty in implementing this method is selecting the value of W_M.

A sliding window which limits the number of requests considered for SPTF scheduling is considered in [53]. Four scheduling algorithms are considered in [53]: (i) Short sighted *optimal access time algorithm - OAT*, which finds the optimal allocation for a fixed number of requests. (ii) Sliding window OAT, which replaces the completed request with a new request. (iii) Short-sighted greedy that considers the best among n requests. (iv) Fully greedy which considers all of the requests in the queue. The relative performance of the four methods are 1.84, 1.75, 1.73, and 1.00, which shows the benefit of considering all of the requests in the queue. Further investigation of the effect of the number of requests considered for scheduling is reported in [93].

The window-based SPTF or WB-SPTF policy is proposed in [92], so that only requests within a time window (0,T) of the arrival time of the oldest request are considered for scheduling. The goal is to improve the standard deviation of response time, hopefully at the cost of a small sacrifice in its mean value. T is adjusted according to the load. Trace-driven simulations show that WB-SPTF outperforms SPTF in overload mode.

A mixed workload consists of requests to continuous and discrete objects. Continuous data requests have an implicit deadline associated with the delivery of the next data block, e.g., to ensure glitch-free viewing of streaming (rather than download) video. In addition to requests to continuous objects (C-requests), media servers also provide access to discrete objects (D-requests). C-requests have to meet performance requirements, while discrete requests should be served with as small a mean response time as possible [9]. One scheduling method serves C-requests according to SCAN and intervening D-requests according to either SPTF or OPT(N), where the latter determines the optimal schedule after enumerating all $N!$ schedules.

The *FAir MIxed-scan ScHeduling - FAMISH method* ensures that all C-requests are served in the current round and that D-requests are served in FCFS order. More specifically this method constructs the SCAN schedule for C-requests, but also incorporates the D-requests in FCFS order in the proper position in the SCAN queue, up to the point where no C-request misses its deadline. D-requests can also be selected according to SPTF. The relative performance of various methods is studied in [9].

There are two categories of timing guarantees or *service levels* for I/O requests: *deterministic* and *statistical*. The latter is meaningful for *variable-bit-rate - VBR* (as opposed to *constant-bit-rate - CBR* streams), since the required I/O bandwidth for VBR varies and provisioning for worst case behavior will degrade performance inordinately. Scheduling with *quality-of-service - QOS* guarantees has been investigated in greater depth in networking than in disk subsystems. According to [136] this requires: (i) controlling and limiting the allocation of resources for different levels; (ii) scheduling of resources to meet performance goals; (iii) determining system design and its parameters; (iv) investigating the tradeoffs between statistical and deterministic guarantees on throughput.

A two level queueing model first proposed in [103] and adopted in [136] for supporting multiple QOS levels, where the first level queue acts as a scheduler, as well as an admission controller to a pool from which requests are scheduled on disks. Three request categories are considered: (i) *periodic requests*, e.g., CBR or VBR videos, which require service at regular intervals of time (deterministic or statistical guarantee); (ii) *interactive requests*, e.g., playing a video game, which requires a best effort mean response time; (iii) *aperiodic requests*. e.g., file transfers, which need a guaranteed minimum level of service or bandwidth.

Anticipatory disk scheduling according to [52] defers the processing of a new request until it has ascertained that there are no further pending requests in the area of disk at which the R/W head resides. This is a demonstration of a non-work-conserving scheduling policy [57], improving performance in some cases.

4 Single Disk Performance Improvement

The following methods to improve disk performance are discussed in this section: (1) disk reorganization and defragmentation; (2) log-structured file systems; (3) active disks and free-block disk scheduling.

File systems attempt to place files on consecutive disk sectors, but this is not always possible due to dynamic file allocation and deletion. Disk defragmentation reorganizes data by placing the "fragments" of a file into one contiguous region, so that the sequential reading of a file can be accomplished with one seek.

One method to improve the disk access efficiency, by minimizing the average seek distance in accessing the set of files on a disk drive is the *organ pipe arrangement* [141], which places the most popular files on the alternate sides of the middle disk cylinder in decreasing access order at points $0, \pm1, \pm2, \ldots$.

Monitored file access frequencies, which vary over time, can be used to adaptively reorganize the files to approximate an organ pipe organization [1]. The average seek distance can be reduced further by allocating disk files by taking into account the frequency with which files are accessed together. Clustering algorithms utilizing the file access pattern have been proposed for this purpose, see e.g., [10].

The *log-structured file systems - LFS* paradigm is useful in an environment where writes dominate reads, e.g., a system with a large cache, where most read requests are satisfied by cache accesses and the disks are mainly used for writing (there is usually a timeout for modified data in the cache to be destaged onto disk). LFS accumulates modified files in a large cache, which is destaged to disk in large segments, e.g., the size of a cylinder, to prorate arm positioning overhead [94]. Space previously allocated to altered files is designated as free space and a background garbage collection process is used to consolidate half-empty segments to create almost full and absolutely empty segments for future destaging.

The write cost in LFS is a (steeply) increasing function of the utilization of disk capacity (u), which is the fraction of live data in segments (see Figure 3 in [94]) and a crossover point with respect to improved UNIX's FFS occurs at $u = 0.5$. The following segment cleaning policies are investigated: (i) when should the segment cleaner be executed? (ii) how many segments should it clean? (iii) the choice of the segment to be cleaned, (iv) the grouping of live blocks into segments. The cost benefit in cleaning is defined as ratio of I/O transfer time and space freed.

Given the high value of disk arm utilization and the fact that each disk is equipped with a relatively powerful microprocessor with local memory, there have been recent proposals for *freeblock scheduling*, which utilizes opportunistic disk accesses as the arm moves in processing regular requests and of course when the disk is idle [66].

5   Raid Concepts

The *redundant array of inexpensive/independent) disks - RAID* paradigm and five RAID levels were introduced in [86]. RAID0 and RAID6 were added later [18]. After introducing the RAID paradigm, we discuss two RAID features: *data striping* and *fault-tolerance*. RAID1 or disk mirroring, RAID5 or rotated parity disk arrays, and RAID6 and related schemes tolerating two disk failures are discussed in the following three sections.

5.1   Motivation for RAID

The motivation for RAID was to cut costs by replacing *SLEDs - single large expensive disks* with an array of inexpensive commodity hard disk drives. There was an incompatibility issue because SLEDs used in conjunction with mainframe computers recorded data according to application specific variable length blocks (multiples of logical record lengths for more efficient storage), while *Fixed Block Architecture - FBA* disks record data as 512 byte blocks or sectors. The translation from the I/O commands generated by OS/390 or z/OS to the FBA disks is carried out by the disk array controller [73].

5.2   Striping

Data access skew or disk load imbalance is a consequence of careless data allocation, such that a few disks have a high utilization, while most other disks are almost idle. Even initially desirable allocations may turn into undesirable allocations due to cyclical variations in the load. The high utilization of the busy disks results in high response times. Tools taking into account file access frequencies and the characteristics of heterogeneous disks and the bandwidth of the associated I/O paths have been developed to determine an initial "optimal" allocation, but more importantly to improve the allocation based on access frequencies observed in an operational system. Local perturbations to the current allocation are then usually sufficient [140].

Striping, which is considered to be a cure to the access skew problem, partitioned large files into *striping units*, which are allocated in a round-robin manner at all disks. The stripe units in a row constitute a *stripe*. The intermixing of stripe units from multiple files on a disk results in a balanced disk load. A counter argument to striping is that without striping the user has direct control over data allocation, e.g., to ensure that two hot files are not placed on the same disk, but this is not guaranteed with striping. Parity striping maintains a regular data layout, but allows enough space on each disk for parity protection [37].

Randomized data allocation is an alternative to striping, which relies on a precomputed distribution [98]. A comparison of this method with striping in a multimedia storage server environment is provided.

The sizes of stripe units should be large enough to satisfy the most commonly occurring accesses, otherwise multiple disks need to be accessed to satisfy one request. There are the following tradeoffs: (i) In a lightly loaded system the parallelism afforded by small stripe unit sizes may be beneficial in reducing response time. This is especially so for accesses to larger blocks, so that the long transfer time will dominate the synchronization delay. (ii) In a highly loaded system the positioning time overhead may result in an increase in the mean overall response time, so that a large stripe unit size is preferable. (iii) Larger stripe units allow sequential accesses to large files, but disallow parallel accesses.

The optimal stripe unit size in kylobytes (KB) is estimated in [17] as the product of a factor (equal to $1/4$), the average positioning time (in milliseconds), data transfer rate (in MB per second), and the degree of concurrency minus one, plus 0.5 KB.

An approximate analytic model for choosing a global near-optimal striping unit is presented in [106]. The data allocation method is based on the concept of heat and temperature, where *heat* is the number of accesses to an extent (a large chunk of data) per unit time. *Temperature* is the ratio between heat and extent size [26]. Using queueing terminology temperature is the arrival rate of requests to a segment and heat is the product of the arrival rate and the time to transfer the file (proportional to file size), which is in fact the disk utilization ensuing from the file access.

A greedy algorithm can assign files to the least utilized device. A *perfectly balanced file assignment - PBFA* is defined as an assignment where all disk utilizations are approximately equal. The Sort Partition (SP) algorithm proposed in [62] sorts the files in decreasing order of their (relative) utilizations and assigns the files in that order over the disks. In fact disks are selected randomly and files with the same size and transfer times are assigned to the same disk. The intuition behind this allocation is that assigning files with equal or relatively close service times to disks results in a reduction of the second moment of service time and hence mean waiting time. SP achieves a good approximation to PBFA when the number of files is large. It can be shown that SP finds the allocation with minimum response time.

A hybrid partition algorithm is also proposed which deals with adding new files to an existing allocation. There are the conflicting goals of clustering same size files and balancing disk utilizations. A heuristic to deal with the problem is proposed in this paper.

A dynamic load balancing step called *disk cooling* uses a greedy policy to determine the best candidate extent to be moved to attain the highest gain while minimizing the required data movement [106].

5.3   RAID Fault-Tolerance and Classification

Fault-tolerance for early RAID designs was required due to the lower reliability of commodity disks replacing SLEDs and that there was an order of magnitude increase in the number of disks (to maintain the same overall capacity) [86], [33]. A single parity disk is added in RAID3 and RAID4 and it provides error correction capability, since the identity of the failed disk is known. RAID3 and RAID4 are different in that RAID3 is used for parallel processing, so that all disk drives are accessed concurrently to process a single request, while RAID4 utilizes a larger stripe unit than RAID3 and the disks may be accessed independently.

RAID5 alleviates a weakness of RAID4, which is due to the fact that the parity disk is a potential bottleneck for write-intensive applications. This is because updating data on any disk requires the updating of the corresponding block on the parity disk. RAID5 rotates the parity at the stripe unit level, so that all disks have the same number of parity blocks [33], [18]. RAID1 or disk mirroring or shadowing is a degenerate case of RAID4 with one disk protecting the other.

The practice of using parity to protect against single disk failures continues, although the previous dichotomy of SLEDs versus inexpensive commodity disks has disappeared, since all disks have the same form factor and capacity. The advantage of parity protection is as follows: (i) there is no data loss when a disk failure occurs and the system continues its operation by recreating requested blocks on demand, although performance is degraded, (ii) the system can automatically rebuild the contents of the failed disk on a spare disk. A *hot spare* is required, since most systems do not allow *hot swapping*, replacing a broken disk with a spare, while the system is running.

6   RAID1 or Mirrored Disks

We first categorize mirrored disk systems according to their queueing structure, as independent and shared disk systems. We also discuss request routing methods in the former case, We next consider a shared queue, part of which is *nonvolatile RAM - NVRAM* (also called *nonvolatile storage - NVS*) and discuss the prioritized processing of read requests with respect to writes to improve their performance. Finally, we discuss different RAID1 configurations from the viewpoint of their reliability and load balancedness, while operating in degraded mode.

6.1   Independent versus Shared Queue Configuration

Mirrored disks have two configurations as far as their queueing structure is concerned [125]: (i) *Independent Queues - IQ:* Read requests are immediately routed to one of the disks according to some routing policy, while write requests are sent to both disks. (ii) *Shared Queue - SQ:* Read requests are held in this queue, so that there are more opportunities for improving performance. Many variations are possible, e.g., deferred forwarding of requests from the shared queue (at the router) to independent queues.

Request routing with IQ can be classified as *static* and *dynamic*. Examples of static routing are uniform or random versus round-robin or cyclic routing. Both methods halve the arrival rate to individual disks, but cyclic routing improves performance by making interarrival times "more regular" [125].

Improved routing is possible if the router can determine other request attributes, e.g., the address of the data block being accessed, so that a read request can be routed to the disk providing the shortest positioning time, but this would be only possible if the detailed disk layout and the current position of the arm is known to the router. An approximation to SSTF can be implemented based on the block numbers, i.e., send a new request to the disk with the closer block number. Request attributes can be used in *affinity-based routing*, so that one disk serves requests to outer disk cylinders, while the other disk serves requests to inner cylinders, i.e,, cylinders $(1, C/2)$ and $(C/2 + 1, C)$. There will be a load imbalance with zoning, since outer tracks have a higher capacity than inner tracks.

A dynamic policy takes advantage of its awareness of the current state of the systems, e.g., the number of requests at each disk, so that it can use the *Join the Shortest Queue - JSQ* policy. However, this policy does not improve performance when requests have a high service time variability [135]. Simulation studies have shown that the routing policy has a negligible effect for random requests, and that performance is dominated by the local scheduling policy [125].

SQ provide more opportunities for improving performance than IQ for two reasons: (i) resource-sharing argument that a single queue for both servers (disks) is better than independent queues for read requests. [57]. (ii) there are more requests available for minimizing the positioning time. For example, the SATF policy with SQ attains a better performance than IQ, since the shared queue is approximately twice the length of individual queues [125].

6.2 Improving Global Performance Measures

Schemes to improve RAID1 performance have the common theme of minimizing the processing required by write requests and giving higher priority to read requests, so as to improve read response times.

A study concerned with caching write requests in NVS and opportunistically writing out the modified blocks when the disk arm is already in the vicinity of the modified block is reported in [108]. In *distorted mirrors* scheme there is no need for NVS. The "secondary disk" stores the data on the "primary disk" in permuted order, so that data is updated in place on the primary disk, but a *write anywhere* policy, which incurs a low positioning time, is used on the secondary disk [109]. This method requires the controller to maintain pointers to empty data blocks on the secondary disk and is only effective for lower disk space utilization. Note that only the primary disk is suited for processing multiblock sequential requests.

In *doubly distorted mirrors* both disks have master and slave partitions [84]. Distorted and doubly distorted mirrors have a higher space overhead due to metadata and a more complicated recovery scheme than traditional mirrored disks. A combination of caching, write through, and write twice schemes is used in *improved traditional mirrors* [85]. Write through is accomplished at low cost by allocating backup data blocks, whose location can be identified quickly by a mathematical formula.

One method to improve RAID1 performance is to always have one disk available for processing read requests [89]. A period T is defined and the first disk processes reads during the first phase $T/2$, while the second disk processes writes and vice-versa. Writes are deferred by being buffered in an NVS cache. Disk updates are carried out more efficiently by sorting them according to disk location to reduce disk access time. This method works best when reads and updates require the same processing time $(T/2)$. However, the amount of time taken for read and update accesses is not necessarily $T/2$ and depends on the frequency of reads and updates and the buffer hit ratio for reads. The assumption that the processing of reads and updates can be overlapped is a major weakness of this method.

The performance of the above method can be improved as follows [124]: (i) eliminating forced idleness by allowing write requests to be processed individually; (ii) instead of CSCAN, using SATF or even exhaustive enumeration, which is feasible for sufficiently small batch sizes, to find an optimal destaging order; (iii) introducing a threshold for the number of read requests, which when exceeded, defers the processing of write batches.

6.3   Improved Data Layouts for RAID1

In the basic RAID configuration the load due to read requests on the surviving disk doubles. This may result in "infinite" disk queues due to disk overload. Several RAID configurations to balance the load after a disk failure are discussed below. On the other hand these configurations tend to be less reliable than basic RAID1, which under the best circumstances does not encounter data loss even when $N/2$ disks from a total number of N disks fail.

Interleaved declustering: There are N disks, n clusters, and $m = N/n$ disks per cluster. The data one each disk in the cluster is distributed evenly over the remaining disks, so that if one of the m disks in a cluster fails, its load is distributed evenly over the remaining $m - 1$ disks.

Chained declustering the data on each disk is replicated on the next disk (modulo the number of disks N) [49]. This data layout is less susceptible to data loss, than interleaved declustering, since it takes two consecutive disk failures for data loss to occur. Consider four disks with primary and secondary allocations: (A,D'), (B,A'), (C,B'), (D,C'). A balanced disk load after a disk failure is achievable for read-only requests. Consider the original loads a, b, c, and d to A, B, C, and D, which are assumed to be equal. A balanced allocation after the failure of the second disk is (a,d/3), (c/3,b), (2d/3,2c/3), so that the loads on surviving disks increase equally (by 1/3).

Group rotate declustering: This method combines mirroring with chained declustering [20]. Data on $N/2$ primary disks is striped and the stripes are duplicated in a rotated manner on an equal number of secondary disks. This layout is less reliable than basic mirroring, since only one failure per group will result in data loss.

7 RAID5 Disk Arrays

We describe the operation of a RAID5 system in normal, degraded, and rebuild modes with emphasis on the scheduling of disk requests.

7.1 RAID5 Operation in Normal Mode

The updating of a single data block in RAID5 requires the updating of the corresponding parity block, so that when the old data and parity blocks are not cached four disk accesses are required, hence the term *small write penalty*. This penalty can be reduced by using *read-modify-writes - RMW* accesses, so that extra seeks are eliminated.

Modified data blocks can be first written onto a duplexed NVRAM cache, which has the same reliability level as disks [68]. Destaging of modified blocks and the updating of associated parity blocks can be deferred and carried out at a lower priority than reads. Destaging is initiated when NVRAM utilization exceeds a certain high-mark and it is stopped when it reaches a low-mark [127]. An alternative method based on the rate at which the cache is getting filled is proposed in [133] and shown to be an improvement over the previous method using a trace-driven simulation study.

The small write penalty can be alleviated by carrying out full-stripe writes, by first caching modified files in the cache. Because of its similarity to LFS [94], this scheme is referred to as the *log-structured array - LSA* [70]). LSA was implemented in StorageTek's Iceberg disk array [18], which is also a RAID6 disk array, so that both parities are computed efficiently. One advantage of LSA, and also LFS, is that data can be compressed, since it is not being written in place. There are many interesting issues associated with the selection of the stripes to be garbage collected, which are beyond the scope of this discussion.

7.2 RAID5 Operation in Degraded Mode

A RAID5 disk array can continue operation in *degraded mode* with one failed disk. since blocks on the failed disks can be reconstructed on demand by accessing all of the corresponding blocks on surviving disks, according to a *fork-join request*, and XORing them to recreate the block. The time taken by a fork-join request is the maximum of the completion times of requests at surviving disks. As far as writes are concerned, if the disk at which the parity block resides is broken, then we simply write the data block, but if the data block is not available, then we read all of the corresponding data blocks on surviving disks and XOR them with the new data block to compute and write the parity.

Since each surviving disk has to process its own requests, in addition to the fork-join requests, we have a doubling of disk loads when all requests are reads. If all disk requests are read requests then the disk utilization would have to be below 50% in normal mode so that the system does not saturate in degraded mode.

The *clustered RAID* or *parity declustering* organization is proposed in [77] to reduce the increase in disk load in degraded mode. This is accomplished by selecting the *parity group size* to be smaller than the number of disks, so that a less intrusive reconstruction is possible. *Balanced Incomplete Block Designs - BIBDs* have been proposed to balance the disk load due to parity updates on finite capacity disks [81], [47]. BIBD layouts are available for certain parameters only [43] and *nearly random permutations* is a more flexible approach [72].

Properties for ideal layouts are given in [47]: (i) Single failure correcting: the stripe units of the same stripe are mapped to different disks. (ii) Balanced load due to parity updates: all disks have the same number of parity stripes mapped onto them. (iii) Balanced load in failed mode: the reconstruction workload should be balanced across all disks. (iv) Large write optimization: each stripe should contain $n - 1$ contiguous stripe units, where n is the parity group size. (v) Maximal read parallelism: reading $n \leq N$ disk blocks entails in accessing n disks. (vi) Efficient mapping: the function that maps physical to logical addresses is easily computable. The *Permutation Development Data Layout - PDDL* is a mapping function described in [101], which has excellent properties and good performance in light load (like the PRIME data layout [3]) and heavy load (like the DATUM data layout [2]).

7.3   RAID5 Operation in Rebuild Mode

In addition to performance degradation, a RAID5 system operating in degraded mode is vulnerable to data loss if a second disk failure occurs. If a hot spare is provided, the rebuild process is initiated immediately after a disk fails. Instead of wasting the bandwidth of the hot spare, which does not contribute to disk performance, the *distributed sparing* scheme distributes a capacity equivalent to the spare disk as spare areas uniformly over $N+2$ disks [118]. On the other hand the hot spare can be shared among several arrays. The *parity sparing* method utilizes two parity groups, one with $N+1$ and the other with $N'+1$ disks, which are combined to form a single group with $N + N' + 1$ disks after one disk fails [16].

The two performance metrics in rebuild processing are the mean response times of application-generated read requests when rebuild is in progress and rebuild time. *Read redirection* can be used to improve performance as rebuild progresses in RAID5 [77]. This amounts to reading reconstructed data from the spare disk, rather than invoking a fork-join request. *Piggybacking* is another option, where a block which was reconstructed on demand is also written on the spare disk [77].

Rebuild methods are classified into two categories in [47]. *1- Disk-oriented rebuild:* This scheme reads successive *rebuild units* from a surviving disks when the disk becomes idle. When all the corresponding rebuild units (say disk tracks) have been read, they are XORed to obtain the track on the failed disk, which is then written onto a spare disk. The performance of this method, when user requests are given a higher priority than rebuild requests, is studied via simulation in [47] and analytically in [114], [118] using a vacationing server model. *2-Stripe-oriented rebuild:* This scheme synchronizes at the level of the rebuild unit, i.e., the next set of rebuild units are read only after the previous one is written to disk. The performance of these method is evaluated via simulation in [47] and the first method is shown to outperform the latter method,

Rebuild requests are best processed at a lower priority than user requests, especially read requests, e.g., processed when there are no remaining user requests and the disk is idle. This corresponds to the *vacationing server model - VSM* in queueing theory [112], which is utilized in the analysis reported in [114], [118]. Using the framework of the M/G/1 queueing model the mean waiting time of user requests is $W_{VSM} = W_{M/G/1} + \bar{r}$, where $\bar{r}$ is the mean residual service time of rebuild requests. This implies that W_{VSM} increases with the size of the rebuild unit. A larger rebuild unit, e.g., several successive tracks, reduces the positioning time overhead, but the same effect is achieved when multiple consecutive tracks are read because the disk is idle.

The *permanent customer rebuild model* introduced in [72] is motivated by the queueing system with the same name [13]. The processing of rebuild requests at the same priority as user requests, degrades the response times of user requests. On the other hand prioritizing the processing of user requests is expected to result in a reduction in rebuild time, but the reverse seems to be true. This is because unless the disk utilization is very low few tracks are read one after the other, so that the positioning overhead is higher [120].

8   Two Disk Failure Tolerant – 2DFT Disk Arrays

RAID5 disk arrays are susceptible to failure until rebuild processing is completed. 2DFTs can be constructed at the cost of the capacity of one disk drive (in addition to that of RAID5), which is a small cost to pay given rapid increases in disk capacities.

The performance of 2DFTs is impacted more than RAID5 (a 1DFT) in that the updating of each block requires the updating of two check blocks, rather than just one. Three representative 2DFTs are considered in [44] and their performance is compared against each other and also RAID0 and RAID5 (with the same number of disks).

8.1   RAID6 Overview and Cost of Operation

RAID6 uses two check disks to tolerate two disk failures, but more generally a disk array with n data and k check disks, with data words (d_i) and checksum words (c_i), as shown below

$$\boxed{d_1 \ | d_2 \ | d_3 \ | d_4 \ | \ldots \ | \ldots \ | d_n \ | c_1 \ | c_2 \ | \ldots \ | c_k}$$

to tolerate k disk failures. If the words are w bits wide then the constraint $2^w \geq n + k$ applies. Let vector $\mathbf{d} = (d_1, d_2, \ldots, d_n)$ and $\mathbf{c} = (c_1, c_2, \ldots, c_k)$. Vector $\mathbf{u}$ is a concatenation of vectors $\mathbf{d}$ and $\mathbf{c}$, and is computed as $\mathbf{u} = \mathbf{dG}$, where $\mathbf{G}$ is a concatenation of two matrices: $G = [I|P]$: I is an $n \times n$ identity matrix and P is a parity matrix. Matrix $\mathbf{G}$ is generated from a Vandermonde matrix, as shown in [63].

It follows that checksum words c_i are linear combination of the data words d_i, and the unique weights are specified by $\mathbf{G}$. For a small write, when d_j' overwrites d_j, all the checksum words need to be changed: $c_i' = c_i + g_{i,j}(d_j' - d_j), 1 \leq i \leq k..$ Note similarity to parity updating in RAID5, except that instead of one, k check disks need to be updated. To recover from failures we express the missing data blocks as a function of check blocks, and since there are k check blocks, up to k data blocks can be recovered.

RAID6 is a special case which allows two disk failures. The two parities in this case are referred to as P and Q parities and they are laid out in parallel diagonals, similarly to RAID5. The operation of RAID6 in normal mode is similar to RAID5 except that two check blocks need to be updated. A data block to be read from a failed disk can be recreated using one of two surviving check blocks. When two disks fail, a missing data block can be reconstructed using one of the two parity blocks. When both failed disks hold data blocks, then to recreate one of them all the corresponding blocks on surviving disks have to be read and we need to solve two equations in two unknowns to recreate the missing block.

8.2   The EVENODD Scheme

EVENODD has two desirable features [12]: (i) The coding is based on the XOR operation, i.e., the same hardware as RAID5. (ii) It utilizes two parity disks to recreate two failed disks, which is the minimum redundancy possible. EVENODD organizes data as symbols in an $(m - 1) \times (m + 2)$ array, which is referred to as a *segment*. The $m + 2$ columns represents disks, where columns $0 : m - 1$ are the data disks, and column m (resp. $m + 1$) holds parity P (resp. Q), i.e., horizontal (resp. diagonal) parities.

The EVENODD scheme works only when m is a prime number [12], but when the number of data disks is not prime, the number of disks is set to the first larger prime number, with the contents of the virtual disks set to zeros. Parity P is the XOR of all the data symbols in the same row, as in RAID5, while Q parities are computed over similarly marked diagonals in Figure 1, but an additional parity over the diagonal (∞) is applied to these parities. The implication is that an update of a block in the diagonal requires the updating of all of the Q parities.

D0	D1	D2	D3	D4	P	Q
◇	♣	♡	♠	∞	(0:4)	◇ ⊕ ∞
♣	♡	♠	∞	◇	(0:4)	♣ ⊕ ∞
♡	♠	∞	◇	♣	(0:4)	♡ ⊕ ∞
♠	∞	◇	♣	♡	(0:4)	♠ ⊕ ∞

Fig. 1. EVENODD data layout with m=5.

The P and Q parities are rotated as in RAID6 to attain a balanced load as far as the updating of parities is concerned.

Although updating parity seems prohibitively expensive, especially when diagonal elements are updated, the EVENODD scheme can be made very efficient by using small symbol sizes [12]. If the minimum block size is 4KB, we can choose the symbol size to be $4096/(m-1)$ so that each column in a segment constitutes one block. Two suitable prime numbers are: $m = 17$ and $m = 257$. We can vertically stack as many segment as required, to make the stripe unit as large as desired. The disk access pattern of EVENODD is exactly the same as RAID6, so that their performance is indistinguishable from the viewpoint of disk accesses. Of course, internally, different computations are required [12].

8.3   The RM2 Scheme

The RM2 2DFT scheme is defined as follows [87]: "Given a redundancy ratio p and the number of disks N, construct N parity groups each of which consists of $2(M-1)$ data blocks and one parity block such that each data block should be included in two groups, where $M = 1/p$." Each disk contains one parity and $M-1$ data units in a segment, a repeating pattern, so that the parity stripe units are distributed evenly. An algorithmic solution to this problem is based on an $N \times N$ *redundancy matrix - RM*, which will be not discussed here for the sake of brevity.

An RM2 data layout is constructed as follows: (i) select the target redundancy ratio p and set $M = 1/p$, (ii) select the total number of disks N that satisfy: $N \geq 3M - 2$ if N is odd or $N \geq 4M - 5$ if N is even. The next two steps specified in [87] deal with the layout of data and parity blocks using the RM2 matrix. Note that the inequalities imply that $p = 1/M \geq 3/(N+2)$, which means that RM2 has a higher redundancy ratio than RAID6, which is always $2/N$.

For example, given $p = 1/3$ and $M = 1/p = 3$, $N = 7$ is the smallest number satisfying the inequalities. Given the seven disks $\mathcal{D}_0, \mathcal{D}_1, \ldots, \mathcal{D}_6$, the parity stripe units laid out on the first row are $P_0, P_1, \ldots, P_6$, row two holds data blocks: $D_{2,3}$, $D_{3,4}, D_{4,5}, D_{5,6}, D_{0,6}, D_{0,1}, D_{1,2}$ and the third row: $D_{1,4}, D_{2,5}, D_{3,6}, D_{4,0}, D_{5,1}$, $D_{6,2}, D_{0,3}$, where $D_{i,j}$ means that the stripe unit is protected by P_i and P_j. Note that each data block is protected by two parity blocks.

In normal mode the cost of reads and writes is similar to RAID6, With one failed disk when the required data block is not available, it can be reconstructed easily using one of the parities, while if a parity block is not available its update can be ignored. When there are two disk failures and the required data block is not available, then we need a recovery path and the number of steps varies from 1 to $2M - 2$. In fact 2.4 disk accesses are required per surviving disk for $N = 7$, $M = 3$.

9   Performance Evaluation Studies

We first review performance studies of single disk systems, followed by RAID1, RAID5, and RAID6 systems.

9.1   Single Disk Performance

Disk simulation studies can be classified into random-number-driven and trace-driven. A trace-driven simulation study concludes that an extremely detailed model is required to achieve high accuracy in estimating disk performance [96]. The main difficulty in estimating disk performance is modeling the effect of the onboard disk cache, since the prefetching and cache replacement policies are not published. DBMSs and OS I/O routines also initiate prefetching for sequential accesses. This is less of a problem when dealing with random disk accesses, which are generated by OLTP applications.

Some other disk parameters, such as the seek time characteristic, which is required for building a disk simulator are not necessarily available in disk manuals. The DIXTRAC tool automates the extraction of disk parameters [100], and makes them available at [30]. These parameters are used by the DiskSim simulation tool [14].

An I/O trace, in addition to being used in trace-driven simulations, can be analyzed to characterize the I/O workload, which is then used in a random-number driven simulation study. The analysis of an I/O trace for an OLTP environment showed that 96% of requests are to 4 KB blocks and 4% to 24 KB blocks [90]. Other notable disk I/O characterization studies are [50], [51], where the second study is concerned with I/O requests at the logical level.

The analysis of a single disk is quite complicated for anything other than a FCFS policy under favorable assumptions, e.g., an M/G/1 queueing model. Zoning introduces complications, which are dealt with in [118]. A review of analytic studies of disk scheduling policies is reported in [25], but most of these studies make unrealistic assumptions to make the analysis tractable. For example, the analysis of the SCAN policy in [24] assumes: (i) Poisson arrivals to each cylinder; (ii) the disk arm seeks cylinder-to-cylinder, even visiting cylinders not holding any requests; (iii) satisfying a request takes the same time at all cylinders. Clearly this analysis cannot be used to predict the performance of the SCAN policy in a realistic environment.

Most early analytic and simulation studies were concerned with the relative performance of various disk scheduling methods, e.g., SSTF has a better performance than FCFS, at high arrival rates SCAN outperforms SSTF [25]. Other studies propose a new scheduling policy and carry out a simulation study to evaluate its performance with respect to standard policies [31], [102]. Two more recent simulation studies of disk scheduling methods, which also review previous work are [142], [121], [122]. There are few analytical studies which take into account the track buffer and the effect of readahead into it [107].

Scheduling policies for a mixed workload consisting of discrete or D-requests and continuous or C-requests is reported in [9]. D-requests access small data blocks, which are located between large datasets accessed by C-requests. C-requests originate from a finite number of sources and read successive blocks of video blocks using a SCAN or C-SCAN policy. while D-requests are processed according to SPTF. There are two performance metrics: the mean response time of D-requests and the fraction of C-requests that miss a deadline. Scheduling techniques for hard real-time computing systems seem to be applicable to this problem [15].

There have been numerous studies of multidisk configurations, where the delays associated with I/O bus contention are taken into account. *Rotational Positing Sensing - RPS* is a technique to detect collisions when two or more disks connected to a single bus are ready to transmit at the same time, in which case only one disk is the winner and additional rotations are incurred at the other disks, which can result in a significant increase in disk response time. There is also a delay in initiating requests at a disk if the I/O bus is busy. Reconnect delays are obviated by onboard disk caches, so that RPS delays are no longer an issue.

9.2 Performance Studies of Mirrored Disks

We first discuss performance studies of mirrored disks dealing with the effect of scheduling on a local performance measure, such as seek distance, followed by global performance measures such as response time.

The expected seek distance for read (resp. write) requests with a degree of replication k is the expected value of the minimum (resp. maximum) of k seeks. A simple expression can be obtained when disk requests are uniformly distributed. For read requests $S_k^{read} \approx C/(2k+1)$, i.e., $C/5$ for $k = 2$ versus $C/3$ for a single disk. For writes $S_k^{write} \approx C(1-(2/3)(4/5)...(2k)/(2k+1))$ for k disks, e.g., $7C/15$ for two disks [11]. In a continuous domain $(0,1)$ the expected value of the seek distances for reads and writes is 5/24 and 11/24, respectively [79], [21].

The analysis for reads is optimistic in that it does not take into account the synchronization effect that writes have on the placement of disk arms, i.e., after processing the write requests on behalf of a single request at both disks, their arms are placed at the same cylinder. The effective degree of replication with k disks is determined via a Markov chain model in [65]. A more detailed Markov chain model takes into account the shortest seek policy in determining which disk arm is to be moved [131]. Numerical results for a large number of cylinders and small k shows that the expected seek distances are almost indistinguishable from the results in [65].

A greedy policy to minimize the average seek time in mirrored disks is proposed in [46]. It is assumed that disk accesses are independent and that both disk arms can be positioned autonomously, but only one transfer at a time is possible. The arm nearer to the target cylinder t in the range [0,1] is chosen for the seek, but in addition the other arm jockeys to a desirable position: $t/3$ if $t \geq 1/2$ and $(2-t)/3$, otherwise. Given independent uniformly distributed requests, the mean seek distance is $5/36$ versus $5/24$ without such an optimization.

Latency is a matter of concern if the arms on all disks are synchronized, i.e., they seek together for reads. It is suggested in [80] that both arms of mirrored disks seek to the target cylinder and the data be accessed via the arm with the lower latency. The mean latency then drops from $1/2$ to $1/3$ of rotation time. The latency for writes with data replication is the maximum of the latency on all disks, i.e., the mean latency to complete a write increases from one-half of rotation time for one disk to two-thirds of rotation time for two disks. More generally the latency for read requests is $1/(k+1)$ and $k/(k+1)$ for writes, when the degree of replication is k disks [11]. Multiple arm disks or data replication on multiple disks has also been considered for improving performance due to latency. The arms on one disk are placed $180/n$ degrees apart or with n disks their rotation is synchronized so that the arms appear to be $180/n$ degrees apart [80].

As far as anticipatory disk arm positioning is concerned, it is best to position the disk arm at $C/2$ in the case of a single non-zoned disk. With mirrored disks it is best to position one arm at $C/4$ and another arm at $3C/4$, when all requests are reads, so that the positioning time is reduced to $C/8$ [55]. When the fraction of write requests is f_w then both arms should be placed at $C/2$ when $f_w > 0.5$ and otherwise at $1/2 \pm s_{opt}$, where $s_{opt} = 0.25(1 - 2f_w)/(1 - f_w)$ [55].

A method to improve RAID1 performance so that one disk is always available for processing read requests, which is based on [89] was already discussed in Section 6.2. A rough performance analysis of the method is provided.

Analytic and simulation studies of various scheduling policies in mirrored disk systems is provided in [126]. A distinction is made between single queue (SQ) and multiple queue (MQ) policies. In the SQ category there are the primary/secondary (PS) and equitable queueing (EQ) disciplines. The following PS disciplines are considered: (a) a request is not served until both disks are idle; (b) a read may proceed as soon as a write is completed. With the EQ disciplines: (i) reads may execute concurrently, but no concurrency is allowed between reads and writes; (ii) as in (i) updates are required to wait until both disks are available, but a read can start as soon as an update is completed; (iii) this policy is a modification of (a) with reads started at both disks, when one read is finished the other read is aborted (no description is provided how this is accomplished).

MQ policies allow updates to proceed although both disks are not idle, which was not the case with SQ policies. The following scheduling policies are considered: (1) Reads are sent to the queue for the first and second disk with equal probabilities. Several variations of policy (1) are considered, one of which routes reads to the shorter queue. A variation in which reads are initiated at both disks is considered and in one subvariation a read is aborted as soon as the corresponding read request completes. (2) In this common queue (CQ) policy requests are dequeued as soon as a disk is idle, and if one of the disks is busy the write request will be enqueued locally. The performance of these policies is determined via simulation as well as Markovian models, which are solved using the matrix-geometric method, As would be expected the common queue (CQ) policy demonstrates the best performance, except at very low utilizations.

Reconstruction in RAID1: Reconstruction policies can be classified into *ordered* and *greedy* in [8]. Ordered rebuild copies the blocks of the surviving disk onto a spare disk in an ordered manner using a vacationing server model, as in [118]. An ordered rebuild policy with a permanent customer model is analyzed in [71]. A greedy rebuild policy is proposed and analyzed in [8], which rebuilds a yet to be constructed track closest to the disk arm. It is shown that the greedy policy has the better performance.

9.3   RAID5 Performance

There have been several RAID prototypes, which are partially described in [18]. Performance (measurement) studies of commercial RAID systems have been published in conjunction with the recently developed SPC-1 benchmark [110]. There is also a great deal interest in evaluating the performance of parallel file systems by benchmarking [42].

There have been numerous analytical and simulation studies of disk arrays. Markovian models have been used successfully to investigate the relative performance of variations of RAID5 disk arrays [69]. This analysis utilizes by an approximation for the mean response time of fork-join requests [78]. The approximation utilizes an exact formula which has been given for the mean response time of a 2-way fork-join queueing system with Poisson arrivals (with rate λ) and exponential service times (with rate μ), i.e., $R_2^{F/J} = (3/2 - \rho/8)R$, where $R = (1/\mu)/(1 - \rho)$ is the mean response time at each one of the two servers ($\rho = \lambda/\mu$ is the utilization factor). Note that $R_2^{max} = 1.5R > R_2^{F/J}$.

Several performance evaluation studies of RAID5 systems have been carried out based on an M/G/1 model. Most notably RAID5 performance is compared with parity striping [37] in [19] and with mirrored disks in [20]. When the striping unit is small, a user request results in accesses to multiple disks. Estimating the mean response time of the resulting fork-join request is carried out by obtaining the first two moments of response times and using them to approximate the individual response times with a Coxian distribution. This is because the expected value of the maximum of N random variables with a Coxian distribution [128] can be evaluated easily, since it can be expressed as the sum of exponentials. It is stated in [78] that the analysis is applicable to the case when there are local requests at the servers. In fact simulation results have shown that this is not so and that the mean response time for fork-join requests is higher in the presence of interfering requests. The analysis in [19] may overestimate the mean response time, since the components of fork-join requests are not independent.

A performance analysis of clustered RAID in all three operating modes is given in [72]. An M/G/1 queueing model is adopted in estimating user response times. To estimate the response time of fork-join requests the first two moments of the response times of read requests are obtained used in estimating the two parameters a and b of the *extreme value distribution* $P[X < x] = exp(-e^{-(x-a)/b})$, with mean $a + \gamma b$ and variance $(\pi b)^2/6$, where $\gamma \approx 0.577$ is Euler's constant. The expected value of the maximum of N random variables with this distribution is given simply as $R_N^{max} = a + b(\gamma + ln(N))$. As discussed in Section 6.3, the analysis in [72] utilizes the *permanent customer queueing model* [13], i.e., there is one rebuild request at each disk at any time, so that a completed request is immediately replaced by a new one. No validation results are presented.

An analysis of RAID5 disk arrays in normal, degraded, and rebuild mode appears in [114] and [118], where the former (resp. latter) study deals with RAID5 systems with dedicated (resp. distributed) sparing. The analysis in [114] and [118] is presented in a unified manner in [117], which also reviews other analytical studies of disk arrays. These analyses consider a disk-oriented rebuild policy and a *vacationing server model* is utilized for this purpose [112].

Based on an observation that the distribution of service times and also response times for the components of fork-join requests has a coefficient of variation smaller than one, the response time distribution is approximated by an Erlang distribution with k stages. The calculation of the maximum of N such random variables requires the summation of $k + 1)^N$ multinomials, which can be very costly. The method reported in [36] is therefore used to reduce the computational cost.

Simulation is used to compare the performance of rebuild strategies, and the effect of parameters such as the rebuild unit, the size of the rebuild buffer on performance [120]. The preemption of rebuild requests can be used to improve disk response time, but this will result in an increase in rebuild time, so that even more disk requests will encounter an increased response time [116].

The analysis in [58] approximates the service time of disk requests with an Erlang distribution and utilizes a technique developed in [54] for computing the distribution of fork-join requests in this case. A limitation of this approach is that it is only applicable when disk service times can be approximated by an Erlang distribution. Finally, a general approximation based on interpolation between light and heavy loads for fork-join queues is developed in [134]. The interarrival and service time can follow the exponential, Erlang, and hyperexponential distributions.

A simulation study of clustered RAID is reported in [47], which compares the effect of disk-oriented and stripe-oriented rebuild. The former method outperforms the latter, since the latter synchronizes rebuild processing at the level of stripe- or rather rebuild- units from all disks, while the former requires more buffer space. The effect of the rebuild unit on performance is investigated in [48].

An analytic throughput model is reported in [129], which includes a cache model, a controller model, and a disk model. The model is validated against a state of the art disk array, showing a 15% prediction error on the average. Mean-value analysis [59] is applied to evaluating the performance of a mirrored disk system, which is then validated against measurement results [132].

A simulation study investigating the performance of a heterogeneous RAID1 system: MEMS-based storage backed-up my a magnetic disk is reported in [130]. Specialized simulation and analytic tools have been developed to evaluate the performance of MEMS-based storage [41].

A comprehensive trace-driven simulation study of a cached RAID5 design is reported in [127], The Pantheon simulator was developed at HP Labs [138] for simulating disk arrays and was used in the evaluation of AutoRAID [137].

RAIDframe is a simulation as well as a rapid prototyping tool for RAID disk arrays [27]. It is superseded by the DiskSim package [14], which in spite of its ability to model I/O subsystem components (buses, controllers), is more tuned for detailed modeling of individual disks rather than disk arrays.

The performance of a RAID system tolerating two and more disk failure is investigated in [2] via simulation. The performance analyses of 2DFT methods in [44], [45], is an extension of the analyses in [114], [118], but only considers operation in normal and degraded modes. A queueing model based on M/G/1 is developed based on cost functions, which are systematically derived for RAID6 and RM2 in normal and degraded modes. The cost function is the sum of the multiples of the number of simple read, write, and read-modify-requests (for updating data and parity blocks). As noted earlier EVENODD has the same performance as RAID6, when the symbol size is small. Another difference is that a modified extreme value distribution is utilized in estimating the mean response time of fork-join requests [45].

RM2 operation in degraded mode with two failed disks involves multiple accesses to related blocks for on demand reconstruction. Due to the difficulty of obtaining disk access times for such requests, a hybrid approach is used in [44] i.e., we use simulation to obtain the moments of service time, which are then used in an M/G/1 queueing model.

10   Conclusions and Recent Developments

In this tutorial we have reviewed scheduling and data allocation methods for individual disks, RAID1 or mirrored disks, RAID5 rotated parity disk arrays, and RAID6, EVENODD, and RM2 disk arrays. We have also covered analytic and simulation methods that can be used to evaluate the performance of disks and disk arrays.

There are several technologies contending to replace disks. The *micro-electro-mechanical system -MEMS based storage* is also nonvolatile and an order of magnitude faster than disks, but it is currently ten times more expensive [130]. *Magnetic RAM - MRAM* has approximately the same speed as DRAM, but is less costly, stores 100's MB/chip and is non-volatile [139].

Storage constitutes 50% of the investment in data centers [35]. More importantly, the explosion in storage capacity in computer installations has led to a dramatic increase in the cost of storage management. In 1980 one data administrator managed 10 GB (gigabytes) of data, while in 2000 this number is 1 TB (terabyte) [40], but given that 5 TB now costs $5000, the cost of administering it is an order of magnitude higher. Consequently automating storage management and data allocation has emerged as a very important area [139].

An "iterative storage management loop" called Hyppodrome was recently proposed at HP Labs [6]: (i) design new system; (ii) implement design; (iii) analyze workload and go back to (i). Hyppodrome uses Ergastulum to generate a storage system design [5]. There is an intermediate step for selecting RAID levels [7], configuring devices, and assigning stores onto devices.

Storage Area Networks- SANs, which are based on Fibre Channel technology allow disks and disk array controllers to be attached to the SAN, rather than to a server. Disk data is accessible directly to the multiple servers attached to the SAN. Appropriate protocols are required to coordinate accesses to these shared data. An interesting discussion of algorithmic approaches relevant to storage networks appears in [97].

NAS embeds the file system into storage and is typically based on Ethernet [35]. Four typical NAS systems are described in [35]: (i) storage appliances such as SNAP! from Network Appliance; (ii) *NASD - network attached secure devices* [34]; (iii) Petal whose goal is to provide easily expandable storage with a block interface [60]; (iv) *internet SCSI - iSCSI* protocol implements the *SCSI - small computer system interface* protocol on top of TCP/IP [99], [75].

The Storage Networking Industry Association (www.snia.org) among its other activities has formed the Object-based Storage Devices (OSD) work group, which is concerned with the creation of self-managed, heterogeneous, shared storage. This entails moving low-level storage functions to the device itself and providing the appropriate interface. A recent paper on this topic is [74]

Finally there is increasing interest in constructing *information storage systems*, which provide availability, confidentiality, and integrity policies against failures and even malicious attacks, see e.g., [143].

References

1. S. Akyurek and K. Salem. "Adaptive block rearrangement", *ACM Trans. Computers 13*(2): 89-121 (May 1995).
2. G. A. Alvarez, W. A. Burkhard, and F. Cristian. "Tolerating multiple failures in RAID architectures with optimal storage and uniform declustering", *Proc. 24th In'l Symp. Comp. Architecture*, 1997, pp. 62-72.
3. G. A. Alvarez, W. A. Burkhard, L. Stockmeyer, and F. Cristian. "Declustered disk array architectures with optimal and near-optimal parallelism", *Proc. 25th Int'l Symp. Computer Architecture*, 1998, pp. 109-120.
4. G. A. Alvarez et al. "Minerva: an automated resource provisioning tool for large-scale storage systems", *ACM Trans. Computer Systems 19*(4): 483-518 (2001).
5. E. Anderson et al. "Ergastulum: quickly finding near-optimal storage system designs", *Technical Report HPL-SSP-2001-05*, HP Labs, 2001.
6. E. Anderson, M. Hobbs, K. Keeton, S. Spence, M. Uysal, A. C. Veitch. "Hyppodrome" Running circles around storage administration", *Proc. File and Storage Technologies Conf. - FAST'02*, USENIX, 2002, 175-188.
7. E. Anderson, R, Swamitham, A. C. Veitch, G. A. Alvarez, and J. Wilkes. "Selecting RAID levels for disk arrays", *Proc. File and Storage Technologies Conf. - FAST'02*, USENIX, 2002, 189-201.
8. E. Bachmat and J. Schindler. "Analysis of methods for scheduling low priority disk drive tasks", *Proc. ACM SIGMETRICS Conf.*, 2002, pp. 55-65.
9. E. Balafoutis et al. "Clustered scheduling algorithms for mixed media disk workloads in a multimedia server", *Cluster Computing 6*(1): 75–86 (2003).
10. B. T. Bennett and P. A. Franaszek. "Permutation clustering: An approach to online storage reorganization", *IBM J. Research and Development 21*(6): 528-533 (1977).

11. D. Bitton and J. Gray. "Disk shadowing", *Proc. 14th Int'l Very Large Database Conf.*, 1988, pp. 331-338.

12. M. Blaum, J. Brady, J. Bruck, and J. Menon. "EVENODD: An optimal scheme for tolerating disk failure in RAID architectures", *IEEE Trans. Computers* 44(2): 192-202 (Feb. 1995).

13. O. J. Boxma and J. W. Cohen. "The M/G/1 queue with permanent customers", *IEEE J. Selected Topics in Communications* 9(2): 179-184 (1991).

14. J. S. Bucy, G. R. Ganger, and Contributors. "The DiskSim Simulation Environment: Version 3.0 Reference Manual", *Technical Report CMU-CS-03-102*, Jan. 2003.

15. G. C. Buttazzo. *Hard Real-Time Computing Systems: Predictable Scheduling Algorithms and Applications*, Kluwer Academic Publishers, 2000.

16. J. Chandy and A. L. Narasimha Reddy. "Failure evaluation of disk array organizations", *Proc. 13th Int'l Conf. Distributed Computing Systems -ICDCS*, 1993, pp. 319-326.

17. P. M. Chen and D. A. Patterson. "Maximizing performance on a striped disk array", *Proc. 17th Int'l Symp. Computer Architecture*, 1990, pp. 322-331.

18. P. M. Chen, E. K. Lee, G. A. Gibson, R. H. Katz, and D. A. Patterson. "RAID: High-performance, reliable secondary storage", *ACM Computing Surveys* 26(2): 145-185 (1994).

19. S.-Z. Chen and D. F. Towsley. "The design and evaluation of RAID5 and parity striping disk array architectures", *J. Parallel and Distributed Computing* 10(1/2): 41-57 (1993).

20. S.-Z. Chen and D. F. Towsley. "A performance evaluation of RAID architectures", *IEEE Trans. Computers* 45(10): 1116 1130 (1996).

21. C. Chien. "Seek distances in disks with dual arms and mirrored disks", *Performance Evaluation 18:* 175-188 (1993).

22. E. G. Coffman Jr. and P. J. Denning. *Operating Systems Theory*, Prentice-Hall, 1972.

23. E. G. Coffman Jr., E. G. Klimko, and B. Ryan. "Analyzing of scanning policies for reducing disk seek times", *SIAM J. Computing* 1(3): 269-279 (1972).

24. E. G. Coffman, Jr. and M. Hofri. "On the expected performance of scanning disks", *SIAM J. Computing* 11(1): 60-70 (1982).

25. E. G. Coffman, Jr. and M. Hofri. "Queueing models of secondary storage devices", in *Stochastic Analysis of Computer and Communication Systems*, H. Takagi (ed.), North-Holland, 1990, pp. 549-588.

26. G. Copeland, W. Alexander, E. Boughter, and T. Keller. "Data placement in Bubba", *Proc. ACM SIGMOD Int'l Conf.*, 1988, pp. 99-108.

27. W. V. Courtright II et al. "RAIDframe: A Rapid Prototyping Tool for Raid Systems", http://www.pdl.cmu.edu/RAIDframe/raidframebook.pdf.

28. P. J Denning. "Effects of scheduling in file memory operations", *Proc. AFIPS Spring Joint Computer Conf.*, 1967, pp. 9-21.

29. Disksim. http://www.pdl.cmu.edu/DiskSim/

30. Diskspecs. http://www.pdl.cmu.edu/Dixtrac/index.html.

31. R. M. Geist and S. Daniel. "A continuum of disk scheduling algorithm", *ACM Trans. Computer Systems* 5(1): 77-92 (1987).

32. E. Gelenbe and I. Mitrani. *Analysis and Synthesis of Computer Systems*, Academic Press, 1980.

33. G. A. Gibson. *Redundant Disk Arrays: Reliable, Parallel Secondary Storage*, The MIT Press, 1992.

34. G. A. Gibson et al. "A cost-effective, high bandwidth storage architecture", *Proc. ASPLOS VIII*, 1998, 92-103.
35. G. A. Gibson and R. Van Meter. "Network attached storage architecture", *Comm. ACM 43*(11): 37-45 (Nov. 2000).
36. A. Gravey. "A simple construction of an upper bound for the mean of the maximum of N identically distributed random variables", *J. Applied Probability 22*: 844-851 (1985).
37. J. Gray, B. Horst, and M. Walker. "Parity striping of disk arrays: Low-cost reliable storage with acceptable throughput", *Proc. 16th Int'l Very Large Database Conf.*, 1990, 148-161.
38. J. Gray and G. Graefe. "The five-minute rule ten years later, and other computer storage rules of thumb", *ACM SIGMOD Record 26*(4): 63-68 (1997).
39. J. Gray and P. J. Shenoy. "Rules of thumb in data engineering", *Proc. 16th ICDE*, 2000, pp. 3-16.
40. J. Gray. "Storage bricks have arrived" (Keynote Speech), *1st Conf. on File and Storage Technologies - FAST '02*, USENIX, 2002.
41. J. L. Griffin, S. W. Shlosser, G. R. Ganger, and D. F. Nagle. "Modeling and performance of MEMS-based storage devices", *Proc. ACM SIGMETRICS Conf.*, 2000, pp. 56-65.
42. National Energy Research Scientific Computing Center. "Global Unified Parallel File System Project - GUPFS", http://www.nersc.gov/projects/gupfs.
43. M. Hall, Jr. *Combinatorial Theory, 2nd ed.*, John Wiley, 1986.
44. C. Han and A. Thomasian. "Performance of two-disk failure tolerant disk arrays", *Proc. Symp. Performance Evaluation of Computer and Telecomm. Systems - SPECTS '03*, 2003.
45. C. Han, A. Thomasian, and G. Fu. "How much more costly is it to tolerate two disk failures?", submitted for publication, Oct 2003.
46. M. Hofri. "Should the two-headed disk be greedy? - Yes it should", *Information Processing Letters - IPL 16*(2): 83-85 (1983).
47. M. C. Holland, G. A. Gibson, and D. P. Siewiorek. "Architectures and algorithms for on-line failure recovery in redundant disk arrays", *Distributed and Parallel Databases 11*(3): 295-335 (1994).
48. R. Y. Hou, J. M. Menon, and Y. N. Patt. "Balancing I/O response time and disk rebuild time in a RAID5 disk array", *Proc. 26th Hawaii Int'l Conf. on System Sciences - HICSS, Vol. I*, 1993, pp. 70-79.
49. H.-I Hsiao and D. J. DeWitt. "A performance study of three high availability data replication strategies", *J. Distributed and Parallel Databases 1*(1): 53-80 (Jan. 1993).
50. W. W. Hsu, A. J. Smith, and H. Young. "I/O reference behavior of production database workloads and the TPC benchmarks - an analysis at the logical level", *ACM Trans. Database Systems 26*(1): 96-143 (2001).
51. W. W. Hsu, H. Young, and A. J. Smith. "Characteristics of production database workloads and the TPC benchmarks", *IBM Systems J. 40*(3): 781-802 (2001).
52. S. Iyer and P. Druschel. "Anticipatory scheduling: A disk scheduling framework to overcome deceptive idleness in synchronous I/O", *Proc. 17th Symp. Operating System Principles*, 2001, pp. 117-130.
53. D. Jacobson and J. Wilkes. "Disk scheduling algorithms based on rotational position", *HP Technical Report HPL-CSP-91-7rev*, 1991.
54. C. Kim and A. Agrawala. "Analysis of the fork-join queue", *IEEE Trans. Computers 38*(2). 250-255 (1989).

55. R. P. King. "Disk arm movement in anticipation of future requests", *ACM Trans. Computer Systems 8*(3): 214-229 (1990).

56. L. Kleinrock. *Queueing Systems Vol. I: Theory*, Wiley-Interscience, 1975.

57. L. Kleinrock. *Queueing Systems Vol. II: Computer Applications*, Wiley-Interscience, 1976.

58. A. Kuratti and W. H. Sanders. "Performance analysis of the RAID5 disk array", *Proc. Int'l Computer Performance and Dependability Symp.*, 1995, pp. 236-245.

59. S. S. Lavenberg (ed.) *Computer Performance Modeling Handbook*, Academic Press, 1983.

60. E. K. Lee and C. Thekkath. "Petal: Distributed virtual disks", *Proc. ASPLOS XII*, 1996, pp. 84-92.

61. E. K. Lee and R. H. Katz. "The performance of parity placements in disk arrays", *IEEE Trans. Computers 42*(6): 651-664 (June 1993).

62. L. Lee, P. Scheuermann, R. Vingralek. "File assignment in parallel I/O systems with minimal variance of response time", *IEEE Trans. Computers 49*(2): 127-140 (2000).

63. W. Litwin and T. Schwarz, S. J. "LH$^*_{RS}$: A high-availability scalable distributed data structure using Reed Solomon codes", *Proc. ACM SIGMOD Conf.* 2000, pp. 237-248.

64. C. L. Liu and J. W. Layland. "Scheduling algorithms for multiprogramming in hard read-time environment", *J. ACM 20*(1): 46-61 (Jan. 1973).

65. R. W.-M. Lo and N. S. Matloff. "Probabilistic limit on the virtual size of replicated data system", *IEEE Trans. Knowledge and Data Eng. - TKDE 4*(1): 99-102 (Jan./Feb. 1992).

66. C. R. Lumb, J. Schindler, G. R. Ganger, and D. F. Nagle. "Towards higher disk head utilization: Extracting free bandwidth from busy disk drives", *Proc. 4th Symp. Operating System Design and Implementation* USENIX, 2000, pp. 87-102.

67. W. C. Lynch. "Do disk arms move?" *Performance Evaluation Review 1*(4): 3-16 (Dec. 1972).

68. J. Menon and J. Cortney. "The architecture of a fault-tolerant cached RAID controller", *Proc. 20th Annual Int'l Symp. Computer Architecture*, 1993, pp. 76–86.

69. J. Menon. "Performance of RAID5 disk arrays with read and write caching", *Distributed and Parallel Databases 11*(3): 261-293 (1994).

70. J. Menon. "A performance comparison of RAID5 and log-structured arrays", *Proc. 4th IEEE High Performance Distributed Computing*, 1995, pp. 167-178.

71. A. Merchant and P. Yu. "An analytical model of reconstruction time in mirrored disks", *Performance Evaluation 20*(1-3): 115-129 (1994).

72. A. Merchant and P. S. Yu. "Analytic modeling of clustered RAID with mapping based on nearly random permutation", *IEEE Trans. Computers 45*(3): 367-373 (1996).

73. A. S. Meritt et al. "z/OS support for IBM TotalStorage enterprise storage server", *IBM Systems J. 42*(2): 280-301 (2003).

74. M. Mesnier, G. R. Ganger, and E. Riedel. "Object-based storage", *IEEE Communications Magazine 41*(8): 84-90 (2003).

75. K. Z. Meth and J. Satran. "Features of the iSCSI protocol", *IEEE Communications Magazine 41*(8): 72-75 (2003).

76. W. Michiels, J. Korst, J. Aerts. "On the guaranteed throughput of multizone disks", *IEEE Trans. Computers 52*(11): 1407-1420 (Nov. 2003).

77. R. Muntz and J. C. S. Lui. "Performance analysis of disk arrays under failure", *Proc. 16th Int'l Very Large Database Conf.*, 1990, pp. 162-173.

78. R. Nelson and A. Tantawi. "Approximate analysis of fork-join synchronization in parallel queues", *IEEE Trans. Computers* 37(6): 739-743 (1988).
79. S. W. Ng. "Reliability, availability, and performance analysis of duplex disk systems", *Reliability and Quality Control*, M. H. Hamza (ed.), Acta Press, 1987, pp. 5-9.
80. S. W. Ng. "Improving disk performance via latency reduction", *IEEE Trans. Computers* 40(1): 22-30 (Jan. 1991).
81. S. W. Ng and R. L. Mattson. "Uniform parity distribution in disk arrays with multiple failures", *IEEE Trans. Computers* 43(4): 501-506 (1994).
82. S. W. Ng. "Crosshatch disk array for improved reliability and performance", *Proc. 21st Int'l Symp. Computer Architecture*, 1994, pp. 255-264.
83. A. W. Ng. "Advances in disk technology: Performance issues", *IEEE Computer* 40(1): 75-81 (May 1998).
84. C. Orji and J. A. Solworth. "Doubly distorted mirrors", *Proc. ACM SIGMOD Int'l Conf.*, 1993, pp. 307-316.
85. C. Orji, M. A. Weiss, and J. A. Solworth. "Improved traditional mirrors", *Proc. Int'l Conf. Found. Data Organization and Algorithms - FODO'93*, 329-344.
86. D. A. Patterson, G. A. Gibson, and R. H. Katz. "A case study for redundant arrays of inexpensive disks", *Proc. ACM SIGMOD Int'l Conf.*, 1988, pp. 109-116.
87. C.-I. Park. "Efficient placement of parity and data to tolerate two disk failures in disk array systems", *IEEE Trans. Parallel and Distributed Systems* 6(11): 1177-1184 (Nov. 1995).
88. J. S. Plank. "A tutorial on Reed-Solomon coding for fault-tolerance in RAID-like systems", *Software Practice & Experience* 27(9): 995-1012 (Sept. 1997), also see "Note: Correction to the 1997 tutorial on Reed-Solomon coding", see http://www.cs.utk.edu/~plank.
89. C. Polyzois, A. Bhide, and D. M. Dias. "Disk mirroring with alternating deferred updates", *Proc. 19th Int'l Very Large Database Conf.*, 1993, 604-617.
90. K. K. Ramakrishnan, P. Biswas, and R. Karedla. "Analysis of file I/O traces in commercial computing environments", *Proc. Joint ACM SIGMETRICS/Performance '92 Conf.*, 1992, pp. 78–90.
91. A. L. Reddy and J. Wyllie. "Disk scheduling in multimedia I/O system", *Proc. ACM Multimedia '93*, 1993, pp. 289-297.
92. A. Riska, E. Riedel, and S. Iren. "Managing overload via adaptive scheduling", *Proc. 1st Workshop on Algorithms and Architectures for Self-Managing systems*, 2003.
93. A. Riska and E. Riedel. "It's not fair - evaluating efficient disk scheduling", *Proc. 11th IEEE/ACM Int'l Symp. on Modeling, Analysis and Simulation of Computer Telecommunication Systems - MASCOTS '03*, 2003, pp. 288-395.
94. M. Rosenblum and J. K. Ousterhout. "The design and implementation of a log-structured file system". *ACM Trans. Computer Systems* 10(1): 26-52 (Feb. 1992).
95. C. Ruemmler and J. Wilkes. "A trace-driven analysis of disk working set sizes", *Technical report HPL-OSR-93-23*, HP Labs, 1993.
96. C. Ruemmler and J. Wilkes. "An introduction to disk drive modeling", *IEEE Computer* 27(3): 17-28 (March 1994).
97. K. A. Salzwedel. "Algorithmic approaches for storage networks", in *Algorithms for Memory Hierarchies, LCNS 2625*, U. Meyer, P. Sanders, J. Sibeyn (eds.), Springer 2003, pp. 251-272.
98. J. R. Santos, R. R. Muntz, and B. Ribeiro-Neto. "Comparing random data allocation and data striping in multimedia servers", *Proc. ACM SIGMETRICS Conf.* 2000, pp. 11 55.

99. P. Sarkar, S. Uttamchandani, and K. Voruganti. "Storage over IP: When does hardware support help?" *Proc. 2nd Conf. on File and Storage Technologies - FAST '03*, USENIX, 2003.
100. J. Schindler and G. R. Ganger. "Automated disk drive characterization", *CMU SCS Technical Report CMU-CS-99-176*, 1999.
101. T. J. E. Schwarz, J. Steinberg, and W. A. Burkhard. "Permutation development data layout (PDDL) disk array declustering", *Proc 5th IEEE Symp. on High Performance Computer Architecture - HPCA*, 1999, pp. 214-217.
102. M. I. Seltzer, P. M. Chen, and J. K. Ousterhout. "Disk scheduling revisited", *Proc. 1990 USENIX Summer Technical Conf.* pp. 307-326.
103. P. J. Shenoy and H. M. Vin "Cello: A disk scheduling framework for next generation operating systems", *Proc. ACM SIGMETRICS Conf.*, 1998, pp. 44-55.
104. P. Scheuermann, G. Weikum, and P. Zabback. "Adaptive load balancing in disk arrays", *Proc 4th Int'l Conf. Found. Data Organization and Algorithms - FODO'93*, pp. 345-360.
105. P. Scheuermann, G. Weikum, and P. Zabback. ""Disk cooling" in parallel disk systems", *IEEE Data Engineering Bulletin 17*(3): 29-40 (Sept. 1994).
106. P. Scheuermann, G. Weikum, P. Zabback. "Data partitioning and load balancing in parallel disk systems", *Very Large Database J. 7*(1): 48-66 (1998).
107. E. Shriver, A. Merchant, and J. Wilkes. "An analytic behavior model for disk drives with readahead caches and request reordering", *Proc. Joint Int'l Conf. on Measurement and Modeling of Computer Systems*, 1998, pp. 182-191.
108. J. A. Solworth and C. U. Orji. "Write-only disk caches", *Proc. ACM SIGMOD Int'l Conf.*, 1990, 123-132.
109. J. A. Solworth and C. U. Orji. "Distorted mirrors", *Proc. 1st Int'l Conf. Parallel and Distributed Information Systems - PDIS*, 1991, pp. 10-17.
110. Storage Performance Council. www.storageperformance.org.
111. D. Stodolsky, M. Holland, W. C. Courtright II, and G. A. Gibson. "Parity logging disk arrays", *ACM Trans. Computer Systems 12*(3): 206-325 (1994).
112. H. Takagi. *Queueing Analysis. Vol. 1: Vacation and Priority Systems*, North-Holland, 1991.
113. T. J. Teorey and T. B. Pinkerton. "A comparative analysis of disk scheduling policies", *Comm. ACM 15*(3): 177-184 (1972).
114. A. Thomasian and J. Menon. "Performance analysis of RAID5 disk arrays with a vacationing server model", *Proc. 10th ICDE Conf.*, 1994, pp. 111-119.
115. A. Thomasian. "Priority queueing in RAID disk arrays", *IBM Research Report RC 19734*, 1994.
116. A. Thomasian. "Rebuild options in RAID5 disk arrays", *Proc. 7th IEEE Symp. Parallel & Distributed Systems - SPDP*, 1995, pp. 511-518.
117. A. Thomasian. "RAID5 Disk Arrays and Their Performance Analysis", Section 37 in *Recovery in Database Management Systems* V. Kumar and M. Hsu (editors), Prentice-Hall, Jan. 1998.
118. A. Thomasian and J. Menon. "RAID5 performance with distributed sparing", *IEEE Trans. Parallel and Distributed Systems 8*(6): 640-657 (June 1997).
119. A. Thomasian. "RAID5 disk arrays and their performance evaluation", in *Recovery Mechanisms in Database Systems*, V. Kumar and M. Hsu (eds.), pp. 807-846.
120. A. Thomasian, G. Fu, and C. Han. "Rebuild strategies for redundant disk arrays", *Proc. 21st NASA/IEEE Conf. on Mass Storage Systems and Technologies*, April 2004.
121. A. Thomasian and C. Liu. "Some new disk scheduling policies and their performance", *Proc. ACM SIGMETRICS Conf.*, 2002, pp. 266-267.

122. A. Thomasian and C. Liu. "Disk scheduling policies with lookahead", *Performance Evaluation Review 30*(2): 31-40 (Sept. 2002).
123. A. Thomasian and C. Liu. "Fairness in SPTF based scheduling", Draft, May 2003.
124. A. Thomasian and C. Liu. "Performance of mirrored disks with a shared NVS cache", Draft, Jan. 2003.
125. A. Thomasian. J. Spirollari, C. Liu, C. Han, and G. Fu. "Mirrored disk scheduling", *Proc. Symp. Performance Evaluation of Computer and Telecomm. Systems - SPECTS '03*, 2003.
126. D. F. Towsley, S.-Z. Chen, and S.-P. Yu "Performance analysis of fault-tolerant mirrored disk systems", in *Performance '90*, Peter J. B. King, Isi Mitrani, and Robert J. Pooley (eds.), North-Holland 1990, pp. 239–253.
127. K. Treiber and J. Menon. "Simulation study of cached RAID5 designs", *Proc. 1st IEEE Symp. on High Performance Computer Architecture - HPCA*, 1995, pp. 186-197.
128. K. S. Trivedi. *Probability and Statistics with Reliability, Queueing and Computer Science Applications, 2nd edition.*, Wiley-Interscience, 2002.
129. M. Uysal, G. A. Alvarez, and M. Merchant. "A modular, analytical model for modern disk drives", *Proc. 9th IEEE/ACM Int'l Symp. Modeling and Simulation of Computer and Telecommunication Systems - MASCOTS'01*, 2001, pp. 183-192.
130. M. Uysal, A. Merchant, and G. A. Alvarez. "Using MEMS-based storage in disk arrays", *Proc. 2nd Conf. File and Storage Technologies - FAST'03*, USENIX, 2003.
131. A. Vakali and Y. Manolopoulos. "An exact analysis of expected seeks in shadowed disks", *Information Processing Letters 61*(6): 323-329 (1997).
132. E. Varki, A. Merchant, J. Xu, and X. Qiu "An integrated performance model of disk arrays", *Proc. 11th IEEE/ACM Int'l Symp. Modeling and Simulation of Computer and Telecommunication Systems - MASCOTS'03*, 2003, pp. 296-305.
133. A. Varma and Q. Jacobson. "Destage algorithms for disk arrays with nonvolatile caches", *IEEE Trans. Computers 47*(2): 228–235 (1998).
134. S. Varma and A. M. Makowski. "Interpolation approximations for symmetric fork-join queues", *Performance Evaluation 20*(1-3): 361-368 (1994).
135. W. Whitt. "Deciding which queue to join: Some counterexamples", *Operation Research 34*(1): 226-244 (Jan. 1986).
136. R. Wijayaratne and A. L. Narasimha Reddy. "Providing QOS guarantees for disk I/O", *Multimedia Systems 8*: 57-68 (2000).
137. J. Wilkes, R. A. Golding, C. Staelin, and T. Sullivan. "The HP AutoRAID hierarchical storage system", *ACM Trans. Computer Systems 14*(1): 108-136 (1996).
138. J. Wilkes. "The Pantheon storage-system simulator", *Report HPL-SSP-95-14*, HP Labs, Palo Alto, CA, revised May 1996.
139. J. Wilkes. "Data services – from data to containers" (keynote speech), *2nd Conf. on File and Storage Technologies - FAST '03*, USENIX, 2003.
140. J. L. Wolf. "The placement optimization program: A practical solution to the disk file assignment problem", *Proc. ACM SIGMETRICS/Performance'89 Int'l Conf.*, 1989, pp. 1-10.
141. C. K. Wong. "Minimizing head movement in one dimensional and two dimensional mass storage systems", *ACM Computing Surveys 12*(2): 167-178 (1980).
142. B. L. Worthington, G. R. Ganger, and Y. L. Patt. "Scheduling for modern disk drivers and non-random workloads", *Proc. ACM SIGMETRICS Conf.* 1994, pp. 241-251.
143. J. J. Wylie et al. "Survivable information storage systems", *IEEE Computer 33*(8): 61-68 (Aug. 2000).

Author Index

Lecture Notes in Computer Science

For information about Vols. 1–2901

please contact your bookseller or Springer-Verlag